Knowledge, Space, Economy

Edited by John R. Bryson,
Peter W. Daniels, Nick Henry
and Jane Pollard

London and New York

First published 2000
by Routledge
11 New Fetter Lane, London EC4P 4EE

Simultaneously published in the USA and Canada
by Routledge
29 West 35th Street, New York, NY 10001

Routledge is an imprint of the Taylor & Francis Group

Typeset in Goudy by Keystroke, Jacaranda Lodge, Wolverhampton
Printed and bound in Great Britain by Clays Ltd, St Ives plc

British Library Cataloguing in Publication Data
A catalogue record for this book is available from the British Library

Library of Congress Cataloging in Publication Data
Knowledge, space, economy / editors, John R. Bryson . . . [et al.].
    p. cm.
    Includes bibliographical references and index.
    1. Economic geography. 2. Space in economics. 3. Industrial location.
  4. Marketing channels. 5. Knowledge management. 6. Information technology
–Economic aspects. 7. Competition. I. Bryson, J. R.

HF1025 .K567 2000
330.9–dc21   00-057636

ISBN 0–415–18970–5 (hbk)
ISBN 0–415–18971–3 (pbk)

Contents

Illustrations

Figures

Tables

Contributors

John Allen is Professor and Head of Geography at the Open University, where he has edited and co-authored such volumes as *A Shrinking World? Global Unevenness and Inequality* (Open University/Oxford University Press, 1995), *Rethinking the Region: Spaces of Neoliberalism* (Routledge, 1988), and *Human Geography Today* (Polity Press, 1999). He is currently writing a book on spatiality and power.

Morag Bell is Professor of Geography at Loughborough University. She has research interests in the relations between concepts of culture, environment and development in Western thought and practice since the late nineteenth century. Much of her recent work has focused on the interactions between institutional knowledge, citizenship and environmental transformation within the context of British imperialism and its legacies. She has published widely and received research funding from a range of sources including the British Academy, the ESRC and the Welcome Trust for the History of Medicine.

John R. Bryson is Senior Lecturer in Economic Geography at the University of Birmingham. His recent work has focused on specialist knowledge-intensive business service firms and their interaction with the internal labour markets of both large and small client organisations. Following from a research interest in private sector flows of knowledge and information he is currently undertaking the first major research project into the geography of charitable or third sector organisations.

David B. Clarke is Lecturer in Human Geography and Affiliate Member of the Institute of Communications Studies at the University of Leeds. He is also an ESRC Research Fellow. He has published widely on consumption, the media, new technologies, modernity, postmodernity, and space. Having recently edited *The Cinematic City* (Routledge, 1997), he is currently working on two books on the spatiality of the consumer society, *Consumption and the City* (Routledge, forthcoming) and *Commodity, Sign and Space* (Blackwell, forthcoming); and, with Marcus Doel and Kate Housiaux, *The Consumer Society Reader* (Routledge, forthcoming).

Ian Cook is Lecturer in Human Geography at the University of Birmingham. He has been researching the connections that commodities make for about ten years, starting with tropical fruits grown in Jamaica and sold in British supermarkets, moving on to the geographies in and of the UK retailing and consumption of pasta, spices, chicken and bread (with Phil Crang and Mark Thorpe), and is now concentrating on Jamaican hot pepper sauces (with Michelle Harrison). He has written about this and also about the connections between academic writing, 'the field' and the classroom. More can be found on his web page at http://www.bham.ac. uk/geography .

Phil Crang is Reader in Human Geography at Royal Holloway, University of London, and was formerly Lecturer in Geography at University College London. He is co-editor of the journal *Ecumene: a journal of cultural geographies*. His research interests centre on the geographies of commodity culture. With colleagues at UCL and Sheffield, he is currently undertaking research on South transnationality and commodity culture in an ESRC funded project focused on the food and fashion sectors.

Peter W. Daniels is Professor of Geography and Director, Service Sector Research Unit at the University of Birmingham. He has published articles and books on the location and development of office activities and on the emergence of service industries as key agents in metropolitan and regional restructuring at the national and international scale. His books include *Services and Metropolitan Development* (Routledge, 1991), Service *Industries in the World Economy* (Blackwell, 1993), *The Global Economy in Transition* (with B. Lever, Longman, 1996), *Services in the Global Economy, Vols I and II* (with J. R. Bryson, Elgar, 1998). He was recently President of the European Research Network on Services and Space (RESER).

Nick Henry is Reader in Urban and Regional Studies in the Centre for Urban and Regional Studies, University of Newcastle upon Tyne and was formerly Lecturer in Geography at the University of Birmingham. His major research interests lie in economic geography and regional development in the 'advanced' economies, with a special emphasis on high technology and service industries. These interests include an understanding of the social and political construction of 'the economic' and the impacts of economic change upon wider society. His most recent book is *The Economic Geography Reader: Producing and Consuming Global Capitalism* (with John R. Bryson, David Keeble and Ron Martin, Wiley, 1999). He is currently working on Birmingham's multicultural economic networks and Britain's so-called 'cool economy'.

Kevin Hetherington is Reader in Sociology at Lancaster University. He researches on issues of spatiality, materiality and identity. He is currently working on research into museums and the spatial politics of access and is co-editor of the journal *New Research in Museum Studies*. Recent books include *The Badlands of Modernity* (Routledge, 1997), *Expressions of Identity* (Sage/TCS, 1998) and *New Age Travellers* (Cassell, 2000) as well as a special issue of *Environment and Planning D: Society and Space* (Vol. 18, 2000), co-edited with John Law, on the theme 'After Networks'.

Jeremy Howells is a Director at Policy Research in Engineering, Science and Technology (PREST) and a Senior Research Fellow at the ESRC Centre for Research in Innovation and Competition (CRIC), University of Manchester and UMIST. He has written, co-authored and edited nine books on research and innovation, two of which are to be shortly published by Edward Elgar publishers.

His research interests centre on research and development, innovation, technology transfer and knowledge.

Bob Jessop is Professor of Sociology at Lancaster University. He previously taught politics and sociology at the University of Essex. He is best known for his contributions to state theory, Marxist political economy, the regulation approach to Fordism and post-Fordism, the study of postwar British politics, governance and governance failure, and recent changes in welfare regimes. His three current research projects are concerned with welfare state restructuring; the discourses and contradictions of the knowledge-driven economy; and the re-scaling of the state in a post-national era. Further information and recent publications can be found at his home page: http://www.comp.lancs.ac.uk/sociology/rjessop.html

David Knights is Professor of Organisational Analysis and Head of the School of Management at Keele University. In 1994 he founded and managed a Financial Services Research Forum of 26 leading corporations to fund strategic research. He still helps run this research activity at the University of Nottingham Business School where he worked briefly prior to moving to Keele. He has supervised over 30 externally funded research grants to the value of around £3 million and currently is involved in five ESRC funded projects on Bank Fraud, Business Reengineering, Innovation, Education and Virtual Markets. A selection of recent books includes *Managers Divided: Organisational Politics and Information Technology Management* (with F. Murray, Wiley, 1994), *Regulation and Deregulation in European Financial Services* (with G. Morgan, Macmillan, 1997), *Financial Service Institutions and Social Transformations: International Studies of a Sector in Transition* (with T. Tinker, Macmillan, 1997) and *Management Lives!: Power and Identity in Work Organisations* (with H. Willmott, Sage, 1999).

John Law is Professor of Science Studies and Sociology at the University of Lancaster. He has written widely on technologies, materialities, spatialities, and organisations, and is currently working on medical and health technologies and subjectivities. His recent publications include *Actor Network Theory and After* (co-edited, Blackwell, 1999) and *Organising Modernity* (Blackwell, 1994) and his new book *Aircraft Stories: Decentering the Object in Technoscience* will appear with Duke University Press in 2001.

Andrew Leyshon is Professor of Economic and Social Geography at the University of Nottingham. He is currently undertaking research in three main areas. First, he is examining the geographical consequences of competition and organisational change in the retail financial services industry, with particular reference to new technologies of information and knowledge. Second, he is researching alternative systems of monetary exchange in the form of Local Exchange and Trading Systems. Third, he is investigating the changing economic geography of the music industry, with particular reference to the impact of software formats and Internet distribution systems. He is the author of *Money/Space* (with Nigel Thrift, Routledge, 1996) and co-editor (with David Matless and George Revill, Guilford Press) of *The Place of Music*.

Edward J. Malecki is Professor of Geography at the University of Florida. He has a BA degree in International Studies and a PhD in Geography, both from The Ohio

State University. He was a Visiting Researcher in the Centre for Urban and Regional Development Studies at the University of Newcastle upon Tyne, and a Fulbright Professor at the University of Economics and Business Administration in Vienna. Malecki is the author of *Technology and Economic Development: The Dynamics of Local, Regional and National Competitiveness*, second edition (Addison Wesley Longman, 1997) and is co-editor of *Making Connections: Technological Learning and Regional Economic Change* (Avebury, 1999) and *The Industrial Enterprise and Its Environment: Spatial Perspectives* (Avebury, 1995). He serves on the editorial advisory boards of several international journals and is Associate Editor of *Entrepreneurship and Regional Development*.

Pamela Odih is a Lecturer in Sociology at Goldsmiths College, University of London. Her research focuses on four main areas: consumer research, gender studies, media and cultural studies and postmodernity. Gender time and its bearing on consumption and wider gender relations is her specialist area. Recent publications include Gendered Time in the Age of Deconstruction, *Time and Society*, 8(1).

Steven Pinch is Reader in Geography at the University of Southampton. His main research interests are in the geography of welfare and spatial aspects of economic restructuring. His principal books are *Cities and Services* (Routledge and Kegan Paul, 1985) and *Worlds of Welfare* (Routledge, 1997). In addition to work on the British motor sport industry, his publications include articles on 'territorial justice' in social service allocations; contracting-out of public services; the geography of corporate philanthropy; changing working practices; social polarisation; and evolving spatial divisions of labour in southern England. He is currently collaborating with Dr Nick Henry on a project on the geography of Britain's music and multimedia industries.

Jane Pollard is Lecturer in Economic Geography at the University of Birmingham. Her research interests focus on urban and regional economic development, and particularly the relationships between money and finance and processes of social, political and economic change. Current research is examining the restructuring of retail financial services in Britain and the USA and the entry of new players into retail banking markets in Britain.

Michael Pryke is Lecturer in Geography at the Open University. His research interests include the geographies of money and finance, issues of time-space and financial power and the sociology of money. He has recently edited *Unsettling Cities* (Routledge, 1999) with John Allen and Doreen Massey.

James Sidaway is Lecturer in Geography at the University of Birmingham. He is the author of *Imagined Regional Communities: The Geopolitics of Integration in the Global South* (Routledge, forthcoming) and over thirty book chapters and papers on themes of political geographies and geopolitics, the sociology and history of geographical knowledges and development.

Mark Thorpe is Associate Director and Head of Qualitative Research at Simons Priest & Associates, a company offering specialist market research services to commercial clients. He previously worked in the Brand Development Unit of The Research Business International. Before becoming a commercial researcher, Mark worked at University College London, the University of North London and the City Literary Institute. At present, much of his time is devoted to working with leading retailers in developing better understandings of shopper and consumer behaviour. Mark is on the advisory committee of the ESRC study 'Environment and Identity – A Cross Setting Study'.

Frank Webster was educated at Durham University and the LSE. He was Professor of Sociology at Oxford Brookes University from 1990 to 1998, and is now at the University of Birmingham. He is the author of several books, including *Information Technology: A Luddite Analysis* (with Kevin Robins, Ablex, New Jersey, 1986), *The Technical Fix: Computers, Industry and Education* (with Kevin Robins, Macmillan, 1989), *Theories of the Information Society* (Routledge, 1995), *The Postmodern University?* (edited with Anthony Smith, Open University Press, 1997), *Times of the Technoculture: From the Information Society to the Virtual Life* (with Kevin Robins, Routledge, 1999) and *Theory and Society: Understanding the Present* (edited with Gary Browning and Abigail Halcli, Sage, 1999).

Jane Wills is Lecturer in Geography at Queen Mary and Westfield College, University of London. Jane has particular research interests in the geography of individual relations, management cultures and trade union organisation. She is currently exploring the development of European works councils, social partnership and new unionism in the UK. Her publications include *Union Retreat and the Regions* (written with Ron Martin and Peter Sunley, Jessica Kingsley, 1996), *Geographies of Economies* (edited with Roger Lee, Arnold, 1997), *Dissident Geographies* (written with Alison Blunt, Longman, 2000) and articles in journals such as *Antipode, Economic Geography, Transactions of the IBG* and the *Human Resource Management Journal*.

Acknowledgements

It will probably seem strange for our acknowledgements to begin with the editors thanking each other. However, the project could not have got off the ground unless there had been the sort of lively discussion that characterises the regular meetings of the Developed Market Economies Research Group in the School of Geography at Birmingham. During 1997 it became apparent that members of the Group had, and indeed continue to have, research interests about aspects of economic geography that are linked by a common thread. This takes the form of the role of knowledge and/or information in the production and consumption of goods and services, and its implications for spatial patterns of development. We were also aware of the developing interest and debates in the social sciences about the nature of knowledge and its impacts on power relations or the role of the state as a 'gatekeeper' in knowledge society. Our pooled interest, the willingness of the group to share ideas and information, and the belief that it would be timely to attempt to produce an interdisciplinary collection reflecting some of the current thinking about the complex interactions between space, knowledge and the economy encouraged us to begin discussions about an edited interdisciplinary collection.

We began by drafting a proposal that could be used to encourage/persuade invited contributors from a range of disciplines that this was a worthwhile exercise. This involved several 'brainstorming' sessions during which one of the editors had the unenviable task of trying to record the thoughts of his colleagues (with faster minds than his) on a laptop computer. A rather absent-minded approach to saving what had been drafted caused much consternation amongst the more computer literate of the editors but the finished product seemed a fair reflection of the pooled thoughts of the group!

Subsequent events depended very much upon the initial encouragement for the project that was received from Manuel Castells. We are indebted to him for his support and encouragement. We are also very grateful to all those contributors who initially signed up to the project and especially to those who, at a later stage and at short notice, agreed to fill gaps created when it proved necessary for some of the initial contributors to withdraw because they were unable to meet their commitments. The balance of the book would have been adversely affected had the latter not been prepared to fill the breach. The reviewers of the initial book proposal made some very helpful suggestions that were incorporated in the final version. We would also like to acknowledge the constructive comments made by the two reviewers, one from the UK and the other from the US, who undertook the task of scrutinising the manuscript prior to its acceptance for publication. Responsibility for the final content of the book does, of course, rest with the editorial team.

Edited books are about teamwork and their production involves numerous individuals. The contributors willingly and constructively responded to feedback on their draft chapters and dealt quickly with various queries along the way. We would particularly like to thank Claire Clarke in the School of Geography and Environmental Science office who assisted with typing, collating and printing various sections of the manuscript. Kevin Burchill in the School's Drawing Office re-drew some of the diagrams. Some of the ideas about knowledge and its implications that are explored in the various contributions by the editors were shared with postgraduate students in the School taking the MSc World Space Economy course; we appreciated their interest and thoughtful comments. Finally, we would like to thank Sarah Carty at Routledge who originally commissioned the book. Andrew Mould has subsequently taken up the reins, ably supported by Ann Michael who has demonstrated much patience as deadlines have slipped past and the excuses have become ever more imaginative. Such is the joy of publishing!

The authors and publishers would like to thank the following for permission to reproduce copyright material:

For figures 8.1, 8.2 and 8.3, Taylor & Francis Ltd (PO Box 25, Abingdon, OX14 3UE), from S. Pinch and N. Henry (1999b) 'Paul Krugman's geographical economics, industrial clustering and the British motor sport industry', *Regional Studies* 33(9): 815–27. For figures 8.4, 8.5, 8.6, 8.7 and 8.8, Elsevier Science, from N. Henry and S. Pinch (2000) 'Spatialising knowledge', *Geoforum* 31(2): 191–209. For figure 9.1, the President and Fellows of Harvard College, from M. Storper and R. Salais (1997) *Worlds of Production: The Action Frameworks of the Economy*, Cambridge, MA: Harvard University Press, p. 33. For table 7.1, Sage Publications, Inc., from R. V. Knight and G. Gappert (eds) (1982) *Cities in the 21st Century*, p. 56. For table 11.1 the Heritage Foundation, Washington DC, at http://www.heritage.org/

John R. Bryson
Peter W. Daniels
Nick Henry
Jane Pollard

Birmingham, May 2000

1 Introduction

John R. Bryson, Peter W. Daniels,
Nick Henry and Jane Pollard

The knowledge-driven economy is not only a new set of high-tech industries such as software and biotechnology, which are built on a science base. Nor is it only a set of new technologies – information technology and the Internet. It is about a set of new sources of competitive advantage that apply to all industries, high-tech and low-tech, manufacturing and services, retailing and agriculture. The key to our competitiveness is how we combine, marshal and commercialise our know-how.

(*Observer*, 30 July 1998)

Capturing some of the more populist representations of current economic transformation, the writer in The *Observer* reflects one of a number of recent appeals to the growing significance of knowledge in contemporary capitalism (see, for example, Coyle 1997; Drucker 1993; Leadbeater 1999). A key argument developed in these literatures is the suggestion that economic competitiveness is now bound up not with new materials *per se* but with new ways of producing, using and combining diverse knowledges; the same ingredients, in essence, can be rearranged in new, and better, recipes. In similar vein, these arguments are to be found in academic commentaries on economic transformation. A selection of these might include, for example, Lash and Urry (1994) on economies of sign and space, Lundvall and Johnson (1994) on the learning economy, Quah (1996) on dematerialisation and Thrift (1998a) on soft capitalism. It is now clear that in a number of disciplines, including geography, sociology, economics, cultural studies, management, psychology and policy studies, there is a growing acceptance that flows and translations of knowledge are integral to understanding contemporary global capitalisms.

The recognition of knowledge as a factor in the production and reproduction of economies and societies can be traced to the influential work of Daniel Bell (1973) which was informed by the perceived shift from manufacturing to services. More recently, the work of Manuel Castells (1996) on 'informational capitalism' has focused further attention on the growing importance of knowledge in contemporary capitalist societies. His 'informational capitalism' signals the importance of innovation, knowledge and learning in a globalising, rapidly evolving economy. Castells has been a key advocate, arguing that while information and knowledge have always been important for economic growth, we have entered an era in which the ability of firms, regions, nations and labour to produce, circulate and apply knowledge are fundamental to their competitiveness. Indeed, information and knowledge are now viewed as components and products of production processes in numerous sectors, not only in so-called 'high-technology' sectors. Thus, 'informational capitalism' is not about the growth of a post-industrial information processing sector (à la Bell's sectoral approach). Rather, all departments, sectors and sub-sectors of the economy are, or are becoming, informational in the sense that information and knowledge are widely embodied in the work process and accessed through the increasing powers of information technologies.

An ubiquitous feature of this literature is the nagging recognition that knowledge is a very slippery concept. What is knowledge? What is it to know? What differentiates knowledge from information? Has the importance of various knowledges altered over the last 20 years, and if so in what ways? All of these are key questions to which it is only possible to construct partial answers.

In this volume, we focus on the nexus of knowledge, space and economy. In so doing, we concentrate on three themes. First, we examine *different forms* of what are construed as 'economic knowledges', that is knowledges that are deemed central to constructing competitive advantage in contemporary capitalism. Such a remit leads into several areas of inquiry: what forms do economic knowledges take?; how are different knowledges – geopolitical, regulatory, technical, consumer and genetic, to give only a few examples – made economic either indirectly (through their production, appropriation and use by firms) or directly through their commodification?; how is the production and circulation of economic knowledges mediated by power?

Second, if economic knowledges are being made, then *they are made in space and place*. How do space and place make knowledges 'economic'? What are the spatialities of different economic knowledges? How are different forms of economic knowledge produced, mobilised and translated? What are the geographies of these processes? What are the geographies of the knowledge-space-economy?

Third, although the editors' disciplinary background is economic geography, we believe it is vital to encourage an *interdisciplinary research agenda* around the notion of knowledge, space, economy. Thus, we have brought together a group of scholars from a range of social science disciplines to explore the meaning and interpretation of knowledge in the contemporary space economy. The sheer diversity of knowledges that are of economic significance, and their contested nature, is conveyed via the different disciplinary backgrounds of the contributors.

The knowledge-space-economy nexus

Although different authors are grappling with different conceptions of knowledge and its production, there are perhaps a number of cognate features of what we might call 'knowledge economies'. Hodgson (1999: 181–2) considers five developments that have heightened the importance attached to knowledge production, use and circulation: (1) processes of production and their products are becoming more complex and sophisticated; (2) increasingly advanced knowledge and skills are being required in many processes of production; (3) there is an increasing reliance on specialist and idiosyncratic skills; (4) the use and transfer of information is becoming more extensive and important for economic activities; and (5) there is increasing uncertainty in contemporary economic life.

Knowledge economies seem to be characterised by the de-materialisation of production, a shift away from dealing with raw materials and machines toward dealing with other minds; there is a shift from 'action-centred to intellectual skills' (Hodgson, 1999: 184). There is also a shift away from physical dexterity and skills to mental processing ability (see also Reich 1991a). The term 'knowledge economies' is also intended to capture a sense of accelerating technological change and, related to this, the need for continuous innovation. This acceleration of technological change, and the associated growth of information available to firms, organisations and other agents, is also, it seems, bound up with growing uncertainty. In a rapidly changing world of increasing complexity, predicting future events becomes ever more difficult (Beck 1992). Thus, a more 'knowledge intensive' economy is one in which the ability to learn is an important aspect of competitiveness (Porter 1990; Drucker 1993).

If it is accepted that the production and translation of different forms of economic knowledges is of increasing importance in contemporary capitalism (for a critique of this approach see Hudson 2000), then a series of key questions arise concerning the production and distribution of knowledge in economic networks. Contrary to neo-classical economic theory, there are seemingly wide variations in the capacities of firms to create, circulate and manage knowledge. Knowledges and the ability to commodify them are unevenly distributed.

Some of the early attempts to examine knowledge-based economies focused on the changing role of information technologies and the possibilities for such technologies to erode the 'friction of distance'. This would, goes the argument, lead to the spatial dispersion of many functions and, in essence, signal the 'end' of geography (see O'Brien 1992). Prioritised, implicitly and explicitly, in such writing was an emphasis on the 'hardware' of information transmission and its diminishing costs in real terms as technologies evolve rapidly. In addition to the technologically determinist bent of such writings, there was also a focus on traded flows of what is now commonly called codified or explicit knowledges, as opposed to tacit, know-how forms of knowledge.

In line with criticisms about globalisation and the end of geography, however, (see, for example, Martin 1994a), the reinvigorated importance of place in contemporary economic times has come to the fore and many have turned to knowledge-based explanations to explain agglomeration or clustering. Invoking high-tech milieu, financial and business services or craft-based industrial districts, these explanations have tended to emphasise the idiosyncratic quality to relationships or transactions between firms (see Scott 1988). More recently, there have been more explicit elaborations of the non-codified, tacit nature of productive knowledges, many of which are untraded (Storper and Salais 1997). Implicit in these arguments is the suggestion that tacit knowledges may not travel well and, significantly, that certain spatialities may help or hinder the production and translation of different forms of economic knowledge.

Many of the chapters in this volume contribute to these debates concerning knowledge-space issues and, most especially, the relationship between forms of knowledge, alternative spatialities of knowledge and different paths of knowledge territorialisation. Several contributions explore the analytical foundations of the knowledge-space-economy through discussion of, for example, the archetypal innovative high technology cluster, drawing on tacit, localised, technologically-based knowledges. Others provide analyses of a variety of economic systems tracing characteristic flows of knowledge and their implications for changing organisational geographies.

A number of distinctive elements emerge from the volume. First, some authors problematise the (still) productivist bias to the identification of knowledge, considering consumer, as well as producer, knowledges, and circuits and networks as well as linear flows. Second, knowledge is often characterised as either tacit or codified, leading to its subsequent 'mapping' on to the grid of global-local space. Whilst this approach is discernible in several contributions, others visualise 'local' knowledges as comprising a mix of near and far, proximate and distanciated knowledge flows. Third, this collection is testimony to the eclecticism of different disciplines as they grapple with the emerging contours of the knowledge-space-economy. Most especially, this broadening of the spatial matrix of economic knowledge contributes to the on-going conceptual reassessment of 'the economic' and 'the economy' (Crang 1997; Thrift and Olds 1996; Amin and Thrift 2000). Increasingly apparent in recent social science literatures about knowledge, is recognition that the metaphors used to think about the economy are changing as its socio-cultural construction is acknowledged. In this vein, the chapters of the book explore the knowledge-space-economy nexus through objects as diverse as the cultural industries, international geo-politics and concepts of the self.

The structure of the book

The themes highlighted around the knowledge-space-economy nexus are developed through three parts. Part I charts the current state of (interdisciplinary) thinking on the changing role of knowledges in capitalism. These chapters, from authors with backgrounds in Geography, Management, Sociology and Economics, exemplify some of the ways and senses in which economic knowledges are being discussed in different disciplines. Recurring themes include acceptance of the significance of knowledge in the economy, privileging of particular forms of knowledge, and the material forms, impacts and implications of different knowledges for objects and actors within contemporary economy and society.

There are two further parts to the book. These draw from and illustrate many of the issues examined in Part I from two different, but related, vantage points. Part II comprises a range of theoretically informed case studies that reveal knowledges at work in space and place. Specifically, these chapters through their focus on *organisations*, explore the diversity of couplings between knowledge, economy and space as diverse forms of knowledge are made 'economic' in time and space.

The focus changes in Part III to consider some of the ways in which individuals contribute to the production, consumption and translation of knowledges. Again, some of the themes highlighted in Part I – such as the growing importance of knowledge and uncertainty in contemporary life – are revisited. With all the uncertainties of the contemporary moment what emerges in this part are some of the difficulties and partialities involved in 'knowing' and, indeed, the difficulties that can be encountered in attempting to change patterns and forms of knowledge exchange.

Part I Knowledge, space, economy

Beginning with a geographer's view of knowledge-space-economy, John Allen addresses current commentaries on the spatiality of economic knowledge. He argues for a variety of configurations, shapes and forms that knowledge, and most particularly economic knowledges, may take. Specifically, he suggests that for all the plurality of forms of economic knowledge, cognitive knowledges have enjoyed a relatively privileged position (see Howells below). Allen draws on Foucault's work on knowledge and power to highlight how cognitive knowledge has become the 'obvious' knowledge of the contemporary knowledge-based economy – highly prized, highly valued and associated with a select set of economic activities and labour. He argues that the acceptance of such cognitive-based accounts of economic knowledge excludes or downgrades the contribution of a variety of other 'knowledge' activities and sectors (broadcasting media or advertising, for example) and impedes efforts to unravel current processes of economic transition. His chapter urges us to acknowledge that there is more than one type of symbolic activity that constitutes economic knowledge and he uses the writing of Cassirer (1979) to highlight the important contribution of aesthetic and expressive, as well as cognitive, knowledge within the economy.

In making this argument, Allen reflects on how it is only cognitive forms of economic knowledge that have become the centre of attention in delineating the contemporary space economy. Furthermore, he articulates the way in which this form of economic knowledge has become enfolded within the widely accepted dichotomy of tacit and explicit knowledge, with its subsequent geographical mapping on to the global–local scale.

Drawing on literatures from the sociology of scientific knowledge, particularly the theoretical perspective of actor network theory, John Law and Kevin Hetherington also focus on the spatialities of the contemporary economy. They do not question the merits or otherwise of the claims about the information economy promulgated by Castells (1996), Bell (1973) and others but, rather, they reflect on how these structures might come into being. Law and Hetherington argue that it may not be wise to consider the transmission of information or global capital flows à la Castells without simultaneously asking specific questions about how the materiality of these relations is achieved and how it is sustained. Actor network theory is used to 'read' how objects, subjects, and processes are bound together relationally to produce knowledge(s). The performance of these networks determines the influence (power) of knowledge (see also the chapters by Allen and Odih and Knights). Further, they argue that knowing the global is always the product of the local, of a place and its particular materiality. Through an account of the materiality of knowing, Law and Hetherington sketch a conception of the economy with its knowing locations, points of passage, and active making of space through performance. As such, they conclude that if proper account is taken of material semiotics and the performed spatialities implied, then the concept of 'globalisation' is a misleading catch-all phrase that implies the presence of one global system or all-embracing space-time box within which 'knowing' is located. This, they suggest, misses the materiality of the 'global order' and the complex spatialities implied in its enactment. To their mind, there are a variety of configurations of knowledge, space, economy produced as part of materially heterogeneous networks of near and far, human and non-human.

Jeremy Howells, by contrast, employs a neo-Schumpeterian perspective to explore the specifics of the key relationship between knowledge and innovation and its centrality to firm performance. He concurs with the argument for a historical process involving the increasing volume, spread and speed of knowledge and highlights the recognition of this trend by those involved with the management and organisation of firms. Knowledge and innovation are viewed today as vital determinants of organisational competence and capability. This is best understood, he argues, by exploring the way in which knowledge activity is situated within the firm – the firm as a knowledge repository/generator – and the ways in which innovation and knowledge are shaped by the internal and external relationships of the firm. The problem for the firm is how to harness the knowledge and information embodied in its individual workers for the corporate benefit and for the good of effective collaboration with external suppliers and clients. The construction of an effective organisational 'memory' is one requirement; although knowledge is a relational, individual-centred concept ('the knowing self'), the ultimate objective is shared understanding.

In essence, Howells narrows the spatial focus to investigate the spatialities of knowledge production in innovative firms. This chapter highlights the geographical context of the innovative firm by drawing on the increasingly common characterisation between tacit, local knowledge and codified, ubiquitous knowledge (see Allen above).

As with previous rounds of technological transformation, the emergence of new information and communication technologies prompts questions about the changing spatialities of the generation and circulation of forms of productive knowledge.

For some commentators, the telecommunications revolution of the knowledge space economy presages 'the end of geography'. By contrast, Castells (1996) was at pains to highlight the geo-politics of the informational economy such that the variable geometry of the informational economy is shaped by the actions and responses of nation states. This changing form and function of the state in a globalising, knowledge-driven economy is the subject of the chapter by Bob Jessop.

For Jessop, writing from a disciplinary background in sociology, knowledge must be characterised, like labour, as a fictitious commodity. Its valorisation of what he terms the 'extra-economic' is merely one contradiction of several which knowledge capitalism will face. This is no different to previous capitalist periods that have also faced similar contradictions. Uncertain of what the after-Fordist regime may entail, Jessop's summation is that there will be an increasing tendency towards a structural coupling of economy and state which can be summarised as the Schumpeterian workfare state. Translated into new state forms, this suggests a multiscalar and networked form of governance that is driven by the imperative of international competition. These forms of governance will be delivered via diverse mechanisms ranging from local 'learning region' initiatives (see Malecki, Chapter 7) to World Trade Organisation initiatives on global intellectual property rights (see Sidaway and Pryke, Chapter 11).

To conclude the first section, writing from the perspective of management theory, Pamela Odih and David Knights continue the theme of dominant and subordinated knowledge time-spaces within the contemporary economy. They show how marketing knowledges treat space and time as 'given' or, at best, 'rational' rather than variable elements in the environment of marketing. Consumers, for example, are often regarded as allocating time within fixed spatial contexts; the idea that the use of time and space incorporates cycles of activity or rhythms of consumer behaviour that vary across time and space is rarely acknowledged within a dominant linear theorisation of time-space. This problem can perhaps be rectified, they argue, by using the notion of 'social time space'. This acknowledges that, for example, consumption by individuals is influenced by the symbolic meaning of different goods and their potential to inform others of their identity and life-style attributes. Yet Odih and Knights illustrate that even these discourses, which appear to contradict ingrained dualistic conceptions of time and space, inadvertently reinforce the hegemony of linear time and spatial solidarity. In so doing, such conceptions inadvertently take for granted what they most need to question.

To sum up, the chapters of Part I set an interdisciplinary tone to debates about the intersections of knowledge, space and economy. United by a recognition of the significance of the knowledge-space-economy nexus, what is striking is the cross-cutting nature of the theoretical bases which inform the writings of authors from different disciplines. Throughout these contributions the role of space and place in the construction of diverse, and particular, forms of economic knowledge is stressed. Part II develops and refines these arguments through case study and other analyses of the knowledge-space-economy nexus.

Part II Knowledge at work in space and place

The chapters in the second part of the book draw from and illustrate many of the issues examined in Part I. In a number of cases, through the development of a diverse range of theoretically informed case studies, the significance of different knowledge(s), as they become 'economic' in space and place, is explored.

Part II begins with contributions from Ed Malecki and from Nick Henry and Steven Pinch on how 'local knowledge' has become a central component structuring the geography of the knowledge economy. Malecki discusses how local knowledge creates 'sticky places' (Markusen 1996), able to both attract and embed the highly prized cognitive, and most especially technologically innovative, knowledges identified earlier by Allen. The desire to create, sustain and attract these knowledge forms, through policy concepts such as that of the 'learning region', is currently hegemonic within local and regional economic policy. Malecki explores the theoretical basis of such policy, derived as it is from the idea that knowledge has become much less specific to organisations and is 'in the air' to be shared amongst some or all of the firms in a locality. On the basis of the assumption that all places are not alike, he suggests that firms in particular localities and regions have a great deal to gain from sharing knowledge (private and public).

The Boston high technology region is used as an example of a learning region where local, 'sticky' knowledge has played a fundamental role in sustaining the innovative capacity and competitiveness of its firms and institutions. This, and other examples, are used to argue that regional economic change in the 1990s, and no doubt beyond, is best understood as a function of learning processes. The tacit knowledge embedded in local or regional cultures is fundamental to success; it is not easily incorporated into economic models and has therefore been overlooked but its significance should not be underestimated. Indeed, regional economic policy should aim at nurturing such 'learning regions' and the web of public and private institutions that enable them.

Complementing the more general discussion of the significance of agglomeration and local/regional networks of production by Malecki and Howells, Henry and Pinch provide a detailed elaboration of a 'sticky place'. Contrasting the theoretical approach of 'geographical economics' (Krugman 1991; *The Economist* 1999; Martin 1999a) to that of 'new economic geography', Henry and Pinch reveal some of the processes that produce local economic knowledges. They use a case study of the British motor sport industry, and its regional agglomeration in Motor Sport Valley, to show how global dominance in the production of racing cars has been attained through an economic system comprising the social and spatial embeddedness of an expert 'knowledge community'. The region has produced a capacity to facilitate rapid exchange and dissemination of knowledge about the best way to design and manufacture high performance racing cars. This knowledge circulates within the regional knowledge community, for example, as key personnel move from one firm to another and has resulted in the creation of a form of 'knowledge pool' on a constant learning trajectory. It is also very much the case that the knowledge upon which Motor Sport Valley is based is not readily transferred to another locality. The knowledge itself may be mobile but its embodiment in people who have certain ways of doing things (social or cultural norms) derived from being part of a particular place makes this prospect more elusive than might be imagined.

The empirical context chosen by Jane Pollard and Andrew Leyshon to analyse the increasing significance of knowledge is that of UK retail banking. Their chapter charts the changing knowledge architectures of different retail banking products and the associated shifts in the geographical anatomy of the retail banking industry. Pollard and Leyshon contrast the dispersed branch, cash point and other activities of retail banks with the much more intense agglomeration of the high technology activities that have typically been at the forefront of any discussion of the geography of the knowledge economy (see Malecki, Chapter 7 and Henry and Pinch, Chapter 8). Using Storper and Salais' (1997) *Worlds of Production* framework, they demonstrate the diversity of strategic, technical and regulatory knowledges that retail banks must now accommodate, often sourced from other industries or from other countries such as the United States. There is a clear trend towards codified and centralised knowledge bases and, in this context, retail banks have found it necessary to adjust their operating geography towards more concentrated forms rather than the more dispersed branch systems that gathered, amongst other things, local, tacit economic knowledge. It is concluded that not only is competition in UK retail banking now very intense, the variety of knowledges that it needs to harness has become very disparate, spanning customers, products, technologies, regulation, marketing and strategy. While each of these, or combinations of them, may be used locally, they are better organised, enhanced and updated via more centralised arrangements that also involve interaction with other places, people and knowledges. New retail banking geographies are emergent, and they are at least partially based on processes involving, in some instances, the explicit reduction of flows of tacit knowledge so prized in other economic contexts.

The previous contributions in this part encapsulate a still tentative, yet growing, analysis of knowledge forms and how their production and translations are spatialised. One set of knowledges closely associated with the contemporary economy are those listed under the rubric 'managerial' created by the symbolic analysts of the management consultancies (see Allen above). John Bryson demonstrates how knowledges constructed around the thinking and ideas of certain key individuals shape the organisation of production and economic geographies in time and space.

In his contribution, the spread of Taylorism is examined to show not only the long-standing impact of this paradigm on the organisation of production but also the variable interpretations that it elicited on the part of managers and/or in different places. Taylorism was by no means readily accepted as the way forward for production. Just like many other innovative ideas the sceptics were at least as numerous as the supporters with many of the concerns linked to the 'real' motives of knowledge as power with respect to impacts on jobs, wages and other working conditions. Bryson argues that the dissemination of the knowledge needed to implement the managerial model of Taylor depended on who was transmitting it and their interpretation. Thus, the diffusion of Taylorist ideas was, at least in part, a function of how these ideas were conveyed to those who might use or be persuaded by them.

This theme is developed by James Sidaway and Michael Pryke who show how the contemporary geopolitical gaze is, with the end of the Cold War and the ascendancy of neo-liberalism, increasingly one based around geo-economics. Through a fascinating piece of deconstruction, they provide an example of the 'economic' (yet also clearly political) labelling of diverse parts of the world. 'Free or unfree' is a question of economics not human rights. They look at how flows of private capital are steered from developed to developing countries using bundles of knowledge couched in a 'language' that is carefully crafted to achieve the objective in hand. On the one hand, private investors need to be able to make sense of the investment opportunities in developing countries while, on the other, the receiving countries need to be persuaded that such investment is a 'good thing'. The language involved in this process engages with notions of freedom (from poverty or unemployment) and the hegemony of the free market. Sidaway and Pryke suggest that these ideas are a central part of the institutional discourses of influential global organisations such as the World Trade Organisation, the World Bank or the Organisation for Economic Co-operation and Development. Such ideas are embodied in sophisticated ways in the regularly updated literature and indices made available to investors and others via publications of the kind produced by the Heritage Foundation in the USA. As a consequence, the geopolitical designation of developing countries is paralleled by, and intertwined with, a powerful geo-economic classification based on their global investment potential.

The final chapter of Part II returns us once more to the constitution of 'useful' knowledge. If the contribution of Jessop in Part I reminds us that knowledge is a fictitious commodity, and Sidaway and Pryke provide an insight into the scripting of useful knowledge, then Morag Bell reminds us that the meaning and application of 'useful' knowledge is far from clear-cut. Through the example of genetic engineering she argues how space, place and time are constitutive of what becomes defined as 'useful' (and invariably commodified and 'economic') knowledge. Bell widens knowledge beyond the immediately 'economic' to look at that associated with human identity and life quality. Part of twentieth-century modernity has encompassed improving human and environmental relations in order to achieve a 'better' life but what are the sources, ownership and authority of the knowledges that can fulfil this objective? Bell characterises the situation as one of great doubt and uncertainty in circumstances where there are moral and material boundaries to what is 'useful' knowledge. As an example of uncertainty the Human Genome Project is examined in some detail. It provokes many debates about the ethical and moral significance of new genetic technologies for individuals, their identity and their place in the social and economic order. At the same time the national and global economic significance of genetic 'engineering' is also a source of much dispute. What precisely is the value of this knowledge, for whom is it intended, how is it sourced and controlled, and what purposes are served in its production, exchange and application?

Part III Becoming in the (k)now: spaces of identity

Part III of the book considers some of the ways in which individuals contribute to the production, consumption and translation of knowledges and how individual identities are constructed, in part, through their knowledge relations. Once again, the themes highlighted in Part I, especially the growing importance of knowledge and uncertainty in contemporary life, rear their head. With all the uncertainties of the contemporary moment, what emerges in this part of the book are some of the difficulties and partialities involved in 'knowing' and indeed the difficulties that can be encountered in attempting to change patterns and forms of knowledge exchange.

The centrality of knowledge for society and for social life is examined by David Clarke. For Clarke, historical transformations in power/knowledge have been paralleled by transformations of identity (see also, Odih and Knights, Chapter 6). The contemporary information revolution signifies a fundamental change in the way that capitalist society reproduces itself. Modern identity, Clarke argues, was particularly translucent when analysed through work – the logic of production – and the stable identity that it provided. Postmodern identity is about keeping your options open. Furthermore, it is much more about a consumerist logic. Objects increasingly define individual identities and provide the means for individual transformation. Commodities are the instruments of social integration so that consumption has become an instrument for social control or compliance. There is a certain self interest in conforming to the rule of the consumerist society even though we delude ourselves that this society also incorporates freedom of choice. This new form of identity is driven by market solutions whereby the information revolution is merely the market-driven redefinition of knowledge under postmodernism. Thus, we should not overlook the contribution of this revolution to social inequality and patterns of inclusion and exclusion.

Frank Webster continues Clarke's theme, concentrating on identity in an era of 'virtual culture' and society. Webster's argument starts from the premise that everything around us is subject to change at an accelerating rate. This marks something of a departure from, say, the mid-nineteenth century when the vestiges of Empire made it seem as if countries like Great Britain were invulnerable and the social and economic order would remain much the same. The virtual society – and its crucial ingredient of knowledge and information – removes the possibilities of foundationalism. The virtual world is more malleable but actually very difficult to reconstruct on the basis of clearly defined goals or objectives (simply because it is so malleable). Knowledge is crucial to these developments in that it has enabled an assault on all 'traditions'. Knowledge has, Webster argues, increased both our capacity to comprehend the world as changeable and to change it. Yet, this has not increased our ability to achieve change with any degree of certainty. The virtual world offers infinite possibilities yet we have less confidence than ever in what we want to create. For Webster, however, this resulting threat to self-fulfilment has reintroduced the value of social relationships as a route to creating a sense of self, underlining the importance of what he terms 'the imperative of sociability'. In some ways this has aided and abetted the continuing rise of capitalism as the engine of world economic development. Its success relies on individual identity becoming subservient to the common 'good' requiring an acceptance of surveillance and conformity to certain societal and economic standards. Unfortunately not everyone, not every society, nor every nation can fulfil these and exclusion is just as important a part of capitalism as inclusion. Thus, for Webster, many individuals and nations are irrelevant to the triumph of global capitalism.

The path to a consumerist logic is also the subject of the chapter by Ian Cook, Phil Crang and Mark Thorpe. Their chapter on category management and circuits of knowledge in the food business reveals the sometimes tortured attempts of this production-led business to incorporate 'consumer knowledge' (see also, Odih and Knights). These authors highlight again the link between power and knowledge and the competing claims of different circuits of knowledge. Using notions of commodity culture they explore food retailing and consumption in and around a number of localities in North London. The chapter examines how knowledges of foods and the diversity of their geographies are used, produced and contested by manufacturers, retailers and consumers. They illustrate how the use of 'Category Management' (CM) in the food trade attempts to structure the food provision system according to the categorical understanding of consumers. Yet, Cook, Crang and Thorpe question whether or not the knowledge that the food industry harnesses for this purpose equates with that of consumers. They demonstrate, along with other contributors to Part III of the book, that knowledge has many guises and roles.

While manufacturers and retailers endeavour to acquire knowledge about consumers that will benefit sales and profits, they are also being encouraged to recognise the knowledgeable subject on the shop floor. Jane Wills examines the new forms of corporate governance and managerial practices founded on fostering the involvement and participation of employees at work. Quality circles, consultative forums and teamwork are just some of the practices that draw on the knowledge embodied in the corporate labour force; practices which allegedly enable modern businesses to confront competition, and to restructure or revise production processes more effectively. Drawing upon her research on information sharing within UK-owned companies involved in European Works Councils (EWCs), Wills highlights the chasm between rhetoric and reality in the management of knowledge at work. Both companies and employees, if for different reasons, are willing to support the development and implementation of structures for improving information exchange but it seems that the net effect has been to reduce rather than increase worker involvement. While a growing number of firms are using EWCs as part of their corporate communications activities, and these are generally welcomed by managers, unions and employees, there remain many doubts about their effectiveness. This may in part be the result of the workplace 'culture' in the UK that has long embraced a deep mistrust between managers and workers. Once again, knowledge and power are closely intertwined, and the implementation of EWCs reflects the political economy and social relations of which they are but a part. It remains the case that employees have few, if any, opportunities to put their knowledge to work in the workplace.

Part I
Knowledge, space, economy

2 Power/economic knowledge
Symbolic and spatial formations

John Allen

Introduction

There are, I think, a number of things that can be said about our current understanding of economic knowledge. The first, and perhaps most obvious, is that there is not one economic knowledge but many. Economic knowledges come in a variety of shapes and forms, ranging from the much prized activities of innovation and R&D which materialize as patented or intellectual objects to the more elusive know-how and ways of doing things that somehow add value to an enterprise. A second observation is the consensus that has arisen around the value of the tacit–explicit knowledge distinction first drawn by Michael Polanyi (1958) and the tendency, certainly among geographers, to map the distinction on to the local–global scale of activity.[1] And a third, rather different kind of observation, is the almost complete absence of any thought given to the fact that power and its relationships may actually shape the production and circulation of economic knowledge. Knowledge linked to power, whether in the guise of a discursive regime of 'truth' or the restrictive practices of an institution, involves domination and influence of one sort or another. The three observations are not unrelated and, when considered together, nor are they without consequence.

One consequence, indeed the one explored in this chapter, is that our understanding of economic knowledge is both too broad *and* too narrow. It is too *broad* in the sense that it is increasingly difficult to distinguish between activities which produce or manifest knowledge and those which do not. The preoccupation with tacit, unstated knowledges, that is, those activities which generate a type of informal know-how which is hard to specify, has arguably led to a proliferation of contextual forms of knowledge which range from conventions of style and interpretation to the deeply practical. The uncertainty generated over what is and what is not knowledge, and the anxiety over the possible dilution of the meaning of the term has not passed without effect, however. When pressed to define knowledge, the fallback position adopted by many, perhaps without thought, entails a *narrower* definition of knowledge based around the language of conception, judgement, innovation, and the pushing back of frontiers of whatever kind, be it in relation to product design, process innovation, or organizational change. What are prized above all are cognitive qualities rather than those relating to aesthetics or expressive characteristics, for instance.[2]

This combination of breadth and narrowness in the definition of economic knowledge, however, represents more than just an uneasy ambivalence about what constitutes this much-valued commodity at a time when a 'knowledge-based economy' is said by many to be on the horizon. Power is also part of the equation, and indeed it is part of what holds together the broad and narrow definitions of economic knowledge in one assessment and allows for the possibility of oscillation between the two poles. Following Michel Foucault (1972, 1980), it can be argued that what we take to be our stock of economic knowledges at any one point in time represents a particular construction which, in turn, serves to limit how we put such knowledges to work and where we expect to unearth them. A loose discursive regime of truth or, more colloquially, a ring of truth is attached to the idea that economic knowledge is broadly cognitive and that it is to be found in previously undisclosed contexts and practices. However, we need not fully endorse Foucault's strictures on the links between power and knowledge to see that what we think of as 'economic knowledge' will have some bearing on how economic agencies make sense of a high value economy. In this chapter, this contention is explored and developed in three ways.

First, in relation to what has been left out, or rather not understood, in various attempts to trace economic knowledge through cognitive practices that appear to owe more to a past 'material' age than to today's less tangible practices. Foucault's account of dispersed yet systematic discursive practices, which connect the thinkable and the sayable, are central in this respect. Following that, the work of Ernst Cassirer, a German philosopher writing on the diversity of symbolic knowledges in the first part of the twentieth century, is used to exemplify the different bases upon which knowledge may be said to rest in the advanced economies. Of particular interest here, is the expressive side to economic knowledge which has yet to register in any broad epistemological sense. The latter point is reinforced through an examination of recent work on symbolic economies and semiotic production. Finally, we turn to the spatial effects of what has rapidly gained the status of an economic 'truth'; namely, the local and contextual basis of economic knowledges. Arguing against the premature mapping of tacit and explicit knowledges along a local/global axis, a relational conception of knowledge based on distanciated relations is proposed. Around both the significance of symbolic knowledges and their spatial translation, the power/knowledge nexus remains central to what it has been possible to count as 'true'.

The 'obviousness' of economic knowledge

As is well known, knowledge as a dynamic component of the advanced economies owes much to the pioneering work of Daniel Bell and others writing in the 1960s and 1970s around the perceived shift from manufacturing towards a less tangible world of services and information. Making sense of any period of economic transition is never an easy affair, not least because more than one possible direction of change is always in evidence. For Bell, though, there was only ever one pathway in view: the path towards a post-industrial economy, and the singular feature driving it was knowledge, in this case codified, abstract knowledge. In *The Coming of Post-Industrial Society* (1973) he set out the basis of what the future had to hold; namely, a knowledge-driven economy based upon an 'abstract system of symbols' which shape the practices of innovation in science, technology and related fields. Widely criticized at the time, elements of this vision have nonetheless remained remarkably persistent over time.

It would, of course, be too much of a caricature to attribute the many different versions of a knowledge-based economy in circulation to the ideas of Bell alone. The widespread impact, within and beyond academia, of such writers as Peter Drucker (1969, 1999) and Alvin Toffler (1980, 1990), both then and now, on the popular understanding of the significance of knowledge and innovation, together with all kinds of electronic media, should not be overlooked. The racy language of globalization, the increased pace of economic activity attributed to the revolutions in information and communications technologies, and the disembedded character of many economic institutions, are all in line with a notion of knowledge that is formal, abstract and codifiable, and thus easily transferable. Indeed, such an account is not that far removed from the more recent assessment of global networks of knowledge and information provided by Manuel Castells in *The Rise of the Networked Society* (1996).

It is not appropriate here to outline Castells' argument for a networked age of knowledge and information, but it is significant to note that, in the above text, Castells draws his definition of knowledge and information directly from Bell's 1970s post-industrial tract.

> Knowledge: a set of organized statements of facts and ideas, presenting a reasoned judgement on an experimental result, which is transmitted to others through some communication medium in some systematic form.
>
> (Castells 1996: 17; Bell 1976: 175)

Castells, in quoting Bell, remarks that he had no compelling reason to improve on this definition. In so far as it is always difficult to erase elements of ambiguity from any definition, this account of knowledge is quite unexceptional. It is nevertheless steadfastly cognitive in outline: formal, systematic and rooted in abstract reason. Above all, it is a definition that it is easy to subscribe to, given its 'obvious' connotations. And, in fact, this is precisely the point.

Knowledge, power and cognition

A cognitive sense of knowledge is obvious and, in many ways, we are only able to make sense of economic knowledge in that way because it is difficult to think outside of the lines of reason, judgement, conception and innovation. Power and economic knowledge exist in a relation of immanence that lays down the lines by which we are able to articulate and make sense of something. As a form of intuitive 'truth', to draw upon Foucault's (1981) paper, 'The order of discourse', the framework of thinking that defines economic knowledge is limited by what it is possible to say about it without running the risk of appearing odd or beyond comprehension. There is an everyday sense to the way in which the 'obviousness' of the discourse works its way through economic institutions and beyond, fixing the norms of discussion and debate, and engaging new lines of practice which, nonetheless, still tie economic knowledge to a broad sense of cognition. So, for example, alongside the abstract coupling of knowledge and innovation that takes place within R&D centres, we have the practical know-how and reason learnt from experience at the workplace; or, together with an abundance of formal knowledges communicated through business and economic journals, we have the production 'experiments' conducted by teams on the shop floor;

or, parallel to the codified knowledges embedded in the global networks identified by Manuel Castells, we have Michael Storper and Robert Salais' (1997) intimate 'worlds' of tacit knowledge production based upon convention and customary use.

What this amounts to in Foucault's terms is that the field of economic knowledge has been delimited largely to that of cognition; not, we should add, because of the legitimating power of some monolithic discourse, but rather because of the variety of discursive practices which 'cross each other, are sometimes juxtaposed with one another, but can just as well exclude or be unaware of each other' (1981: 67). Thus, a practical discourse on the shop floor may draw upon elements from other discourses, for instance in relation to R&D, and bind them into its own system of meaning, yet remain institutionally detached. What holds such dispersed practices together are the regular relationships between them which provide a constant and unproblematic way of talking about economic knowledge. Such regularities are the very stuff of discursive power. However, it is not that we are all trapped within a unifying discourse of cognitive knowledge, unable to think outside of its parameters; rather it is just that it is easier to think *with* such a discourse than *about* it (Foucault 1972).

For Bell, too, it was simply a straightforward proposition that the abstract system of symbols which underpinned a knowledge-driven economy should be cognitive. Given the obvious significance of knowledge to the economy, as was deemed evident at the time by the proliferation of new developments and innovations in science and technology, it was difficult for Bell to think otherwise. Having said that, his thinking, or rather the cognitive discourse of knowledge of which he was a part, was not without consequence or effect for the way in which the emerging economy of services, information and knowledge has been understood. In particular, Bell's insistence in *The Coming of Post Industrial Society* (1974), and reaffirmed in *The Cultural Contradictions of Capitalism* (1976) that the economic order represented a realm separate from that of culture or politics, each with its own set of dynamics and rhythms of change, served to produce a view of economic knowledge that was distinct from culture, meaning and the imagination. If knowledge was indeed composed of symbols, they were formal and abstract in design, not meaningful and expressive.

It is not hard to follow where this line of thinking can lead, especially in relation to those economic activities which are likely to be considered as knowledge-intensive and which are not. Engineers and scientists arguably would be among the most valued occupations, which today would include software engineers as well as civil engineers, financial 'rocket' scientists as well as laboratory-based scientists. Equally, the high value sectors of a knowledge-based economy would not be difficult to identify, from the technologically innovative sectors such as telecommunications, engineering and computing, to many of the finance and information-based services such as banking, insurance, accountancy, law and consultancy. This is not to suggest that a sharp line can be drawn across occupations and industries on the basis of their 'knowledge' content. Rather it is to suggest that once a cognitive-based account of economic knowledge is in play, it has the power to make itself 'true' by association with those activities which legitimize it. Outside of the scope of knowledge, on this reckoning, are those activities which appear to possess few, if any, of the qualities of reason, analysis and judgement.

Let me spell out the nature of my argument a little further. I am not suggesting that a cognitive-based account of economic knowledge is the only account possible.

On the contrary, I am suggesting that a broad sense of economic knowledge, one which encompasses the expressive as well as the analytical, is hard to sustain given the propensity to formalize and codify innovation and creativity. Thus images, sounds, emotions, moods and sensualities appear difficult to gauge in economic terms and, when evaluated commercially, tend to be measured by abstract criteria drawn from the outside. Non-cognitive activities are recognized less for what they are and more for how they may be reproduced or replicated instrumentally.[3] It is in this sense that cognitive knowledge is linked to power, and through its 'obviousness' certain discursive practices restrict what it is possible to say and think about economic knowledge. Like Bell, then, no one is denying that economic activities encompass more than the cognitive. It is merely that those activities which do not fit easily into a schema of abstract symbolism are not immediately considered as part of the driving force of a 'knowledge-based economy'. Quite simply, they are not part of the 'cutting edge'.

Broadening the symbolic

Before we look at the work of Cassirer, it is worth drawing attention to the fact that Bell, in separating the economic from the cultural in *The Cultural Contradictions of Capital*, drew explicitly upon the writings of Cassirer to justify the divide. Abstract and expressive symbolism were held apart by Bell, and only the former was thought to produce *economic* knowledge.[4] Yet for Cassirer, as will be shown, the main point of his writings was to establish a *generalized* epistemological basis for *all* forms of knowledge, not just for those forms which rest upon formal, abstract reason.

In hindsight, it is perhaps ironic that Bell, of all people, should have reduced Cassirer's philosophy of symbolic forms to that of culture and expressive symbolism. In an economic moment when the less tangible, creative side to the production and circulation of goods has been recognized as a potential source of added value, if not the main source of value, or at a time when much service production relies heavily upon the input of signs and symbols to differentiate products and make them meaningful, an understanding of the diversity of symbolic knowledges outlined by Cassirer arguably has much to offer.

Writing in the first half of the twentieth century, Cassirer became convinced early on that a cognitive, science-based view of knowledge was unjustifiably limited. As someone interested in the cultural sciences of the time as much as the natural sciences, Cassirer's work reflected to some extent the heady intellectual and cultural atmosphere of Berlin at the beginning of the century filtered through his more formal grounding in matters of logic, metaphysics and the 'exact' sciences (Verene 1979). In particular, his interrogation of the mathematically minded, 'exact' sciences like physics led him to question the privileged position that such thinking occupied at that time as the benchmark of knowledge (Krois 1987). Above all, it led him to consider the various ways – conceptual, perceptual, affective – in which it is possible to 'know' the world and how it is rendered *meaningful*.[5]

The nub of the issue for Cassirer was that the formal reasoning of natural scientific knowledge, and the abstract judgements of mathematics in particular, whilst valid as a form of theoretical meaning, did not amount to a prototype for all knowledge. It was only *one* form of meaning amongst others; *one* conceptual system of meaning embedded within a shared cultural framework of signs and symbols. Other ways of symbolically apprehending the world which do not coincide with abstract conceptual signs, such as the expressive qualities of art or the referential qualities of language, also invoke meaning, although within quite different frameworks of knowledge and knowing. On this view, it is simply misleading to reserve the accolade, 'knowledge', for one kind of symbolic formation when others of a different nature open up a quite different type of understanding and access to the world.[6]

We should perhaps be clear about what is meant by 'access' here, as the term gives us a clue to the nature of Cassirer's symbolic knowledges. Symbolic knowledges, for Cassirer, do not simply reflect a world 'out there'; rather they provide a means of apprehending and comprehending it (Krois 1987). The validity of a symbolic formation does not rest upon its ability to provide a copy of the material world. On the contrary, it derives its validity from within; that is, from within an organized system of relations between signs which produce meanings that are fixed by convention. The numerical symbols of mathematics are one such system, the aesthetic symbols of light and shade and harmony are another, and so on. The relationship between symbolic knowledges and the material world is a mediated one, therefore, in which meaning is dependent not upon particular numbers, musical sounds, specific images or the marks we place on a piece of paper, but upon their *symbolic function*: what they express, represent, or signify.

Much of this argument is conducted in Cassirer's three-volume, *The Philosophy of Symbolic Forms*, although it is the final volume, published in 1929 and subtitled 'The Phenomenology of Knowledge', which develops the broad standpoint that sensation alongside imagination and understanding may all be placed on an equivalent epistemological footing. With a nod in the direction of Hegel's first major philosophical work, the reference to a 'phenomenology' of knowledge was intended to convey the point that 'knowledge must convey the totality of cultural forms' (1957: xiv), not just those of abstract cognition. Whilst, today, the rigorous systematic thinking of Cassirer and his allusion to totalities would draw breath in some quarters as well as looks of disdain, the reference to *cultural* modes of knowing and experience in this volume would not be out-of-place in most social science circles.

Expressive, representational and abstract symbolism

The third volume of Cassirer's cultural philosophy is given over to the exposition of three symbolic functions: those of expression, representation and signification.[7] Broadly speaking, the first of these functions, symbolic expression, has close affinities to contemporary concerns with embodied or experiential forms of knowledge which stress their non-representational nature (see Shusterman 1997; Thrift 1999). Here the stress is upon an immediate, non-discursive mode of experience which, according to Shusterman, has bodily feeling as its locus. The intense, often vivid, nature of immediate experience recalled by Shusterman and its significance for somatic aesthetics runs in parallel to Cassirer's observation that expressive meaning is related directly to sense perception and bodily awareness. As John Krois summarizes, 'insofar as perceived phenomena appear to us as agitating, soothing, gloomy, joyful, pacifying or otherwise exhibiting a mood, they exemplify what Cassirer calls expressive symbolism' (1957: 86).

Expressive symbolism is perhaps best understood as a structure of feeling where, for example, a stylish piece of fashion or the lyrics of a new musical composition move us in some way that is unrelated to, say, the latest 'language' of fashion or the technical competence by which the music is reproduced. For Cassirer, what stimulates us in relation to design or lyrical composition is not only their respective popular or linguistic appeal; it is, in the case of the former, the harmony of the design, its colour and form, or in relation to music, the unison of sounds, the specific rhythm, tone and pitch which brings a certain satisfaction to the ear.[8] In common with all forms of aesthetic knowledge, an appreciation of film, art, design, music, display and others rests upon their sensuous form – the feelings they express – not simply upon their technical or analytical excellence. In short, there is a creative content to such affects that cannot readily be measured by any cognitive yardstick.

The novel point here, as developed by Cassirer, is that whilst such aesthetic appreciation may be viewed as the result of a series of unformed feelings, the process of understanding involved is far from passive. For Cassirer, meaning cannot be reduced to some kind of psychological or physiological reaction (Krois 1987). On the contrary, the very production of expressive meaning is itself dependent upon the symbolic codes which make the experience an objective cultural moment. As Cassirer argues,

> light, colour, mass, weight are not experienced in the same way in a work of art as in our common experience. For the cultural producer, the words, the colours, the lines, the spatial forms and designs, the musical sounds are not only technical means of reproduction; they are the very conditions, they are the essential moments of the productive artistic process itself.
>
> (1979: 160–1)

In other words, when formulating aesthetic judgements there are specific operations of knowledge involved which take place on the basis of a constitutive symbolic domain. It is this domain, where words, sounds, and images function as an expressive system of signs, which makes possible cultural understanding, regardless of whether such features can be fully articulated by those involved in the appreciation of a particular art form like film or music.

Or rather, it makes possible an understanding that is expressive, not representational. The second of Cassirer's symbolic functions, representation, is perhaps the most familiar means by which shared cultural meanings become established. Once a word or image stands in for or depicts something else, the expressive side diminishes and the referential dimension takes centre stage. Language is the most obvious system of representation, although in the broad semiological approach of Roland Barthes any object or activity can function 'like a language' in the production of cultural meaning. In his classic text, *Mythologies* (1973), Barthes opened up a rich symbolic seam which, for example, made it possible to talk of a 'language' of fashion where clothes function as identity codes for particular social and cultural groupings. Provided that dress codes are understood and that the difference between items of fashion are marked symbolically, a wider realm of signs is communicated that is open to manipulation through advertising, styling, branding and marketing.

This takes us closer to what it is that enables Cassirer to speak about the 'mythical' or imaginative properties of representational systems such as language. At the core of this understanding is the now widely accepted view that language is fundamentally ambiguous in its relation to the material world. In fact, it is this very ambiguity which reveals the extent to which representation, as a form of knowledge, may stand for little that is actually deemed the 'real' world. Or as Cassirer succinctly expressed it, language '*begins* only where our immediate relation to sensory impression and sensory affectivity *ceases*' (1957: 189).

In this sense, anything that functions like a language, as a system or representation, may involve an imaginative play of signs which nonetheless provide 'access' to a particular 'world', be it cultural, political or economic. The contemporary neo-liberal system of thought, for example, and its various coded signifiers – globalization, transnationals, free trade, entrepreneurship, flexibility and the like – is in many ways an imaginative construction (or a virtual vision, in the words of James Carrier and Daniel Miller (1998)) which functions both to make sense of the world's economies and as the basis upon which judgements are brought to bear. As a representational system of meaning, neo-liberalism, like many other knowledge constructions, can be understood as a skilful manipulation of symbols.

The third of Cassirer's symbolic functions, signification, amounts to what was for him the most developed knowledge accomplishment: the systematic manipulation of abstract symbols. This takes us back firmly to the ground of formal reasoning, cognition and abstract judgement. If symbolic expression is at one end of the knowledge spectrum then symbolic abstraction is to be found at the other. This, for Cassirer, was principally the world of mathematics, geometry and physics, where reason has progressed far beyond representation with the introduction of numerical concepts and notations that corresponded to earlier, non-numerical forms. The ability to simplify reality in this way, to analyse and extend reasoning, so that for example the world may be grasped by fewer axioms and principles, held out the prospect of being able to think about new possibilities which have yet to be encountered. This, unmistakably, is the landscape of innovation, where cognitive frontiers are pushed back and knowledge extended through experimentation and 'pure' thought.

Whilst Cassirer's account here dramatically over-emphasizes the significance of numerical symbols within a small range of sciences, for example, by neglecting the role of other notational devices in conceptual advancement (most notably models, diagrams, graphs and maps of whatever kind), the stress overall remains one of reason's advancement through abstract judgement. Creative scientific work, in short, involves the manipulation of abstract symbols.

But, of course, for Cassirer, as we have seen, there is more than one kind of creative work and more to knowledge than simply abstract symbolism. What is useful about Cassirer's account of symbolic knowledges in relation to our predominant sense of economic knowledge is that it enables us to make *judgements* on the economy, and economic value more generally, which reflect expressive and imaginative aspects as well as those of analytical reason. That there is more to judgement than reason, however, has yet to be fully grasped by those already asking searching questions about the relationship between knowledge, economic signs and symbols.

Symbolic economies?

The term, symbolic economies, is both an accurate description of much of what goes on in the advanced economies and a misleading indicator. The symbolic content of economies in terms of both products and processes has only recently been understood as something integral to the way in which service-dominated economies are organized. Yet the manner in which that recognition and understanding has come about has seemed either to restrict symbolic activity to one part of the economy or, in what mainly concerns us here, to see its importance in terms of knowledge only in the context of analytical reason and cognitive know-how. With regard to the former, it is simply misleading to equate the symbolic with the cultural and then proceed to speak about a range of 'cultural industries' – from broadcasting and multi media to the fine arts and the music industries – as if they exercised a monopoly over the creative deployment and extension of symbolic forms.[9] No industry or sector of the economy has a monopoly on culture or symbolism, and if as much was claimed it would come as something of a surprise to the likes of design engineers and 'creative' accountants.

This point, at least, is well understood by writers such as Scott Lash and John Urry, and Robert Reich, who have helped to establish the importance of symbolic practices within the realm of the economic. Having said that, these writers have found it difficult to think outside the discourse of cognitive knowledge or to be able to construct arguments that display an awareness of the very different symbolic schemas upon which judgements may be based. The immanent nature of the power/economic knowledge relation is also evident in the limits of what it is that they have to say and voice.

In *The Work of Nations* (1993), for instance, Reich asserts that the manipulation of symbols, the analysis of data, words, oral and visual representations, forms a critical part of what most professions do to add value to goods and services. The manipulation of symbols is the process by which new knowledge comes into play, leading, for example, to the introduction of new software technologies, inventive legal arguments, innovative financial instruments, new advertising techniques or a breakthrough in architectural design. Whilst the output of such activities is by no means always radical in departure, the work with numbers, sounds, words and images presupposes an appreciation of the various symbolic codes which make meaning and innovation possible.

Reich refers to those who perform this kind of work as symbolic analysts.

> Symbolic analysts solve, identify, and broker problems by manipulating symbols. They simplify reality into abstract images that can be rearranged, juggled, experimented with, communicated to other specialists, and then, eventually, transformed back into reality. The manipulations are done with analytic tools, sharpened by experience. The tools may be mathematical algorithms, legal arguments, financial gimmicks, scientific principles, psychological insights about how to persuade or to amuse, systems of induction or deduction, or any other set of techniques for doing conceptual puzzles.
>
> (1993: 178)

As the foregoing indicates, the creative play of symbolic work is not limited to a particular sector of the economy; symbolic analysis is a practice which cuts across industries involving engineers as much as financiers, production designers as well as marketing strategists. Interestingly, though, the *process* of manipulation seems to be remarkably similar regardless of the symbolic form in question. Abstraction and analysis, reason and conceptualization, are the mainstay of the cerebral activities involved, with little or no attention paid to what it might mean to work with symbols that are not overtly cognitive.

In this respect, Reich's understanding of symbolic function is closer to that performed by signification as outlined by Cassirer. The stress placed upon the systematic manipulation of abstract symbols involving the exercise of judgement on the basis of reason rather than representation or expression amounts to a form of knowledge that would not be out of place in an R&D technical centre. Although Reich would probably not accept this assessment of his treatment of the symbolic, it is nonetheless the case that his account of symbolic form is the same whether it be concerned with something sensuous, metaphorical or cognitive in style.

The principle here, it should be stressed, is *not* that fashion and film occupy the symbolic realm of the expressive, whereas engineering and finance are confined to a different kind of symbolic logic. If this were the case it would be a relatively simple matter of dividing up industry and output between the different forms of symbolic knowledge. That, however, would effectively deny the various forms of symbolic production involved in the design, styling, fabrication, promotion and sale of any particular good or service. In the case of film, for instance, an appreciation and judgement of its aesthetic qualities would be an obvious consideration, in respect of both its visual impact and its narrative construction, but any overall judgement would also entail a technical assessment of its sound, lighting and editing quality, as well as the extent to which a film was marketed and promoted imaginatively. Even something as hard-nosed as the business of credit, and the use of money more generally, has an expressive meaning, the knowledge of which cannot be deciphered through a series of numerical manipulations.[10]

The loss of aesthetic judgement

Reich is not alone, however, in restricting the process of judgement to the realm of cognitive reason. Such a background assumption is also at work in Lash and Urry's account of the proliferation of sign-values in the production and circulation of commodities. In *Economies of Signs and Space* (1994), they stress the importance today of the sign or symbolic content of commodities over their material content. In part, this line of thinking takes its cue from Jean Baudrillard's (1983, 1993) overstretched assumption that the fundamental ambiguity of representational systems has been realized to the extent that, in a symbolically saturated environment, we no longer expect signs to represent anything in particular. The arbitrary nature of the sign has itself become the site of playful representation where economic meaning is now produced in such a knowingly contrived manner that its very inauthenticity is anticipated and understood. People are reflexive. They know a trick or two when they encounter such, but equally they acknowledge the 'reality' of such symbolic play.

It is, of course, unnecessary to wholly embrace this position to accept that economic meaning may owe less to the tangible nature of production and rather more to its semiotic character than has hitherto been the case. According to Lash and Urry, this production of signs has predominantly taken two forms: a cognitive form, which is exemplified through the flow of information, digital codes and other abstract symbols, and an aesthetic form, which in the broadest of genres engages the expressive side to economic life. The latter, in common with Cassirer's account of expressive symbolism, is directed at the world of affects, although widely interpreted to include much of what Cassirer subsumes under the play of representation – the mix of images in advertising, the sign-value of material objects, the semiotic work of branding, and so forth.[11] Overall, this symbolic activity adds up to an aestheticization of the economic, which takes place within the sphere of production as well as in the circuits of exchange and consumption.

As with Reich, then, Lash and Urry do not restrict symbolic work to a particular sector of the economy. They recognize, for example, that the design process is as much an integral part of manufacturing as it is of fashion or any number of consumer services. In contrast to Reich, however, they make a strong case for considering cognitive signs separately from aesthetic signs, in so far as the process of manipulation involves two distinct forms of reflexivity and knowledge. Knowledge on the basis of cognitive reflexivity operates on the understanding that the analytic principles – be they concerned with legal principles, financial calculations or forms of insurance risk assessment – are themselves open to question and subject to re-negotiation. Knowledge via aesthetic reflexivity meanwhile operates on a hermeneutic basis whereby subjects – say in the sphere of consumption, retail and fashion – are actively involved in the construction of their own identities through their engagement with lifestyle and consumer choices. The symbolic interplay that constructs consumer codes is not something that is handed down through a marketing tradition but is itself open to manipulation by active consumers.

Again, we do not have to agree with these examples of reflexivity to admit that the cognitive and the aesthetic, the realm of analysis and abstraction and that of imagination and affect, are two different ways of apprehending and knowing the world. Even though the aesthetic is defined by Lash and Urry in an excessively broad fashion to cover much more than the expressive side to economic meaning, the difference in symbolic function performed by the cognitive and the aesthetic modes of access are sufficient to warrant separate epistemological treatment. If they are considered as separate domains of symbolic activity, each with its own specific mode of operation, that does not mean that Lash and Urry consider them to be on an equivalent epistemological footing.

If knowledge amounts to more than streams of data, flows of images or reams of information, then the faculty of judgement, exercised on the basis of something other than social convention or subjective taste, is central. Know-how of whatever kind involves a process of judging which necessarily appeals to a kind of universality; otherwise knowledge shades into culture more generally and its specificity is lost. If that were the case, then all forms of knowledge become relative and the term, knowledge, as mentioned in the introduction, is rendered elastic and devoid of meaning.

Lash and Urry are only too aware of the significance that judgement holds from the standpoint of what is and what is not considered knowledge. They stop short, however, of adopting the view that it is possible to make judgements which entail aesthetic forms of reflexivity. They accept that cognitive reflexivity presupposes judgement but restrict aesthetic reflexivity to the realm of 'pre-judgement', that is to the social conventions of 'taste'.

Now if (aesthetic) judgement is strictly understood as the application of a universal to a particular, this would not be judgement at all. It would instead be the sort of '*pre*-judgement' of Gadamer's theory of hermeneutics . . . What is at issue is less aesthetic judgement than taste in more prosaic everyday life. And if aesthetic judgements are in important respects already pre-judgements, then so much the more so would be everyday 'judgements' of taste . . . Such (pre-) judgements are already the background practices, the shared assumptions of taste communities. They are not judgements at all, but *Sitten*, habit. They are not 'rules' or structures about which we can be objective, because we already dwell among them.

(Lash and Urry 1994: 316–17)

In divesting themselves of the need for judgement in the aesthetic realm, therefore, Lash and Urry are able only to embrace a benchmark of aesthetic knowledge that takes its standards from intellectual habit and social convention. The predispositions of a taste community, those who know the codes and conventions and use them to act together, thus become the arbiters of the aesthetic realm. This is a well-trodden path, one that reflects the difficulties that Immanuel Kant above all found himself in when attempting to subsume a particular, subjective judgement based on a feeling under something more universal.[12] Faced with the (im)possibility of justifying a universally valid aesthetic judgement, the way forward is often to stress its relative, conventional character. Yet this serves merely to displace the problem. For if aesthetic knowledge falls short of the standards set by absolute judgement, it is nonetheless more than merely social convention.

For Cassirer, as we have seen, the three symbolic domains of knowledge each operate under definite epistemological conditions, none of which can simply be reduced to habit and convention. If that were the case, the foregoing would amount to little more than detailed descriptions about how one type of knowledge differs from another. As defined symbolic formations, however, their judgmental schemas provide the context within which 'new' knowledge, or more accurately, symbolic innovation may be considered and judged. Rather than viewing aesthetic innovation, for example, as the result of established cultural group recognition, developments, say, in visual design or computer generated imagery, would be subject to judgement within specific symbolic schemas.[13]

This position, although somewhat questionable given the contemporary stress placed upon the relative nature of judgement, does serve to broaden the basis upon which economic knowledge and the manipulation of symbols may be assessed.[14] Its relevance, perhaps, lies less with the fact that it is possible to think beyond the dominant discourse of cognitive knowledge and more with the realization that imagination, expression and the practice of judgement occupy the same epistemological realm. They are all part of the symbolic content of economies.

Stretching the symbolic

The dominant discourse of cognitive knowledge also entails certain spatial consequences, one of which presents a particular problem. The problem, put simply, is the relative ease by which a whole array of terms fall into one another without, it would appear, differentiation. Thus, tacit, culture-bound, embedded forms of knowledge merge seamlessly with spatial notions of proximity, face-to-face interaction and being-there, to give the distinct impression that knowledge, as a competitive asset, is a predominantly localized affair. The corollary of this, as Michael Storper (1997a) has alluded, is that the kind of knowledge that does not generate understanding or involve judgements, namely, information, occupies a different spatial realm, that of the codified global networks outlined by Castells (1997: 238–41). In short, tacit, innovative knowledges do not travel well, whereas formal, codified knowledges do, yet their very universality renders their status as economic assets problematic. This understanding has rapidly gained the status of 'the truth' and, indeed, this account of economic knowledge has the power to render itself true.

It is this often taken-for-granted tendency to map the tacit-codified knowledge distinction on to the local-global scale of economic activity which is argued against here, on two grounds. First, such a mapping serves to limit our understanding of knowledge as a process of translation which involves stretched knowledges, that is *both* proximate *and* distanciated forms of social interaction. And, second, the act of conflation lends itself to a conception of knowledge that is largely cognitive in style. Once a broader sense of economic knowledge is entertained, to include expressive as well as cognitive dimensions, the value of the distinctions drawn are less obvious.

One Polanyi or two?

Perhaps the first issue that needs to be addressed is the association frequently drawn between local, embedded knowledges, evident in the literature on learning regions and neo-Marshallian districts, and the unarticulated, tacit knowledges which are prized for their innovative know-how.[15] The association itself is not in dispute, for there is ample evidence to substantiate the view that ideas which are known yet cannot be told, ways of doing things which can be shown yet not explicitly stored, favour close relations of proximity in which context is all important. In the face of economic uncertainty and ambiguity, being-there has considerable merit. However, the notion that such knowledges are *solely* the creation of territorially specific actions and assets – restricted to organizations, regions, places or other such spatial confines – is highly questionable and reflects the delimiting vision of a powerful set of discourses. Tacit knowledges are indeed poorly articulated, but to assume that they do not mutate through dispersed relationships or that they are not dependent on others who are not present is to confuse the transmission of knowledge with its translation.

The confusion, in large measure, stems from an inability to keep apart the ideas of the two Polanyi brothers, Karl and Michael. To Karl Polanyi (1944) we owe a significant debt for the stress that he placed on the 'embeddedness' of economic actions in social, political and cultural institutions and the subsequent adoption of the term by Mark Granovetter (1985) and Granovetter and Swedburg (1992) in his economic sociology. The latter's use of the term to highlight the significance of purposive action within dynamic networks of social relations has been notably influential in the economic geography literature, especially in relation to accounts of regional trust-based networks, economic performance and adaptability.[16] To his brother, Michael Polanyi (1958, 1966, 1967), on the other hand, we owe a debt for his explication of the tacit–explicit knowledge distinction. Much of his work on this distinction revolves around an attempt to develop a theory of non-explicit thought, a kind of informal logic of discovery in which the process of subception, learning without awareness, performs a central role. It is relatively easy to follow how this work lends itself to studies of economic innovation and creativity. What is harder to understand is why the two insights – the embedded nature of economic action and tacit learning – should be folded into the other at the expense of all other possibilities.

Certainly Michael Polyani made no explicit connection between the two, although it is fair to point out that one of the frequent examples he used to demonstrate the phenomenon of tacit knowledge did stress the interiorization of understanding through the practice of learning-by-doing. But it is one thing to refer to the manner in which an observer may skilfully adopt the performance and notions of an artisan 'by seeking to dwell in them from outside' (1967: 30) and quite another to suggest that face-to-face presence and proximity are paramount. Indeed, among the many other examples of tacit understanding that make up Polyani's writings, cognitive, visual, as well as practical, he was at pains to point out the importance of seeking out 'clues' – wherever they may be – which enable us to make sense of something or some action. Such 'clues', as in any mode of inquiry, involve a form of encoding by which we are able to grasp, say, the latest innovation in engineering design or an intricate piece of legal reasoning or perhaps a novel form of service R&D.

What matters in such situations is not the fact of local embeddedness, but the existence of relationships in which people are able to internalize shared understandings or are able to translate particular performances on the basis of their own tacit and codified understandings. Some such understandings and their symbolic system of meaning are likely to be bound up with particular artefacts and ways of doing things that are firm or place specific. But many other relationships may involve learning-by-detection, where unformed ideas are picked up through distanciated contacts and translated in new and novel ways. Such 'thick' relationships may span organizational and industry boundaries, as can the puzzles and performances which constitute them. The translation of ideas and practices, as opposed to their transmission, are likely to involve people moving to and through 'local' contexts, to which they bring their own blend of tacit and codified knowledges, ways of doing things and ways of judging things. There is no one spatial template through which associational understanding or active comprehension takes place. Rather, knowledge translation involves mobile, distanciated forms of information as much as it does proximate relationships.[17]

To think otherwise is to experience a kind of 'lock in' situation that is not altogether different from that in the path-dependent literature. Once tacit knowledge is conceived as a characteristic of the 'local', be it a firm or economic community, it is easily locked into a static geography which, in a rather 'obvious' manner distinguishes between embedded, localized, know-how of the hard-to-communicate kind and ubiquitous, global, codified knowledge, which is accessible to all and understood by most. This portrayal may be something of a caricature, yet it is not difficult to find variations on this theme, where ubiquitous knowledge parallels a process of globalization which is said to be slowly but surely encoding localized, tacit knowledges and capabilities. Peter Maskell and Anders Malmberg (1999), for example, argue that

The size and composition of the tacit knowledge-base of a region or country do perhaps constitute fundamental ingredients in its ability to perceive and absorb any valuable innovation generated outside its borders . . . Any codification of a piece of knowledge will eventually lead to its diffusion, thereby undermining the present possessor's possibility of using it as an ingredient in sustaining competitiveness. When formally tacit knowledge is converted into a fully codified form, a process is initiated which will sooner or later – usually sooner – turn it into a ubiquity by making it accessible on the global market.

(Maskell and Malmberg 1999: 16)

The implication of this view, replete as it is with the language of knowledge absorption and knowledge diffusion, is that as the latest innovations and creative techniques become codified and available world-wide, so the competitive edge of the advanced regions is undermined leaving them potentially exposed to cost-cutting practices abroad. In knowledge-based economies, the choice, it would appear, is either to innovate faster or to shield their tacit understanding from the global pressures to codify (1999: 14–16).

At the extreme, it seems that something happens to 'knowledge' once it passes beyond the spatial confines of the locality. Divested of their embedded character, the once uncoded understandings are subject to a process of encoding which enables others to share the recently standardized knowledge transmitted. The implication is that previously territorially specific assets and practices have suddenly become tradable. Of course, this does happen, but to generalize the process in this manner and to render the geography static is to put to work a particular conception of economic knowledge. It is to adopt a transmission model of knowledge rather than one of spatial translation. This has the effect of creating a series of bounded knowledge spaces, when in fact there are primarily knowledge networks which in Bruno Latour's terms 'are more or less long and more or less connected' (1993: 122).

Codifying the expressive

A spatially confined view of tacit knowledges also lends itself to the wrongful impression that codification is merely a process which serves to formalize and distribute somewhat hazy knowledges. The manner in which symbolic communication takes place, however, cannot be reduced to simplistic formulae whereby the encoding and decoding of symbols occurs within a rigid framework which is then passed on, imitated and shared. Encoding and decoding, as a mode of symbolic exchange, as Stuart Hall (1980) has shown, is a much more ambiguous process, open to negotiation and a variety of meaningful 'effects'. It cannot be understood in any straightforward behaviourist manner, although if we are to follow Cassirer there are broad cultural schemas which make meaningful symbolic interpretation and understanding possible.

Much of what has been discussed in terms of the codification of economic knowledge tends to revolve around the need to abstract, simplify or tie down the knowledge in people's heads or the way that they do things. Rather than dwell upon the affective or expressive side to economic action, or focus upon the perceptual and visual dimension to economic activity, the stress, as indicated earlier, has been upon abstracting ideas and practices so that they may be communicated in a replicable fashion. The attempt to name that which has not been named, to encode that which has not been spoken or written, has driven the process of codification largely along cognitive lines. It is perhaps not altogether surprising, therefore, that the spatial counterpart to this process has been an attempt to make explicit precisely those 'embedded' forms of tacit, 'localized' knowledge which appear to lend themselves to abstraction and formalization.[18] Encoding visual, aural or expressive cues does not quite follow that pattern, however.

In fact, much of what takes place in the exchange of expressive or representational forms of meaning is tacit in style. Entering 'into' a work of design, art, music, film, or fashion, for instance, is to adopt Polyani's interiorization mode for much of the time, regardless of any spatial concerns. The ability to appreciate certain musical compositions without being able to describe their features, for example, or the ability to recognize and react to works of art and design without being able to articulate their reasoning, or the ability to judge the mood of a visual composition without being entirely aware of the cues involved, are all instances of the use of symbolism which can be translated into action, yet without the aid of any *explicit* codification.

It is not simply the case that aesthetic or expressive symbolic schemes are poorly shared in comparison to cognitive schemas, but rather the fact that symbolically they are more replete and ambiguous. This is a contentious point, but one adopted by Nelson Goodman in his *Languages of Art* (1969) where, on the basis of a distinction between linguistic and non-linguistic symbols, he argues that knowing and understanding in the arts are symbolically of a different, more ambiguous order. Perhaps, another way to express this difference without wholly endorsing the fuller nature of aesthetic symbols is to stress their broadly negotiable character when it comes to decoding them. There is, as it were, a greater opportunity for symbolic interplay with a variety of codes available to draw upon. This takes us closer to Lash and Urry's position that the 'judgements' of taste employed in the realm of aesthetics are, in contrast to cognitive judgements, more open to changes in habit and convention. In short, the practice of knowing and judging is more arbitrary.

Having said that, it is important not to lose sight of the fact that despite a more tacit, ambiguous regime, the world of affects has not escaped the pressures for a more rigid framework of codification. Where this has happened, however, it has largely involved a shift from aesthetic to cognitive coding, with the more expressive elements subjected to a more rational coding; as for instance in the technical standards set for lighting and sound in relation to film or the formal coding of line and perspective in relation to design. In many ways, such a shift is a testament to the evident difficulty in formally codifying expressive knowledges.

It is also testament to the fact that it is only through the language of cognition and abstraction that it is really possible to make sense of the local/tacit versus global/codified distinction. And even then, as we have seen, this taken-for-granted mapping which reflects a particular combination of knowledge and power is largely a flawed, if not spurious, exercise.

Conclusion

So far as economic knowledge is concerned, the argument outlined in the chapter is straightforward. It is that any consideration of the nature of economic knowledge must go beyond the cognitive qualities of reason, abstraction and technical innovation to embrace the expressive side to economic activity. Expressive, that is, in the broadest sense of the term, so that sounds, images, words, emotions, feelings, and other kinds of performed meaning are considered alongside the more formal systematic qualities of analytical know-how and reason. When stated in this rather forthright fashion, one could be forgiven for wondering what all the fuss was about. After all, is it not self-evident? If that were the case, then we would find it less easy to slip into the language of analysis and abstraction, the vocabulary of technical and practical mastery, or the parlance of science and innovation, when the topic of economic knowledge was under discussion.

The sense in which it is easier to talk *within* a cognitive discourse of knowledge rather than *about* it, is the key to the relations of power involved. Its very 'obviousness', its taken-for-granted character, make it difficult to step outside the discourse. It is not that we cannot stand outside such a discourse, but rather the fact that in order to engage in a debate about the nature of economic knowledge it is easier to position oneself within the cognitive discourse. This was precisely the reason why Bell could smoothly separate the economic from the cultural, why he could readily hold apart abstract and expressive symbolism, in his writings. It would have been problematic to do otherwise.

Indeed, the hold that a cognitive discourse of knowledge exercises can also be seen through the largely understated role that it performs in accounts of tacit and codified knowledges. It is perhaps not altogether surprising to find that the coupling of knowledge and practical know-how that is organizationally or locally-based often reflects the values of technical mastery, rather than those of sensuous form or affect. The spatial sleight of hand which leads to the conclusion that economic innovation and competitiveness are necessarily rooted in place can thus also be seen as merely an extension of this dominant discourse. A cognitive account of economic knowledge reinforces the tendency to equate tacit understanding with local 'embeddedness' and codified knowledge with the wider, ubiquitous arena.

In that sense, when we consider knowledge, space and economy together, it is possible to see how geographers have put to work a particular conception of economic knowledge. Cognitive in outline and territorial in shape, such a conception has produced specific effects, both for which economic activities are valued and where they are likely to be found. In consequence, it is that much harder to recognize and to voice the relational networks which make up mobile, stretched knowledges – whether abstract or expressive in form.

Notes

1 The tendency to map the tacit–explicit distinction on to the local–global scale manifests itself in a variety of ways in the economic geography literature. The 'learning economy' and 'learning regions' literature has probably done most to consolidate this impression, in particular Lundvall and Johnson (1994), Maskell and Malmberg (1999), Malmberg *et al.* (1996), Morgan (1995, 1997), and relatedly Storper (1997a) and Storper and Salais (1997). In other respects, this impression has been reinforced by much of the literature on socio-economic networks, in particular those studies inspired by Granovetter and Swedburg (1992), and aspects of the debate over the significance of Marshallian industrial districts. A closer look at this body of work and its effects is taken up in the final part of the chapter.

2 See, for example, the work of Nonaka (1994), Nonaka and Takenchi (1995) and, more recently, Leadbeater (1999). Leadbeater, in particular, is a good example of someone who moves between broad and narrow definitions of knowledge, yet when pressed falls back on the language of science, technology, and innovation to make a general case for knowledge capitalism.

3 The 'creative industries' as a term has recently gained credence in the UK as a collective name for the likes of advertising, design, film, publishing, interactive software and the performing arts. In a recent mapping exercise conducted by the *Creative Industries Task Force*, however, it is apparent that the driving force rests upon the ability to exploit intellectual property rights and facilitate the global distribution of 'creative' products (Department of Media, Culture and Sport 1998).

4 'I mean by culture – and here I follow Ernst Cassirer – the realm of symbolic forms and, in the context of the argument of this book, more narrowly the arena of *expressive symbolism*: those efforts, in painting, poetry, and fiction, or within the religious forms of litany, liturgy, and ritual, which seek to explore and express the meanings of human existence in some imaginative form.' (Bell 1976: 12)

5 The stress placed upon *cultural meaning* by Cassirer in his work on symbolic forms is intended to convey the objective rather than the subjective or psychological nature of cultural forms. Each of the three symbolic functions – expression, representation and signification – outlined below convey different 'dimensions of meaning' (Cassirer 1957: 448–9). See also Krois 1987.

6 This position is outlined fully by Cassirer in his introduction to Volume III of *The Philosophy of Symbolic Forms*.

7 Krois (1987) prefers to use the term signification, probably to avoid misunderstanding over the general use of signifying practices, following Saussure (1974). For Cassirer, the use of the term, signification, is restricted largely to abstract symbolism and the 'world' beyond reference.

8 This particular example is spelt out in Cassirer's lecture, *Language and Art I*, delivered in 1942 and published in Verene (1979).

9 Sharon Zukin's work, for instance, can give the misleading impression that the 'symbolic economy' is restricted to a small range of cultural sectors. See, for example, Zukin (1995, 1998).

10 See Zelizer's (1989, 1994) account of the personalization of money and its attributed social meanings.

11 More accurately, there is little of the non-representational forms of economic activity in Lash and Urry's account. The expressive dimension to experience where an art form 'moves' us emotionally or 'strikes' us in a particular way is virtually absent from their representational treatment of aesthetics.

12 In the *Critique of Judgement* (1987), Kant wrestles with the fact that aesthetic judgements, if they are to be considered as something more than subjective preference, must claim some kind of universal status. Yet, clearly, judgements of taste in relation to art or music or design rely largely on a subject's 'feeling' for what is in harmony. To avoid an unmediated subjectivism, Kant moves closer to the idea of a universal convention and the possibility of individuals transcending their own particular needs and desires. As Terry Eagleton (1990) outlines it, for Kant,

> Aesthetic judgements are thus, as it were, impersonally personal, a kind of subjectivity without a subject or, as Kant has it, a 'universal subjectivity'. To judge aesthetically is implicitly to declare that a wholly subjective response is of the kind that every individual must necessarily experience, one that must elicit spontaneous agreement from all.
>
> (1990: 93)

This view runs into difficulty, however, once the possibility of radically different artistic appropriations are entertained.

13 Further to Kant, Cassirer argued that expressive forms of knowledge were not simply given, but apprehended through particular symbolic schemas. Artistic appropriations

which result in alternative ways of seeing or hearing, say in relation to cubism or jazz respectively, pose a challenge to the universal status of such schemas, however, as indeed do many non-Western visions.

14 In a variety of ways, the position outlined resonates with that argued by the pragmatist philosopher, Nelson Goodman, in *Ways of Worldmaking* (1978) and his later works, which posit a multiplicity of worlds, each known in different ways, and impossible to combine into a single frame of symbolic reference. Perhaps more than Cassirer, however, Goodman insists on the absolute inseparability of symbolic forms, acknowledging the embodied nature of cognition and the cognitive role of feeling, which perhaps should alert us to the inseparability of different forms of symbolic knowledge in each and every area of the economy.

15 See n. 1 above.

16 Gernot Grabher's regional economic work, for example, would fall largely in this category. See Grabher (1993, 1995, 1996).

17 Nigel Thrift's (1994) consideration of financial centres as arenas of activity, where knowledge is tightly bound within networks not places, is an example of such forms of association. Similarly, Nick Henry and Steven Pinch's (2000) work on the spatialization of knowledge in the UK motor sport industry addresses aspects of the mobile, relational character of economic knowledges.

18 Again, much of the 'learning economy' and 'learning region' literature, errs in this direction. See n. 1 above.

3 Materialities, spatialities, globalities

John Law and Kevin Hetherington

Global visions, local materialities

Six hundred years ago the world was divided into a series of different regions. Europe, the Arab world, China, Japan, the civilisations of the Indus, the Mayas and the Incas, various sub-Saharan civilisations; these and others existed apart from one another. Yes, there were some contacts. Arabs and Christians were engaged in a sustained trial of strength around the Mediterranean. The Chinese made periodic forays far from home. And there was a trade in luxuries between Europe and Asia. But there was no 'world-system'.[1] Economic, social and cultural life subsisted almost independently in the separate regions of the world. Indeed, one might say that those different regions existed in different worlds.

Between the years 1400 and 1900 this all changed. A single world-system emerged as Europe colonised and came to dominate most of these other regions. The world entered a period of sustained economic growth which included revolutions in agricultural production, the harnessing of new energy sources, the growth of manufacture and a world division of labour which depended on immeasurable improvements in transport and communications. At the same time, and as an inseparable part of this, a capitalist world order emerged. This was associated with huge increases in wealth and productivity. It was also characterised by massively unequal distributions in wealth, both within regions and to an even more marked extent, between core and peripheral regions. It was associated with the development of the European (subsequently the world-wide) nation state. And finally it was linked to the ever increasing importance of knowledge as a resource closely related to economic production – and more recently to consumption and cultural change.

Many of the formal trappings of the imperialist world order have now disappeared. The past 40 years has seen the virtual end of political colonialisation, and in certain respects the nation state appears to be under threat. But there is as much continuity as discontinuity. The nexus of capitalist enterprise, world trade, world division of labour, unequal division of resources, and growth in knowledge and communications has continued to develop apace. And it is clear, as we reach the year 2000, that in terms of the flow of goods, information and people we live in many respects in an era that is both mobile and global.[2] Networks of information, of sociation, span the world.

Marx notoriously observed – following Shakespeare's Prospero in *The Tempest* – that in capitalism all that is solid melts into air. He was thinking of old feudal loyalties – and more generally any forms of life which were irrelevant to the logic of capitalist economic accumulation. His aphorism still applies. Economic and cultural stabilities are more than ever elusive and ungraspable. The global economy with its information and capital flows is dominatory, generating asymmetries and distributing and redistributing opportunities and miseries ever more rapidly. Social relations are disembedded from local contexts and stretched across time and space. The world is compressed and our links are distanciated at the same time.[3] And, as a part of all this, cultural production is also more rapid than ever. Fragmented, its diasporic and hybrid character can be taken as a sign for the totality of a cultural shift.

In social science this story has been told in a number of different ways: as capitalist accumulation and world-domination; as a process of industrialisation; and, more recently, as a story about the networks of globalisation, and a shift from production to culture and consumption. Our brief sketch indeed reflects all of these, and this is a necessary context for what follows. But what we are most concerned with in this chapter is the nexus of knowledge, space and economy as seen from one particular point of view: a concern with what we will call *materiality*. So what does this mean? And where does it come from?

The most straightforward answer is that materiality is about stuff, the stuff of the world. Straightforwardly, we can imagine three kinds of stuff. First there are objects. Here, then, a concern with materiality is a concern with machines, houses and supermarkets. It is about satellite communications, military technologies, motor-cars, the growth, the distribution and the consumption of tea and coffee. It is about the fancy corporate headquarters of the multinationals – or the *favelas*, the slums, of Rio de Janeiro. It is about the water supply in a Zimbabwe village, or the cable networks beneath the streets of London.

So stuff is about objects. But it is also about bodies too – for bodies are material. So it is about how bodies display themselves in clothes and cosmetics as objects of the gaze, come to embody their conditions of work, are added to or repaired by prostheses. It is about the conditions of childbirth or the embodiments of child-rearing. It is about blind bodies as they find their way around museums or try to get on and off the bus. It is about ability and disability.

So objects and bodies are stuff. They are material. But so too are information and media, and this is our third category of materiality. Texts such as this, newspapers, the pictures on the television at night, books in libraries, CD roms, maps, films, statistical tables, spreadsheets, musical scores, architect's drawings, engineering designs, all of these are information – but information in material form.

Until recently social science has had problems in thinking about materiality. Materials have usually been present in what's written because it's so *obvious* that the world and its relations are made of materials. But, at the same time, they have also been strangely absent from it – perhaps because it *is* so obvious that the world is made of materials that they've been taken for granted. And when they haven't been taken for granted sometimes the role of materials have been hyped up into some kind of drama in which we learn that technological changes determine how we live. The current candidate for this is the World Wide Web, though the same was said about the printing press, electricity and the electric telegraph. But this 'technological determinism' is too simple. This is because technologies are shaped by social circumstances. The Web is a case in point. Its origins lie in the US military concern to create robust communication networks which would withstand Soviet nuclear attack.[4] Had electronic communications developed under some other regime there is every reason to suppose that they would have been different in character. So instead of saying that technologies determine social life we need to say something more complicated, like: technologies-and-knowledge-about-technologies-and-a-good-deal-of-hard-work-and-capitalist-economic-relations together determine (parts of?) social life. Which catches fewer headlines, but is more realistic. And also reflects the way in which different materials – objects and technologies, bodies and texts are produced by and simultaneously produce social and economic relations.

So materials come in different shapes, forms, and kinds, and they interact together to reshape one another and produce effects. In the first part of this chapter we draw on a discipline called Science, Technology and Society (STS) to show how they interact to produce *knowledge or information*. The implication of this argument is that if we want to understand phenomena such as global capital flows, the transmission of information, cultural hybridity, or economic inequality, it is also important to ask *how the relations that produce these are materially brought into being and sustained in particular locations*. This takes us back to the point about the invisibility of materiality. Thus for all the talk about globalisation, this is a phenomenon that also takes material form and does so in particular locations. And these are worthy of study. Indeed, if we want to understand how globalisation is *achieved* we have no choice: we have to look at the ways in which it is materially produced.

This takes us into questions to do with space. As is obvious, globalisation or world systems are spatial phenomena. They are made by materials which are in space – but which also have spatial effects. Some of those spatial effects have to do with inequality and domination. For instance, the literature on economics tells us that information is costly and that profit – indeed good decision-making – depends upon, is often almost indistinguishable from, superior information, quicker information, less distorted information.[5] The better telegraph, the faster steamboat, the more powerful intranet, these are key tools in achieving advantage. So material arrangements generate information. They also generate rapidly moving information, which is why we say that they have spatial effects. In important respects the City of London is closer to Wall Street than it is to inner-city Salford. And this leads us to reflect on the character of spatiality itself. So we'll make the argument that spatiality isn't just about the Euclidean space of the globe, the space dealt with in physical geography. We'll argue that it is also about material networks which imply a different *form* of space.[6] And then we'll go on to argue that the asymmetries of global capitalism, of information, may be understood in terms of the interaction between Euclidean and network spaces. That they are a consequence of what one might think of as *spatial non-conformities*.[7]

Material heterogeneity and knowing locations

To address global concerns it is often best to be local, specific and material. That is the assumption with which we start.

The place in which we might start is a managing director's office. It might be anywhere in a medium-sized enterprise. Actually it is the office of the director of Daresbury SERC Laboratory in the UK.[8] It's furnished as one might expect. At one end, the end away from the door, there is a large desk and an office chair, a computer, a telephone and various other pieces of equipment. Then, at the other end of the table, nearer the door, there is a modest boardroom table, a table for meetings. It seats six, perhaps eight, people in comfortable upright chairs. Then there is a third area to one side, an informal area, with a coffee table, three or four easy chairs, a few magazines and scientific publications. This is where the director relaxes with high status guests. Where they may drink coffee and eat biscuits.

So where does the coffee come from? We might respond to this by talking about the global and link it with the local. By talking, say, of Andrew's office as the end point of a network associated with coffee beans produced in Columbia or Kenya. And this is not, of course, incorrect. However, for us this move already makes too many assumptions about the materiality of connections, and of how the global and the local are different in character. So we want to remain for the time being in Andrew's office without moving to sub-tropical plantations. In which case to find out where the coffee comes from we need to move through one of the doors into a large room where the secretaries work, typing, fielding phone calls, emails and visitors, keeping diaries, ordering up tickets, reports, and, yes, making coffee. Two rooms, then, with doors that lead also on to a corridor where people may wait to visit the managing director. A corridor where the trappings of power – the pile carpet, the décor – are suddenly absent.

The details don't matter. And in one way they are trivial. Everyone knows that power attracts trappings. But this is the first lesson from STS – though it also comes from the writing of empirical philosopher Michel Foucault. For in the analysis of materiality these are *not just trappings*. They are not idle. They are also performative. That is they act. And as they form part of a materially heterogeneous network of bits and pieces of all kinds, that participate in the generation of information, of power relations, of subjectivities and objectivities.[9]

This is more obvious for some trappings than others. For instance Andrew, the managing director, is frowning at his computer. This is because he's discovered that the biggest project in the laboratory is seriously behind schedule, though it's only been going for a few months. But how does he know this? How has this been made visible? How has this information come into being? The answer is that he's got a spreadsheet up on the PC which tells him how much work time (they call it 'manpower') has gone into the project so far. And he's comparing this with what they planned – and the two are very different. The project is a number of months behind schedule. Indeed, though this isn't obvious in any other way, it's used up most of its contingency time already.

We may think of Andrew, then, not just as a man but more specifically as a *knowing location*. Or a point of surveillance. But he's only a point of surveillance – he only knows – because *he is at the right place in a network of materially heterogeneous elements*. This is the argument, then, about material heterogeneity. We might number: his computer; its software; the figures typed into the spreadsheet; the process of collating those figures carried out by people in the finance department; the work of filling in the time sheets that is done (or supposedly done) on a monthly basis by all employees; the decisions that those employees have made about how to allocate their time (for in practice most work doesn't come in half-day blocks which is all the time sheets allow). And then we can extend the network: into the power company (no electricity, no surveillance), the work of the programmers both locally and at Microsoft, the decisions by previous directors to implement a time-booking system, the production of the time sheets; and then the car that Andrew drove to work; the fact that he and the other employees are paid; the telephone and the email that allow him to summon the other senior managers to an emergency meeting. For, yes, the point of this STS analysis is that the relations that produce knowing locations, information, are endless. That they are materially heterogeneous. And, one way or another, they all have to be in more or less working condition if there is to be such a thing as a 'knowing location'. We're saying, then, that *knowing is a relational effect*.

Let's state this more formally. In approaching knowledge in this way we're using what one might think of as a *semiotics of materiality*. It is about *materiality* for the reasons we have discussed: because knowledge, power, and subjectivities are all produced in circumstances that are materially heterogeneous. This means, inter alia, that the distinctions between human and non-human, between ideas and objects, between knowledge and infrastructure – that all of these are seriously overdrawn. And it is a *semiotics* because it assumes that what is produced, together with whatever goes to produce it, secures its significance, meaning, or status not because it is essentially this way or that, but rather because of how everything interacts together. So Andrew is a managing director not because this is given in the order of things, but because he is at the centre of a network. The spreadsheet is a spreadsheet because it relates to him, his computer, the power supply and everything else in a particular way. If something goes wrong then Andrew isn't a managing director any more – and the spreadsheet is similarly no longer a spreadsheet. A semiotics of materiality suggests that objects, materials, information, people and (one might add) the divisions between big and small or global and local, are all relational effects. They are nothing more than relational effects. Which is why it is so important to study how they are produced.[10]

Knowing at a distance, acting at a distance

Here is another story about knowledge. It is more obviously about globalisation than the events in a laboratory, but it too is about material specificities. It's about the early stages of the imperialist expansion that we mentioned above, the early stages of the growth of the world system in the sixteenth century. It is about the Portuguese route to India.[11]

Though there are a few exceptions, most of the histories of the Portuguese expansion mention their ships and navigational tools as important but essentially infrastructural items, means to the Portuguese end of seizing the spice trade from the Venetians and the Arabs, indulging in holy war, or discovering previously unknown sources of gold.[12] Like the props for managing directors, powerful people and information-gatherers there is a division between social actors on the one hand and important but essentially uninteresting furniture on the other. But as we have just seen, an STS semiotics of materiality refuses this division and prior judgement about what is important or not, and says that if we want to understand how knowledge is produced we need to look at the whole set of heterogeneous elements, human and social on the one hand, and non-human and technical on the other. So how does this work for the Portuguese?

The quick answer is that the ship, its crew and its surroundings (or the navigator, his tables and instruments, and the sun or the stars) need to be seen as a *continuous network*. If the different parts stay in place, if their relations with their neighbours hold them in role, then the network as a whole generates knowledges. For instance, the Portuguese navigator together with his instruments, astronomical tables, and appropriate sightings of (say) the North Star, could determine the latitude of the vessel.[13] The whole network of elements, arrayed together, produced that (vital) knowledge. Other physical effects might also result. The vessel itself, its equipment, its provisions and stores, its crew, knowledge of how to catch the winds, to take advantage of the currents, how to steer a course, knowledge of location, plus charts – these were parts of a network which helped (if all the parts successfully held one another in place) to sustain a watertight and seaworthy ship rather than (for instance) a collection of drowning mariners and a mess of wood splintered somewhere on a reef.

The argument once again is that knowledge, objects and people (or 'subjects') are *relational effects* or *emergent phenomena*. STS writer and philosopher Bruno Latour has a very particular way of saying this. He talks of *immutable mobiles*. In this way of talking, the immutable mobile is a *network of elements that holds its shape as it moves*.[14] Indeed like a ship. Or, one might add, in cybernetic mode, like the electronic symbols, the bits and bytes of contemporary communication. So in this kind of account the vessel or the electronic symbol is a network that holds its shape and moves through Euclidean space.[15] But, we could add, so too is the navigator-chart-instrument-table network (or the electronic network). Or, indeed, the chart all by itself.

Do networks of relations hold their shape as they pass through geographical space? This is the crucial (if oversimplified) question which links knowledge with space. Or, restated, do (sub)networks insert themselves into larger networks of relations which are sufficiently stable so that they hold their shape and may pass through geographical space? These questions are ways of talking both about *action at a distance* or domination, and about *knowledge at a distance* or surveillance. For if the Portuguese were able to control the spice trade for nearly a century, if they were able to bombard the inhabitants of Calicut into submission, if they were able to get to India and get back, then this is because they succeeded by luck or good judgement in generating an array, a global network, within which immutable mobiles might circulate. Such that if a command was given in Lisbon, then war might be fought in India. Such that if a command was given in Lisbon it was both heard and enacted in India.

'Action at a distance.' 'Knowledge at a distance.' A note is needed here about distance and about space. For this, as we noted in the introduction, is an important, indeed a vital, twist to the argument. We want to suggest that making action and knowledge at a distance not only makes action, knowledge and global asymmetry – though it certainly does all of these things. In addition we want, and somewhat counter-intuitively, to suggest that it also *makes distance or space*,[16] performs these into being. Which means that distances and space don't exist by themselves as part of the order of things. But rather that they are created.

That's a simple statement of a counterintuitive notion. But what does it mean? Let's start to answer by thinking empirically. Here the story is that before the Portuguese got to work, Lisbon and Calicut (in India) simply didn't exist for one another. They were in separate worlds. They existed (as we are saying) in different spaces. So it was through their efforts that the Portuguese turned Lisbon and India into places that, though they were distant from one another, were nevertheless in the same world, in the same space. Yes, it took many months to make the passage between the two in one of their vessels. Yes, it also took a lot of effort, time, skill and bravery to move from the Tagus to Calicut and back. It is because of this effort and the work involved in displacement that they were indeed distant from one another. But they were also distant *because they were connected together* in a single world rather than belonging to separate worlds.

We're saying, then, that locations which don't communicate with one another, which know nothing of one another, don't exist for one another exist in entirely *different worlds* or spaces. Like the Incas and the Arabs who, so far as is known, never communicated, never knew of one another. The argument is that *distance demands communication and interaction*. Its very possibility, *depends* on communication or interaction. It depends on joining things up within – and thereby *making* – a single space. And if this is difficult to see – if, for instance, it seems that the Incas and the Arabs really belonged to a single world, existed within a single geographical space – this is because geographical space has somehow come to seem natural. As if it were given. And because (for the case of the Incas and the Arabs) we have chosen to ignore the work of more recent historical geographers who have drawn them on to regions in a single world map. And because we have got so used to the work of the geographers together with the networks of trade, of air traffic control, of electronic links and all the rest, we have come to experience the geographical space that it makes as if it were natural, something given in the order of things. Something that has to be that way. But we're saying that *it isn't natural*. Rather, geographical space, global space, is a material semiotic effect. It is something that is made.

Let's note that the same logic works for Andrew's office. It is linked to other locations on the globe, to be sure. It is located in a world-geographical space. But – and – this is because of the work involved in making and maintaining all the email, telephone and transport links which join it to other offices and laboratories around the globe. The work of keeping up the materially heterogeneous links which maintain the mobilities between places, and define their distances. The materially heterogeneous enactments and performances which create a global geographical space on the one hand and locations in that space such as Daresbury Laboratory on the other. Again, then, we want to say that the possibility of globality – and location in globality – is sustained in that work.

Capitalisation 1

In this semiotics of materiality knowing, knowing at a distance, acting, acting at a distance, and the making of space, are all *relational effects*. And they are *materially heterogeneous* effects. Materials of all kinds are being disciplined, constituted, organised, and/or organising themselves to produce knowledges, subjects, objects, distances and locations. We might, with Foucault, note that this is the effect of a strategic ordering of elements. They could be ordered otherwise in which case knowing, location, and all the rest would be different. And then we'd need to add, again like Foucault, that strategy does not necessarily imply the presence of a self-conscious strategist. But this does not mean that there are not centres of accumulation. Places where surplus accrues. Places of profit. It does not mean, in other words, that what we are calling 'capitalisation' does not take place. So, crucial questions in the context of globalisation are: what can be said about accumulation? and how are asymmetries between the centres of accumulation and the rest generated? There are several responses. Responses that have to do with the configuration of the heterogeneous material elements which make up the network of relations.

One has to do with *delegation*. 'Will you act as my agent at a distance? Will you stay reliable? Will you hold together? Or will you turn traitor, turn turtle, or go native?' In terms of a network logic of material relations these are the same questions. And they have the same logic as the immutable mobile. The issue is, will the configuration of bits and pieces that allow me to profit stay the same, or not? If the king issues an order to bombard Calicut, will it be followed through? Will the ships get there? Will the gunpowder stay dry? Will the crews follow their orders? Will they have avoided disease? If the answer to these questions is yes, then we are in the realm of immutable mobiles. If not, then not. And the same logic works for the laboratory too, albeit on a less dramatic geographical scale. Will the employees do as they are asked? Will their instruments, their computers, bend themselves to the project? Or will they not?

Delegation, then, may be understood in a semiotics of materiality as a way of talking about the immutable mobile. Delegation is sending something out which will hold its shape – so that the centre does not have to do the dirty work itself. Which is, to be sure, not simply a moral but also a practical matter. If the King of Portugal or Vasco da Gama had been obliged to subdue the Indians alone and with their bare hands they would not have been up to the task. Delegation, then, is also something which works through a series of tiers. It is an arrangement in which you push the levers and something happens, something that magnifies itself in the next stage, and then again. (Think of the tiers of simplification and delegation implied in building a spreadsheet.) And, crucially, it is also something that happens in a play between *different material forms*. For delegation into non-human materials – cannon, prison walls, marching orders – is often particularly effective (though there are no guarantees, and the integrity of physical materials is, itself, a relational effect).[17]

But successful delegation, the successful creation of immutable mobiles, the capacity to know and act at a distance, has other asymmetry-relevant effects. Or it may be thought of in different ways. For instance, it may be thought of as the creation of what STS scholar Michel Callon calls an *obligatory point of passage*. For the obligatory point of passage is the central node in a network of delegation, so to speak its panopticon. The place of privilege.[18] This, then, is a second feature of material relations which creates asymmetries.

In what we have written we have already come across two obvious obligatory points of passage. On the one hand there is Andrew-and-his-spreadsheet. And on the other, there is the Portuguese-state-and-some-of-its-officials-and-traders. Here is the argument. Those caught up in one or other of these networks of relations have no choice: if they want to move, if they want to achieve their goals, then they have to do so by making a detour. A detour via Andrew-and-his spreadsheet. Or via Lisbon-and-its-spice-markets. So the pepper growers in India can't sell their crop to the Arabs any more. The network of the Portuguese – their guns, their money – have cut the old links. If they want to make money then they are necessarily enrolled into the Portuguese network. They, or more precisely, their crops, make the long detour via the Cape of Good Hope and Lisbon to get to the European market. They then become faithful delegates of the (newly distant) Portuguese centre, tributaries no doubt held in place by fear and need rather than love or affection. But this makes little difference from a semiotic point of view. For held in place they are. Contributing their ha'pennyworth to the network, buttressing it, and at the same time adding to, further performing, its centre as an obligatory point of passage. As a place of privilege. A centre of accumulation.

But the same is happening at Daresbury Laboratory. Employees do not, for the most part, turn up in Andrew's office in person to receive their orders. Instead immutable mobiles emerge from this obligatory point of passage, delegates that faithfully perform themselves across the space of the laboratory. Such that the elements which make up the network of the laboratory find that they are being displaced, moved to work on new projects, acting in ways that they would not otherwise have done. Being enrolled to act as clients of (what has therefore become) a centre, an obligatory point of passage, a privileged location that can see and act at a distance. That makes the distance and masters it, all at the same time. So Andrew does not bend the workings of the laboratory by himself. He delegates to (what he hopes will be) faithful emissaries. And into other material forms – for instance in the shape of minutes, memoranda and pay cheques. Just like the Portuguese monarch. Which tells us, as we already noted, that 'Andrew' (and the Portuguese monarch) is a heterogeneous relational effect rather than someone whose powers are given in his body.

Delegation and obligatory points of passage are crucial to capitalisation and its asymmetries. But these are also a play around *scale effects*. We've noted that distance is a product, an effect. Made and mastered in the creation of immutable mobiles. But delegation also makes spatial effects. For as we've hinted above, immutable mobiles passing to and from (and thereby creating) a centre also play havoc with scale. We will need to return to and revise the notion of scale below. But for the moment let's note that knowledge of distant events, distant actors, also implies that these are rendered small and simple. This is a version of the argument about power and delegation. Just as Andrew and the Portuguese monarch cannot do all the dirty work themselves, so they cannot know all about everything that goes on within their networks, know all about the dirty work. But, nevertheless, and this is one of the features of power, in some general sense they *need* to know about it. Knowing at a distance, then, necessarily implies pretty heroic simplifications and reductions. And it therefore also implies pretty heroic manipulations of *scale*. This means that that which is large in the geographical sense, spread out over time and over space, gets reduced to a report, to a map (and the development of mariners' maps counts as an exemplary case here) or, in the case of Andrew, to a set of figures in a spreadsheet. Everything – or representatives of everything – is being brought to one place, all at one time. That which was big is thereby being rendered small. And, as it is being rendered small, it generates a capacity to see far for the privileged centre. And, crucially, it also generates a capacity so see what would otherwise not have been visible – indeed what would in some sense not otherwise have existed. Which is, to be sure, where we came in: with Andrew-and-his-spreadsheet and the discovery/creation of a delay that would otherwise not have been visible. A Foucauldian point, one that derives from attention to a semiotics of materiality.[19]

Delegation, the making of obligatory points of passage, and scale reversals – all these are configurational features of the asymmetrical networks of capitalisation which grow out of and produce immutable mobiles. Now we want to mention a fourth and final feature. This has to do with the production and concentration of *discretion*. To say it quickly: with the growth of action and representation at a distance there also grows discretion. To act, or not to act. To act in this way or, alternatively, to act in that. Empirically this is easiest seen for Andrew-and-his-spreadsheet. For he can see far enough – and he can successfully act in enough ways – that there are a *variety of courses of action* open to him. But how might we think of this in terms of the configurations of materially heterogeneous networks?

The STS suggestion is quite simple, and it has to do with the asymmetry generated between the centre (which becomes a centre because it is an obligatory point of passage for a series of tributaries) and those peripheral tributaries which are indeed peripheral precisely because they have no options, no choice. But, stood on its head, what this tells us is that it is probable (not certain) that *because there are many tributaries to the centre, the centre correspondingly has many options*. It has many alternative possibilities for acting at a distance, mobilising this rather than that tributary. The argument, then, has to do with redundancy. The centre enjoys the luxury of redundancy. For it, there are no obligatory points of passage in its heterogeneous networks. If one 'circuit', if one set of immutable mobiles gets choked off, goes native, is turned into matchwood on a reef with drowned mariners, then it can always act through another. Send another vessel (which, since the shipping losses on the Portuguese route to India were heroic, was a very common occurrence). Which is not, to be sure, a recourse that is open to the client who is forced to pass through an obligatory point of passage. Like the unfortunate ruler of Calicut and his spice traders.

Our topic is knowledge and globalisation. But it is also capitalisation and power. We will return to the issue of capitalisation and spatiality below. As we have noted above, spatiality needs to be rethought. We have offered some suggestions about this – to do with scale and the making of distance. In this section, however, we have particularly attended to features of the logic of capitalisation or accumulation as seen from the point of view of such a material semiotics. In insisting on *how* it is that knowledges and actions get generated and distributed to particular locations in the social world, and noting how these may be understood as relational strategies or features of the shape of self-sustaining heterogeneous networks, we have identified four crucial moments: *delegation* (which may take material forms); the creation of *obligatory points of passage*; play with *scale and size*; and finally the far from even distribution of *discretion*.

Spatial enactment

Distance, we have asserted, is made – and putatively mastered – all in the same moment. Lisbon and Calicut become places in a single space only when immutable mobiles such as ships shuttle between them – or, to bring the example up to date, with the growth of cartography, GIS, or the financial networks of the world. Until that moment they simply exist in different worlds. This is the crucial move if we are to understand spatiality – and the phenomena of globalisation – from the standpoint of a material semiotics. As we have argued above, *space is made*. It is a creation. It is a material outcome. Like objects, places, or obligatory points of passage it is an *effect*. It does not exist outside its performance.

This step is at least as radical as the STS argument that materials may be understood as relational effects. It's a radical step because as we've noted above, notwithstanding twentieth-century excitements about the relativity of space-time, in six-hundred years of surveying, cartography, nation-building and GIS, the idea that there is (a single) geographical space has been naturalised for Euro-Americans. This means that it is very difficult to imagine space as anything other than some kind of a neutral container, a medium, within which places – like Lisbon and Calicut – may be located. And this in turn means that any attempt to challenge this picture is very hard work and runs against the grain of common sense. As is indeed suggested by the tropes about global space-time compression which, though they index a sense of variability in distance and speed, tend at the same time to re-enact this naturalised geographical view of space.

There are, to be sure, straws in the wind. The idea that space might be treated as a *performance* – and that geographical space as a neutral container or surface is likewise to be understood as an enactment – is being explored in parts of cultural geography as well as STS. One easy way of opening up the subject is to remember – as geographer David Harvey shows – that it takes a great deal of effort to create a map. Setting up triangulation points, trudging around France, educating the necessary surveyors, defining the length of the basic measure, assembling the appropriate instruments, making the actual measurements, transcribing them on to the flat surface of a sheet of paper – all of this is far from a trivial exercise. From an STS point of view it is an exercise in building a materially heterogeneous network – and of generating representations or immutable mobiles that may be brought together to make the depiction of a Euclidean space. For our earlier description of the heterogeneous engineering involved in knowing and acting in a laboratory – or at the centre of the Portuguese empire – applies just as much here. A privileged centre comes to represent what had never previously been brought together – or at least not in the form of a set of consistent spatial co-ordinates. To generate what (in the case of many cartographic conventions) is aptly called a 'view from nowhere'.[20] Which has then shown a progressive tendency to naturalise itself as some kind of 'objective space' within which we are all located.

Unsurprisingly, all of this is costly. It is a finding of STS that metrology – the making of metrication, of mensurability – is not a trivial exercise. And in the context of spatiality a number of writers – most notably David Turnbull as well as David Harvey – have noted the symbiotic link between that effort and the process of (European) nation-building.[21] Precise geographical maps are important for state power in various ways – for instance they define frontiers and create measures for (taxable) plots of land. They also, as we saw earlier, allow global domination. But the effort put into creating a mensurable geographical space demands huge resources and, historically, this came in large measure from the early-modern European state. Knowledge and power as usual are associated with one another, sustaining and performing one another. But behind this important point hides one that is even more crucial if we want to understand the asymmetries of global knowledge and power. It is the proposal, rehearsed above, that rather than being given in the order of things, space, however naturalised it may seem, is always an effect or an outcome of materially heterogeneous relations. It is sustained and enacted in those heterogeneous relations. It involves a lot of work by all the elements bound up in and producing the network. It is precisely an outcome of a '*performance*' that is not given in advance.[22]

Multiple spaces, interacting spaces, capitalisation 2

So spaces are *made*. But if they are made, then how do we know that they are all the same? How do we know that they are all, for instance, geographical? That they all map on to one another in a nice neat way?

The answer is: *we don't*. And, indeed, there is no particular reason for assuming that they will map nicely on to one another.

Annemarie Mol has explored the possibility of such spatial complexities in a quite different context.[23] The argument runs so. If there were indeed a single space with objects located within it, then (as geographical common-sense usually imagines) cartography would be a struggle to discover the coordinates of the different objects within that single space. If one looked at the process of mapping over time, one might find that mistakes were made, the process might be painful and slow, moving around might be difficult and expensive, but in the end some kind of consensus about the location of objects within that single space would emerge. The series of triangulations would, so to speak, be convergent.

This is an appealing story. First, it fits with our notions of common sense. And second, much of the history of cartography is indeed written in this mode: the discovery of continents, their location, the eradication of error, the definition of a single spatial set of coordinates (ending, no doubt, with the satellite global navigational system which can determine location to a within a few centimetres), and the elucidation of the relations between different systems of cartographic projection – it is a history of progress, a history which uncovers spatial reality. But it is a history that doesn't work so well if space is understood as an effect of material relations, of the performance of heterogeneous elements defined and linked together, rather than something which was already there waiting to be discovered.

So if space is being *made* and the history of globalisation from the early cartographers onwards is about the making of a single geographical space, then what needs to be said? The answer is that looked at in this way, a great deal of energy and effort has been put into the *creation* of a network of heterogeneous materials which *perform one particular version* of spatiality. This is a version of spatiality that has become so important that it has become an obligatory point of passage for many. As it has, indeed, for late twentieth century mariners who are no longer obliged to wrestle with sextants, star charts and tables of stellar and solar positions, but use satellite-based global positioning systems instead.

So what we're saying is that geographical space may have become an obligatory point of passage for many mariners (and others). But if the STS story about the performance of space is right, then it is nothing more than the effect of a particular well-elaborated, well-delegated, well-embedded set of heterogeneous material relations. This in turn means that it (and they) could be otherwise. Or (we suggest more radically) perhaps actually *are* otherwise. For notwithstanding the triumph of geographical space there are, we want to say, *alternative kinds of material relations and alternative kinds of spaces* which exist alongside geographical space.

Let's look at this empirically. Andrew sits at his computer and frowns at his spreadsheet. This is the product, as we have seen, of a materially heterogeneous set of relations. He can see far – some distance across the site of the laboratory to objects and events that are otherwise dispersed or invisible, but also some distance into time, into the future, perhaps a year or more. These are the variations in scale which we described above. Space – and time too – are being scaled down. There is 'space-time distanciation'. Andrew, we might say, travels very fast through space and time, with his immutable but heterogeneous mobiles. He is disembedded.

But, though we used the language earlier, there is a problem if we put it in this way. For what are we doing if we say that he 'travels very fast through space and time'? The answer is: if we talk in these terms then we are naturalising what we have been seeking to de-naturalise in this chapter. In other words, we are buying into a specific version of space with its system of spatial (and temporal) coordinates. We are assuming that 'real distances' are given (for instance) in metres or kilometres. And (by analogy) that 'real time' is given in minutes or months. This is a way of talking which thus assumes that what is happening is a process of speeding up *within* a pre-existing space-time box whose coordinates have been set in the order of things. Obviously this is one possibility, but if we are serious in arguing that spatiality (we need to add temporality) are effects of materially heterogeneous enactments or performances, then we need instead to say something different. We need to say that the privilege of Andrew's location is a double effect. It is

- *the effect of the performance of a particular version of spatiality and temporality*, in which proximity or distance take a network form that has to do with the rapid transmission of immutable mobiles.

And it is also

- *the result of the intersection of that network form of proximity with other and different spatial and temporal forms* – and in particular geographical distance and elapsed time.

Our argument is that it is the interaction between these two forms of space and time (the way in which what is close in one is distant in another) which generates the privileges which accrue to Andrew's location, his ability to capitalise. This means that what we're proposing is a reversal. Instead of saying that messages or information or action speed up within a single space and time frame, we are saying that several intersecting spatialities and temporalities get created. The fidelity of the immutable mobile – its immutability – is a network phenomenon, while its speed – its mobility – is an effect of network immobility within geographical space and chronological time. Can Andrew *see* things that those working on the project have not, cannot? Can he *act* 'at a distance' upon them, and bend their actions? If the answer to these two questions is yes, then they are tributaries to him – because he is able to take a short-cut through network space. While the others do not see 'so far' so fast.

The same logic of spatial and temporal complexity applies to phenomena which are more conventionally global in character. The Portuguese, it is true, displaced the Arab merchants from the Arabian sea by means of military might as well as geographical speed. But they did this because their vessels, their crews, their cannon, had been assembled together into a network, a set of immutable mobiles, which made it possible for the Portuguese state to act at a distance. This located Lisbon and Calicut as distant points on the same world map, but meant that that distance was irrelevant. This was the creation, in other words, of a new form of spatiality that was 'global' in scope, while the Arab merchants and sailors were being left to their own devices, with no state intervening on their behalf. Immutable mobiles their vessels may have been – but only up to the point where they met the first ruthless European navigators, with *their* own networks of immutable mobiles and their own complex versions of spatiality. At which point the Arabs turned out to be infinitely distant from their home ports, infinitely distant from the support they might have wished – and therefore no longer immutable mobiles at all.

Conclusion

We have made an argument about the links between materiality, information, spatiality and capitalisation which runs, in summary form, like this:

- It is a mistake to talk of knowledge, global networks and flows, or sociality, without at the same time noting that these are always *materially produced* in specific and local circumstances.
- It also is a mistake to imagine that materials are passive while people are active. Instead materials (human, textual and technological or artefactual) define one another and hold one another in place. All, in other words, contribute to the performance, *human and non-human* alike.
- If this 'semiotics of materiality' is accepted, then there are no fixed distinctions between (say) humans and non-humans, or between subjects and objects. Instead, effects – including objects, subjects and knowledge – are all produced within *heterogeneous relations*.
- Material relations of various kinds are enacted and performed and take various forms. Here we have concentrated on *networks* which have their own (often implicit) strategic logic.
- This logic displays various features. One is that knowledge or information depends on, indeed is close to being co-terminous with, the existence of *immutable mobiles*.
- Other features of networks that tend to produce and enact economic and informational asymmetries include *delegation* into more durable materials, the creation of *obligatory passage points*, the generation of *scale effects*, and the production and variable concentration of *discretion*.
- As the mention of scale suggests, the relations produced in materially heterogeneous performances also have *spatial and temporal effects*. Networks produce geographical spaces and the distances that make this up on the one hand, and chronological time and the metrics for measuring this on the other.
- If spaces and times are created there is no particular reason why they should be consistent with one another. Indeed, though we have only talked about network and geographical spaces here, it seems likely that *spaces and times are multiple*.
- Finally, important economic and informational asymmetries are also generated in the *interaction and interference between different spaces*. In particular, networks not only produce geographical space but they also allow rapid movement through geographical space – with consequent control and competitive advantage. This advantage may precisely be understood as an intersection between network and geographical space.

Afterword: capitalisation and spatial interference

Capitalisation, the ability to make and sustain an obligatory point of passage is an effect of heterogeneous materiality. It is an effect of the relations built up between different elements. Indeed as we earlier noted, there is a whole literature on what one might dub the cybernetics of capitalisation,[24] competitive advantage, and the circulation of capital. This celebrates the importance of speed, or the immutable mobile in the form of the turnpike, the clipper, the steam packet, the telegraph, the printing press, the telephone and now, to be sure, the internet. Or then again, it attends to the capacity of the obligatory point of passage to process the information that it gathers, in the form of the invention of bureaucracy, of double entry book-keeping, of files, of mainframe computers, of post-its, of networked PCs, of data-bases and spreadsheets.

But to this literature we now need to add the ways in which different cybernetic systems, different obligatory points of passage *intersect* with one another. And, in particular, the ways in which the different spatialities and temporalities which they perform intersect with one another. For if capitalisation is about making obligatory points of passage, it is also about how networks draw from the efforts of neighbouring networks – while in turn protecting themselves from the depredations of their neighbours.

The gist of our argument is that capitalisation has to do with interactive effects – new and complex relations between networks which cannot necessarily be conceptualised as networks themselves. And there are various ways of imagining these. Feminist STS scholar Donna Haraway talks of diffraction effects and 'interferences'.[25] STS philosopher Annemarie Mol talks topologically (in which case Cartesian regions and networks become two topological possibilities to which others – for instance fluids – may be added).[26] One of the present authors, Kevin Hetherington, has imagined these interactions in terms of the generative function of a 'blank figure', like the zero point in the grid of Cartesian space (or an arithmetical series), in which case that zero is both within and outside the space in question.[27] One may think, with philosopher Gilles Deleuze, of the fold and the shortcuts that folding makes possible.[28]

Such theoretical registers take us beyond the scope of the present chapter. But however it is conducted, the thrust of our argument is clear. If we attend to a material semiotics and the performed spatialities that this implies, to talk of 'globalisation' is at best a risky short cut and at worst seriously misleading. It is a risky short-cut because it implies some kind of totality, some kind of global system, and some kind of overall space-time box within which the phenomena which we touched on at the beginning of this chapter are located. A 'global society', a 'global order'. Even a global disorder. But this misses out, or so we have suggested, both on the enacted materiality of that order and also the complex spatialities implied in that enactment. These, or so we suggest, need to be understood if we are to make sense of the asymmetries involved in making obligatory points of passage and the process of capitalisation.

Notes

1 The term derives from Wallerstein (1974a).
2 See, for instance, Urry (2000).

3 The term 'distanciation' is drawn from the work of Giddens and 'disembedding' from Harvey. See Giddens (1990, 1991) and Harvey (1989).

4 For a brief account see Castells (1996).

5 See, for instance, Beniger (1986).

6 See Thrift (1996a).

7 See Hetherington (1999).

8 This case study is explored at greater length in Law (1994).

9 Since the term 'network' has wide currency in social science (in the context of globalisation see, for instance, Castells (1996)), it is important to emphasise the semiotic specificity of the way in which we use the term here. It derives from a body of work sometimes called 'actor-network theory' elaborated in the first instance in the sociology of science and technology. For an introduction to the approach see Latour (1987) or Law (1992) and the annotated bibliography maintained at http: //www.comp.lancs.ac.uk/ sociology/antres.html. For a current assessment of the approach see the papers in Law and Hassard (1999).

10 Semiotics is a branch of linguistics which says (to put it quickly) that the meaning of words depends on their relation to other words. That (for instance) 'man' acquires its meaning in relation to such other words as 'woman', 'boy', 'wimp' or 'ape'. STS – and some other similar approaches including the work of Michel Foucault, parts of feminism, areas of cultural studies concerned with the built environment, and parts of cultural geography extend this beyond words to say that objects and subjects (including people) – that is, all materials – have the attributes that they do as a result of their relations with other materials.

11 See Law (1986, 1987).

12 There is a corresponding literature on those means in the beautiful ghetto of maritime history. For an example of a book that considers *both* the exploration *and* the maritime technologies see Diffie and Winius (1977).

13 The longitude was beyond the reckoning of the Portuguese depending, as it does, on very accurate timekeeping.

14 See Latour (1990).

15 Putting it this way is a convenience. As we note below, space needs to be imagined as a relational effect.

16 Makes it. And vanquishes it. In the same breath. Unless, of course, something goes wrong, in which case distance, which has been made, is indeed not vanquished.

17 For a review of materiality in this mode see Law and Mol (1995).

18 For discussion of the notion of obligatory point of passage see Callon (1986).

19 The relevant reference is obviously *Discipline and Punish* (Foucault 1979), but see also Rabinow (1989) and Harvey (1989).

20 This argument has been developed within feminist STS studies. See Haraway (1991), and (in a different and art-history context) Alpers (1989).

21 See Harvey and Turnbull (1999).

22 See Hetherington (1997b).

23 See Mol (1998, 2000).

24 For an enthusiastic example of the genre see Beniger (1986) and also, to a lesser extent, Hughes (1983).

25 See Haraway (1997).

26 On fluids and networks as different topological metaphors see Mol and Law (1994), and de Laet and Mol (2000).

27 See Serres (1991), Taylor (1993), Appelbaum (1995), Hetherington (1997a), and Hetherington and Lee (2000).

28 See Deleuze (1993a) and Serres (1998).

4 Knowledge, innovation and location

Jeremy Howells

Introduction

Knowledge is a fundamental part of our human existence and being. Knowledge also appears to be growing, although it is difficult (if not impossible) to adequately measure or quantify it. Surrogate indicators, such as journal titles and the number of abstracts, used for example by de Solla Price (1963) suggest a strong growth in knowledge, having doubled every fifteen years since the Scientific Revolution and this has been mirrored by the growth of knowledge industries since the 1950s (Foray and Lundvall 1996: 15–16). The growth of knowledge has been accompanied by the spread of knowledge. This is not necessarily new; the Arabs acquired knowledge about paper production late in the first millennium from the Chinese who then transmitted it during the Crusades to Europe (Cover 1971: 368). However, the absolute volume and speed of growth in knowledge has been especially significant since the Second World War. Its production, diffusion and absorption has also moved from highly localised centres of production and application to much wider patterns of national and international generation and consumption. There appear to be several stages to this growth in knowledge. Burns and Stalker (1994: 26) describe the process of techno-logical innovation and its knowledge base up until 1825 as being a largely singular and 'heroic' process, centred on the 'lonely inventive genius'. From 1825 until 1875 there was a transition in the innovative process, facilitated by localised coteries of scientists and business partners, as 'information about scientific discoveries became available to a wide variety of people. Personal communication was replaced by mass communi-cation' (Burns and Stalker 1994: 26).

More recently, Gibbons et al. (1994) have noted a further epoch change in knowledge activity with a shift from Mode 1 to Mode 2 knowledge production. Mode 1, traditional knowledge production based on single disciplines, is homogeneous and is primarily cognitive knowledge generation contexts set within largely academic paradigms. Mode 2 knowledge production, by contrast, is created in broader, hetero-geneous transdisciplinary social and economic contexts within an applied setting. Mode 1 is primarily hierarchical and tends to preserve its form, Mode 2 is more heterarchical and transient in nature (Gibbons et al. 1994: 1–3). One of the key contrasts between the two modes is that in Mode 1 problem solving is carried out following the codes of practice relevant to a particular discipline and problem solving whilst under Mode 2 knowledge activity is organised around a particular application and is more diffuse in nature. Gibbons et al. (1994: 14) are, however, careful to note that Mode 1 production still continues alongside Mode 2 production: 'Mode 2 is not supplanting but rather supplementing Mode 1.' Indeed certain types of genuinely new forms of knowledge, such as aeronautical and chemical engineering, associated with application-oriented Mode 2 knowledge production, may actually revert back to becoming more like disciplinary-based knowledge production under the style of Mode 1 (Gibbons *et al.* 1994: 4).

Many now see the knowledge-based economy (variously described; Lamberton 1997) as having arrived. Here the view is that we have entered a new historical era where the economy is more strongly and more directly rooted in the production, distribution and use of knowledge than ever before (Foray and Lundvall 1996: 12). Underpinning the knowledge economy is the increasing use of information and communication technologies (ICTs) which is having a profound effect on the generation, distribution and exploitation of knowledge (David and Foray 1995). One of the problems of charting the growth and appearance of the knowledge-based economy is the lack of adequate indicators and data on knowledge and knowledge generation within the economy (Foray and Lundvall 1996: 18–19).

The specific relationship between knowledge and innovation has been a relatively under-researched area. What visions there were of the role of knowledge in the innovation process were in a restricted framework. At best it involved the acknowledgement of the role of formal, *scientific* knowledge in the generation of new technologies. The framework here was of the linear model of innovation, with scientific knowledge acting as some kind of ephemeral reservoir which scientists would 'dip into' to help them invent and discover new products (artefacts). What may be termed the 'process of knowledge' and how it effected innovation was largely ignored. A notable exception to this was Chris Freeman (1982: 4–5) who highlighted the inadequacies of economists via their treatment of knowledge *and* innovation as 'exogenous variables'. In particular Chris Freeman (1982: 4) made the distinction, and highlighted the importance of, what might be termed knowledge-related innovation compared with physically-embodied innovation by noting 'the investment process is as much one of the production and distribution of *knowledge* [original italics] as the production and use of capital goods, which merely embody the advance of science and technology.'

As our understanding of the innovation process has improved and as the scope of the term 'innovation' has widened, academics have become increasingly interested in the role that knowledge plays in the innovation process. Interest in knowledge in relation to the study of innovation also grew in the role of knowledge and learning in the management and organisation of firms. There are a whole range of studies high-lighting the importance of knowledge (see, for example, Winter 1987; Prahalad and Hamel 1990; Nonaka 1991; Nonaka and Takeuchi 1995) and learning (Hedberg 1981; Nelson and Winter 1982; Senge 1990; Slaughter 1993; Bessant and Buckingham 1993) as a vital aspect of a firm's competence or capability (Teece 1982; Nelson and Winter 1982; Prahalad and Hamel 1990). One aspect of this issue obviously relates back to the role of innovation as a competence or capability enhancer. This, in turn, has opened up the issue of the relationship between knowledge and innovation in the process of competence and capability enhancing (or destroying).

Another strand of work has been involved in analysing and categorising the different types of knowledge as it relates to technological innovation. Vincenti

(1984: 575; see also 1990) has developed a schema covering different types of technical knowledge; whilst a study by Fleck and Tierney (1991) distinguishes between seven knowledge types ranging from 'metaknowledge' through to 'formal' and 'informal' knowledge and 'instrumentalities'. Faulkner (1994), after providing a review of earlier attempts at categorisation including that by Winter (1987) and Dosi (1988), goes on to develop a detailed, composite typology of some fifteen types, grouped according to the 'object' of knowledge (p. 447). All these typologies provide important insights into the range of knowledge sources used in the innovation process (Faulkner 1994: 452), but are less appropriate when seeking to explore the *relationship* between knowledge and innovation in industrial economies.

This review seeks to analyse this relationship in terms of knowledge in the context of innovation. It then considers knowledge activity situated within the firm, the primary agent of technological innovation within advanced industrial economies. The chapter discusses knowledge and innovative activity within a wider geographical context, within and outside the firm, highlighting the continuing importance of location in the dynamics of industrial innovation and in the knowledge economy overall.

Innovation and knowledge

Definition and context

Part of the neglect of the knowledge–innovation relationship has undoubtedly been the issue (and often quagmire) of what knowledge is. Turning first to what we mean by innovation, the issue has largely been centred on extending what we mean by innovation in terms of the activities it encompasses. Innovation most broadly is 'the whole technological change process representing a shorthand for doing something new' (Stoneman 1995: 3). From being seen as a 'hard' (artefact based), R&D driven process centred on manufacturing and product and process innovations it has now become accepted that innovation includes 'softer' disembodied innovations, involving organisational and social change increasingly in the service field. Moreover innovation is no longer viewed as a linear, step-change process, rather an activity which involves complex interactive and feedback processes often initiated by the user ('pull') rather than led by research ('push') (Kline and Rosenberg 1986).

By contrast, our understanding and definition of knowledge is far more complex and problematic (Sparrow 1996: 24). A simple answer is that knowledge is what we know. However, more centrally knowledge is 'a mental state that bears a specific relationship to some feature of the world' (Plotkin 1994: 40). Crucially knowledge has a relational characteristic, as it involves a 'knowing self' and something, that 'something' being an event or an entity. Knowing is an active process that is mediated, situated, provisional, pragmatic and contested (Blackler 1995). A final element in knowledge is the need for some kind of memory, an enduring brain state that must exist in the case of knowing by the mind, and allows the bridging of the time gap between events that have occurred and any claim to know about them. It is important to note here that memory about events in the past in turn undergoes change and therefore memory forms an unconscious, altering form of knowing (Plotkin 1994: 8).

There is also an important distinction to be made between knowledge and information. Information relates to individual bits of data or data strands, whilst knowledge involves a much wider process that involves cognitive structures which can assimilate information and put it into a wider context, allowing actions to be undertaken from it. Thus knowledge in turn combines the process of learning (Polanyi 1958: 369). The take-up of learned behaviour and procedures is a critical element within knowledge acquisition, both in terms of capturing and moving it from the individual to the organisational level (Kim 1993), but also in more widely diffusing such competence throughout the whole organisation (Urlich *et al.* 1993). Knowledge cannot be said to 'flow'; although, via information flows and mutual learning experiences which then are assembled or absorbed within a cognitive structure or framework, can be said to be 'shared' or 'transferred'.

To be able to innovate, invent and discover, involves both using existing knowledge but often also requires generating and acquiring new knowledge which in turn involves learning. Innovation also involves sharing learned knowledge. The process of innovation by moving from existing knowledge and learning patterns to new ones through invention and discovery can be termed a heuristic (Plotkin 1994: 152–4).[1]

Thus, although we may share many characteristics in our knowledge frameworks and intelligence, and in the way we learn and perceive, it is still an individual-centred phenomenon.

The tacit and codified knowledge dimension

Although other categorisations exist (see above), knowledge has been divided into two main types in terms of degree of formalisation and presence: explicit and tacit (Polanyi 1967). Explicit or codified knowledge involves know-how that is transmittable in formal, systematic language and does not require direct experience of the knowledge that is being acquired and it can be transferred in such formats as a blueprint or operating manual (what Polanyi (1958: 69–131) originally described as 'articulated' knowing; although articulation and explicitness are not directly equivalent). By contrast tacit knowledge cannot be communicated in any direct or codified way. Tacit knowledge concerns direct experience and is not codifiable via artefacts. As such it represents disembodied know-how that is acquired via the informal take-up of learned behaviour and procedures. It is hard to conceive of situations where tacit knowledge can be acquired indirectly as this would involve some kind of codification and lack of direct experience (Howells 1996: 95).

However, much of the recent discussion surrounding knowledge within economics and social science more generally has centred around the bi-polar distinction between explicit or codified knowledge and tacit knowledge, much of which has been based on interpretations of Michael Polanyi's work. This bi-polar dichotomy represents however a crude characterisation of knowledge as an activity. It in particular misrepresents Polanyi's own thinking which stressed that tacit and explicit knowledge were not divided and that explicit or codified knowledge required tacit knowledge for its interpretation. Polanyi (1966: 7) notes 'While tacit knowledge can be possessed by itself, explicit knowledge must rely on being tactile understood and applied. Hence all knowledge is *either tacit or rooted in tacit* knowledge. A *wholly* explicit knowledge is unthinkable.' This is also emphasised in Schon's (1991: 50) view of the practitioner who makes use of research-based theories and techniques but who is also dependent on tacit recognitions, judgements and skilled performances. Knowledge is therefore much more complex than this dichotomy portrays; particularly as one moves from knowledge being an individual phenomenon through to group, firm or organisation wide process. At the very least, knowledge should be better conceived as involving a spectrum of processes ranging from what might be described as 'tacit' and 'explicit'. Although even adopting this framework Polyani (1958: 70) stressed that articulation would always remain incomplete and therefore on his basis one would never fully reach the 'explicit' knowledge end of the spectrum. However, the distinction remains important in terms of recent discussion and has a strong geographical dimension which is explored more fully later in this chapter.

Knowledge, innovation and the firm: moving from the individual to the collective

If we are to learn anything from Polanyi, knowledge is a much more holistic, but individualistic process than generally conceived. This is not to suggest that we are all individual 'islands of knowledge'. We share information and data that flows between us, we go through similar learning processes with similar teachers in similar environments. Thus although we can only have individual knowledge, we can possess collective understanding. However, this shared understanding is always going to be eventually filtered, perceived, stored and reconverted in our own individual 'knowledge frames'[2] which are going to be slightly different (or in some cases very different; for example between a Wall Street dealer and an Australian aborigine).

The problem with harnessing knowledge within the firm, and especially in relation to new and difficult knowledge associated with innovation, is precisely because it is so hard to share a common knowledge frame between different individuals (even between, for example, members of a research laboratory or more generally between employees within the same firm; see below). This is because we all have different knowledge frames. These differences between knowledge frames are likely to be not that great within a small, single-site firm staffed by people from a similar class and cultural background, but will widen as firm size (numbers of employees) and geographical spread increases. Thus, time and geography will further heighten and exacerbate these differences both directly and indirectly through time and place-based social, cultural, political, institutional and economic processes (for example, via the formation of separate, identifiable communities). The codification and abstraction of information and data and the existence of established 'arenas' (such as the factory shopfloor or a laboratory) where individuals can go through similar learning processes with a high tacit element can all help reduce the gap between the knowledge frames of individuals, at least for certain specific tasks or experience sets. However, such practices and learning fora are never going to replicate exactly the same knowledge frame subset in individuals. On this basis, how can knowledge be viewed within the context of the firm? And what role more specifically does the firm play in the knowledge process if we recognise that knowledge is an individually-based phenomenon? It is suggested here that the firm has a number of different, but overlapping roles and effects in relation to knowledge which it can harness to create its own competitive advantage over other firms. These are described below.

Creation of a common knowledge context

In relation to knowledge, a firm is composed of individual people who each possess their own knowledge frame. These individuals share information and data and share common (similar) learning experience through their work activity. However, the firm does not only provide an arena for sharing knowledge. The sharing of knowledge in a firm also leads to the development of a unique 'knowledge context', associated with the emergence of a community of practice in a knowledge context through which a firm seeks to co-ordinate contextual knowledge (Barley 1996).

What individuals within the firm know and can achieve depends in part on the nature and characteristics of the firm's context. Over time, therefore, each individual develops a subset of their knowledge frame (relating to work and the firm) which in many respects is similar to the knowledge frame subset of other workers within the firm (as such individuals working within the same firm can develop similar path-dependent knowledge contexts). The extent to which they are equivalent knowledge subsets (between any two workers) depends on a number of parameters; for example, on how long they have worked in the firm or whether they work in the same team, group or site with that individual. However, it also depends on the degree of similarity of their knowledge frames before they started work with that firm; in relation to, for example, coming from the same community, having similar educational backgrounds and social history and class status (e.g. between workers in a local factory). The firm also has a set of codified knowledge contexts involving shared information and data sets (personnel files, financial data and technical information and blueprints); shared routines and patterns of working; and common decision-making procedures. All these commonalities can be said to form a common organisational information and routine base which many would equate, at least in part, with an organisation's 'culture' (Hampden-Turner 1994). It is also important to acknowledge that knowledge sharing may not always take place. Thus, Cowan and Foray (1997: 605) have noted that indeed R&D workers (in part because they were more sophisticated ICT users) may form a 'clique'; a restricted community or closed network with little knowledge being shared outside this group with the rest of the firm.

The shared knowledge framework described above, equates with Dosi and Marengo's (1993) term 'common knowledge basis' where people co-ordinate their actions and base such co-ordination on a common framework. A 'common knowledge basis' is simply defined by Dosi and Marengo (1993: 169) as 'a common language that enables communication and co-ordination' and thereby provides a shared cognitive framework for workers.[3] This fits in with the wider notion of knowledge outlined above as it acknowledges that knowledge resides in the individuals working within the firm rather than representing some kind of 'meta' knowledge possessed by the firm.[4] Without such a common knowledge basis, there is a danger that individuals within the firm can, when sending pieces of information between each other, lead to a misunderstanding of such information because the receiver has a different information-processing capability than the sender. Such a common knowledge coding format therefore helps establish a firm's distinctive set of decision rules or routines. Such routines in turn help shape a tradition of practice within a firm. More importantly, it can enable similar practice to continue within a firm despite losses of personnel. It also allows the formation of collective induction whereby individuals from the same firm can develop a form of reasoning by which a number of observed peculiarities or facts become the basis for a general assertion, which in turn leads to common forms of action.

This individual–collective knowledge tension within the firm relates to the original work of Penrose (1959) and Richardson (1972) who were the first to see the firm as a collection of individuals, each with their own knowledge and aptitude, brought together in teams which are focused around specific activities. This has been taken up by Metcalfe and De Liso (1998: 19–21) who stress in the context of knowledge that the firm is much more than simply the sum of the individuals contained within it and the knowledge and skills that they possess as individuals. They note that the organisation of the firm creates its own distinctive 'signature' by co-ordinating the internal division of labour and by transforming the knowledge and skills of the individual members into a collective capability; what McKelvey (1982) has termed a 'competence pool'. The tacitness of much knowledge, its indivisibility in use, the uncertainty of its values in different contexts, its proprietary nature, and the fact that much of what is known is jointly produced by the firm's activities (and indeed decays if the activity ceases; Metcalfe and De Liso 1998: 21) means that the firm provides a central contextual role in harnessing knowledge to produce new innovative capabilities. For this role to be successful though, a firm must have in place an effective organisational memory to act as a shared repository for the information and data it has collected and used (Lewis 1997: 97).

Forum for knowledge absorption, sharing and transformation

Behind this general contextual function the firm has more specific knowledge roles. The ability of firms' to establish an effective knowledge-sharing framework is at the core of a firm's strategic competence or capability. Thus the competitive value of a firm's capability crucially depends on its ability to absorb, configure and transform information. Firms in this way act as knowledge filters and condition the way that information and data are absorbed, shared and transferred. A firm's set of routines provide special mechanisms for searching, gathering and processing information and data.[5] In so doing it shapes the scope and direction of the search for knowledge (Penrose 1959: 77). These organisational routines 'transcend individual contributions, being superindividual and not reducible to individual memory' (Teece 1987: 44).

Firms thus provide knowledge frameworks for individuals working within the firm:

1 allowing workers to determine where and when to find new information and data ('searching routine');
2 providing a decision-framework for individuals for filtering this information in terms of its usefulness to the firm ('decision routine'); and
3 once this information has been accepted (or rejected) providing a context for how that information is absorbed ('absorptive framework and capacity').

In this latter respect, Cohen and Levinthal (1990) have emphasised the role of the firm's ability to absorb new technology and knowledge, although the searching and decision routines are equally important to a firm's success in providing a knowledge context. The searching, accessing and ability to absorb new information provide the 'front end' functions in relation to knowledge and innovative capacity. To be able to effectively utilise (and generate; see below) and apply (Grant 1996) these new knowledge sources by its individuals, a firm needs successful organisational routines and mechanisms that allow the sharing and transformation of this information and knowledge iteratively.

Forum for new knowledge generation and learning

In relation to innovation, the firm critically has a pivotal role to play in providing a forum for the generation of new knowledge which forms part of the innovation process. The firm did not always hold this central role in innovative activity. Up until the end of the nineteenth century the 'heroic' model of innovation was current with individuals working on their own as lone inventors and these individuals formed the main generators of technological innovation (subsequently taken up by firms for manufacture). Individuals no longer hold this role and the main arena for techno-logical innovation is within the firm. The firm allows individuals to come together to work on complex phenomena leading to the development of new products, processes or 'ways of doing things'.

The firm, or more specific knowledge arenas within the firm, such as a research laboratory, pilot plant or factory floor, allow individuals to come together to share existing knowledge but more particularly to allow them to generate new knowledge through teamwork and joint learning. This is particularly important where new knowledge is especially 'sticky' and tacit in nature and is not readily shared over long distances. Such joint teamwork can lead to the generation and development of new products and processes, but it can also lead to new organisational routines and work methods which will allow improvements in productivity, cost reduction and improved work quality. Learning is an important element in the generation of new knowledge and although this process may lead to new routines it is also dependent on existing routines to provide them. Indeed Dosi *et al.* (1995) identify a set of routines, 'dynamic routines', associated with learning which are explicitly directed at developing new knowledge and in turn new products and processes. These routines help individuals within the firm 'to ascertain where to probe, how to probe and how much to probe' in relation to the knowledge process (Dosi *et al.* 1995: 192).

The world within and beyond the firm: the role of geography in knowledge and innovation

This analysis has so far stressed the importance of the individual in the knowledge process and has then moved on to consider the collective situation of the firm as a key forum where knowledge is generated and utilised in the innovation process. However, geographical location is important in this process as it helps condition many of the operations of the individual firm in terms of knowledge and innovative activity as well as shaping and conditioning the inter-relationship of firms between each other. This has taken on an increasing significance as firms are having to rely not only on in-house knowledge and technical resources in relation to technological innovation, but also on resources external to the firm. Many significant innovations are now taking place through the combination of the capabilities of several firms and research organisations. These 'Distributed Innovation Processes' (DIPs; Coombs and Metcalfe 1998) create cross-firm capabilities in technological innovation. Firms can collaborate in various ways to produce new knowledge, one of the oldest in the UK is from industrial research associations (Johnson 1973), but there have been a wide variety of new distributed mechanisms that have emerged to allow such inter-firm and inter-organisational knowledge collaboration (Coombs and Metcalfe 1998: 23). The following section therefore considers a number of ways in which geographical location influences these knowledge and innovation processes at both an individual firm level and also in relation to these distributed inter-organisational settings.[6]

Localised patterns of communication: A key issue where geographical location influences knowledge processes is in the field of communication. Geographical distance has a profound influence on the likelihood of individuals within and between organisations sharing knowledge and information links (Allen and Cohen 1969; Allen 1977). These proximal knowledge and information links can also be seen within particular relations such as between user-producers (Lundvall 1988: 354–5). The problem of communication and information transfer is therefore not just an inter-firm issue, but remains a central barrier to effective knowledge sharing in multi-site (particularly multi-national) firms (Howells 1998). Undoubtedly advances are being made in ICTs which are helping to reduce the impact of distance in certain instances (see below), but nonetheless the key issue here remains the importance of face-to-face contact in research and technical communication and the essential tacit nature of much that is being communicated. This makes geographical nearness a crucial factor in innovation. Indeed von Hippel (1994) terms this pattern of close, informal knowledge links that are difficult to recreate over wider geographical spans, as 'locational stickiness', which is in turn associated with the cost of communicating and transferring knowledge over space (Hu 1995).

Localised innovation search and scanning patterns: Geographical proximity also has a fundamental influence on how a firm searches for a collaborative partner in terms of research or in identifying a new, more technically sophisticated component or piece of equipment. This is particularly true of smaller firms which have spatially much smaller scanning fields than larger, multi-site companies (Taylor 1975). The ability of smaller firms, therefore, to be aware of, and identify, collaborative partners or technology sources will be due to much more geographically restricted scanning fields. Hence the total number of potential contacts will also tend to be smaller (MacPherson 1991). Some of these problems may be overcome by locating in more 'information rich' and contact-intensive innovation agglomerations where firms can improve their chances of making more effective technical linkages. Indeed this is a key underlying thesis of the advantage and growth of Marshallian industrial districts (Marshall 1899).

Localised invention and learning patterns: Innovation is often undertaken to resolve specific technical problems experienced in producing goods and these advances in manufacturing technology often occur as a response to particular local contexts (Pred 1966: 91). Studies have also stressed the highly localised nature of many forms of learning experience both in terms of what may be called psychological distance and technological proximity (Nelson 1987), but also geographical distance as well (Arcangeli 1993). Cleeremans and McCellard (1991) have similarly shown that learning from events is difficult when those events are separated too widely, spatially or temporally. 'Learning by doing' first highlighted by Arrow (1962) and 'learning by using' (Rosenberg 1963) have been increasingly acknowledged as key components in successful innovation and are linked to direct and continuous physical proximity to the production process. The rise of the 'learning economy' with its emphasis on know-how sharing in the innovation process rather than knowing how to do things in isolation (Lundvall and Johnson 1994) has heightened the importance of spatial proximity and local relationships in technological and industrial performance (Camagni 1991). It has also become part of a more conscious process by firms of shared learning experiences (Alder 1990).

Localised knowledge sharing: It was stressed earlier that the firm acts as a localised knowledge sharing mechanism not only in a cultural and social sense, but also in a geographical sense. A key element in this geographical sense is the localised nature of tacit knowledge acquisition and transfer (Howells 1996). This is centred around complex and loosely structured personal contact and discussions (Pavitt 1991: 47) and informal information flows (MacDonald 1997) both within and between the firm. Localities can therefore form 'knowledge pools' for particular sectors or technologies (Henry and Pinch 2000). Thus, although many parts of the innovation process can be codified and easily transferred over long distances, many elements of technological innovation remain tacit in form and indeed these may be the elements that have the most impact on corporate performance for the very reason that they are so difficult to learn off-site and to transfer to a different location.

Localised patterns of innovation capabilities and performance: Lastly, localisation and agglomeration can also help reduce the risks and uncertainty of innovation and increase its likelihood of success. Dislocation of risk and information, whereby decision-makers are dislocated from the information to manage and reduce these risks, can form significant barriers to innovation (More 1985: 206). Technologies tend to go through a shake-down process in their early stages and go through a period of high levels of uncertainty associated with the technical and economic performance of a new product or technique (Gertler 1995: 3). It is the reduction in uncertainty following use by early risk-takers that helps build and sustain a 'bandwagon effect' in innovation diffusion (Mansfield 1961). A concentration of early adopters provides a local economy with an advantage in terms of technological innovation more generally. High concentrations of information flows produce lower perceived and actual risks and uncertainty associated with innovation. Similarly some routines and competences of firms are attributable to local forces that shape a firm's capability at the early stages in their lives (Teece and Pisano 1994). This is often supported in turn by high levels of trust and reciprocity (Hansen 1992: 103) and enables local and regional economies to improve their chances of success leading to improved innovative performance through the creation of such 'commensualistic environments' (Ring 1992).

Declining significance of geography?

The above has stressed the significance of geographical localisation in knowledge and innovation processes, but is its role and significance waning? The arguments for the declining role of geography in knowledge and innovation production centre on the increasing ease of information flows via improvements in communication. This in turn centrally involves the increasing codification and abstraction of knowledge. Codified and abstract knowledge is easier to transfer using new forms of communication technology (although this is not such a novel suggestion; the rise of the telegraph in the United States was also seen to have such an effect; Pred 1966).

Proponents of this view argue that because knowledge is increasingly susceptible to being abstracted and codified, this will have major consequences for the dissemination, transfer and retention of knowledge, and in turn the innovation process. In short, codified knowledge is seen as being more easily transferable (Nonaka 1994; Boisot 1995a, 1998; Hu 1995; Antonelli 1995; Cowan and Foray 1997). In relation to communication, the historic trend has clearly been towards decreased 'friction of distance' in relation to information and knowledge links (Abler 1971) and this has gained particular momentum in the late 1980s and early 1990s with the decline in telecommunication costs and the introduction and spread of new ICTs around the world.

Scientists and engineers down through the ages have tried to make their results and applications more widely available to their respective communities and this has been dependent on making their tacit knowledge and skills more formal and codifiable (in Nonaka's model of 'knowledge conversion' this involves the third mode of knowledge conversion from tacit knowledge to explicit knowledge; Nonaka 1994; see also Nonaka *et al.* 1994). Indeed, Arora and Gambardella (1994) argue that technological information is already cast in frameworks and categories that are universal, allowing a clear division (and dispersion) of innovative labour. Moreover, codification and formalisation of knowledge and skills has been applied widely to the manufacturing and research environment. The 1950s and 1960s witnessed the increasing use of automation as an attempt to capture tacit skills and codify them for replication by machine. This was most marked in the rise of early programmable machine tools which reproduced the skills (if not the knowledge) of the machinist by recording key actions on magnetic tape so that replaying the tape would reproduce the machines' motions under the operator's control. More recently, towards the end of the 1970s, major corporations have sought to improve transborder information flows in the research and design process through the use of new ICTs. A number of multinational companies have integrated their global R&D units via the use of such ICTs and have reduced the need for face-to-face contacts and spatial proximity (Howells 1995).

In contrast to this increasing codification of knowledge and ease of transferability are those who have emphasised the role and competitive advantage of competences in tacit capabilities and 'soft technologies' (Morgan 1990; Grosse 1996), often seen to be displayed by Japanese firms. This view stresses that tacit knowledge will remain a fundamental component of the innovation process and more particularly the innovative advantage that firm's possess is established because of the very reason that it is not easily appropriated by other firms. Indeed Gibbons *et al.* (1994: 26) go further by stressing that the competitiveness of firms is now centred on their tacit knowledge base rather than their pool of proprietary knowledge. Many of these advances and developed competences are contingent on local conditions and much of the learning and transferring associated with them is highly constrained by spatial distance. Thus, in the case of the learning aspect of tacit knowledge, much of it involves 'situated' learning where information that is being transferred or where the problems to be overcome depend on being present in the same physical location (Lave and Wenger 1991; von Hippel 1994; Fleck 1994). Moreover much of this type of knowledge can only be accessed and transferred through intimate social interaction (Kogut and Zander 1996). Moreover, Gibbons *et al.* (1994: 17) envisage tacit components of knowledge 'taking precedence over the codified components under Mode 2 conditions of knowledge production.'

There are therefore two contrasting views about whether tacit (versus codified or explicit) knowledge is becoming less significant over time and related to whether the 'tyranny of distance' in relation to knowledge and innovative activity is also becoming less important (Howells 1998). These are crucial issues in how we might expect the international knowledge and innovation system to be shaped in the future and where the competitive balance of firms, industries and nations will lie. However such bi-polar visions are arguably misleading for two reasons. First, because as noted earlier, Polanyi repeatedly stressed that the split between tacit knowledge and codified/ explicit knowledge was artificial. Tacit knowledge remains an essential prerequisite in *all* knowledge activities. Thus we still rely on elements and frameworks of tacit knowledge when dealing with explicit knowledge. Explicit knowledge depends on tacit knowledge and we as individuals use tacit knowledge in our framework of understanding any part of our whole knowledge construct (Polanyi 1966; Tsoukas 1996). Second, there may be a kind of knowledge life-cycle operating here in relation to innovation and technological change. Thus knowledge (and innovation) at the 'leading edge' will always have a high tacit component, but this becomes less significant with knowledge becoming more susceptible to codification over time. Indeed, advances in knowledge and techniques still tend to be associated with new tacit knowledge (Senker 1995). Thus the relative importance of tacit knowledge *vis-à-vis* codified knowledge in relation to technological innovation will change over time in relation to specific technologies and their trajectories.

Conclusions

In conclusion to this review and analysis, knowledge is centrally about the individual although the firm represents a complex mechanism for the collective gathering, sharing and processing of this individually-centred knowledge. To be effective therefore the firm needs to be able to harness and co-ordinate the knowledge that individual workers possess and be able to exchange information with its external collaborators and suppliers. This harnessing and co-ordination of know-how in turn depends on successful mechanisms and routines involved in shared knowledge activities, such as joint learning, research, design and testing associated with innovative activity. Storage of shared information and mechanisms to create an effective organisational 'memory' of successful routines and repertoires of action is also required.

Location is important to all these knowledge activities, centrally influencing the process of know-how generation, interaction and sharing. However, this influence is not constant and some observers believe that the rise of new ICTs, which facilitate both the transfer but also abstraction and codification of knowledge, will entail a diminishing role for geography and distance in relation to knowledge and innovation. By contrast, the view that tacit and informal knowledge will remain significant especially in relation to innovation and the early stages of new technology suggests that location will still remain profoundly important. Undoubtedly, the impact of geography on knowledge and innovation will change as it has done under previous rounds of technological innovation which have fed back into how knowledge, research and other innovative activities have been organised and structured. Moreover, just as each firm's competence and capabilities regarding knowledge are unique, so will the impacts of (and their responses to) such changes be unique and diverse.

Notes

1 Basically a procedure or strategy for solving a problem, or moving towards a solution of a problem. This infers in more general use, the process or activity which leads to a discovery or an invention Plotkin (1994: 250).
2 This has parallels with Johnson-Laird's (1983) notion of 'mental models'.
3 This echoes Arrow's (1974) notion of an organisation's 'coding'.
4 Although Dosi and Marengo (1993: 169) go on to discuss the issue of 'organizational' knowledge which is more problematic to the conceptualisation of knowledge presented here.
5 The wealth of research on industrial location decision-making provides a direct example of this.
6 See Howells (1999) for a more detailed discussion of this process in relation to a firm's regional innovation context.

5 The state and the contradictions of the knowledge-driven economy

Bob Jessop

The modern world is swept by change. New technologies emerge constantly, new markets are opening up. There are new competitors but also great new opportunities. Our success depends on how well we exploit our most valuable assets: our knowledge, skills and creativity. These are the key to designing high-value goods and services and advanced business practices. They are at the heart of a modern, knowledge-driven economy. This new world challenges business to be innovative and creative, to improve performance continuously, to build new alliances and ventures. But it also challenges governments: to create and execute a new industrial policy.

> The Government must promote competition, stimulating enterprise, flexibility and innovation by opening markets. But we must also invest in British capabilities when companies alone cannot: in education, in science and in the creation of a culture of enterprise. And we must promote creative partnerships which help companies: to collaborate for competitive advantage; to promote a long term vision in a world of short term pressures; to benchmark their performance against the best in the world; and to forge alliances with other businesses and with employees.
>
> (Blair 1998)

This quotation from Tony Blair's foreword to the Department of Trade and Industry's White Paper on *Building the Knowledge Driven Economy* encapsulates several key themes in this chapter.[1] But my approach to them is very different, inspired as it is by regulation school and neo-Marxist state theory (see Boyer and Saillard 1995 and Jessop 1990). In particular I address four issues: the general relationship between technological change and capitalist development; the changing articulation of the structural forms and contradictions of capital accumulation; the state's changing roles in developing the information economy and society; and the repercussions of these changes upon the state's institutional architecture.

Technological change and capitalist development

Discussions of the information revolution or informational capitalism often treat knowledge as a factor of production similar to land, capital, enterprise, or labour. This is linked with a periodisation of modes of development based on changes in the primary factor of production for wealth creation. A common periodisation sees a transition from agriculture (land) through industrialism (capital and manual labour) to 'informationalism' (information and communication technologies and intellectual labour). Such analyses tend to naturalise factors of production, obscuring the conditions under which they enter the economic process and get combined to produce goods and services. They thereby reproduce the fallacy, criticised by Marx, that value is rooted in immanent, eternal qualities of things rather than in social relations (Marx 1976: 993; Schiller 1988: 32).

Focusing on social relations is important not only for a general understanding of the capital–labour relation but also for analysing the role of information, knowledge, and intelligence in so-called post-industrial economies. For labour and knowledge are both fictitious commodities. One must ask under what conditions each gains the form of a commodity. On the one hand, since labour-power is not produced by capitalists for profit, it acquires a commodity form only insofar as it enters labour markets and is employed in the labour process. On the other hand, as knowledge is collectively produced and is not inherently scarce (in economic terms, it is a 'non-rival' good), it only acquires a commodity form insofar as it is made artificially scarce and access thereto depends on payment of rent (Kundnani 1998–9: 54–5; Frow 1996: 89). Hence, instead of naturalising knowledge, one should assume that 'information is not inherently valuable but that a profound social reorganisation is required to turn it into something valuable' (Schiller 1988: 32). It is the state's role in this regard that concerns us below.

Marx (1859) suggested that the most fundamental contradiction in *class-based* modes of production is that between the increasing socialisation of productive forces and private control of the means of production. A key issue today is whether the move from industrialism to informationalism changes this contradiction enough to justify the claim that capitalism has given way to a new mode of production. This claim is advanced in Castells' analysis of the information economy. But his account is ambivalent in three respects. First, although he notes that information and communication technologies (ICTs) have specific historical roots in the military-industrial complexes of advanced capitalism, he also tends to naturalise knowledge as a factor of production and thus locates its origins outside specific class relations. Second, although he emphasises that informationalism (or at least the production of knowledge) involves a new kind of social organisation, namely, a *networking* logic, he also claims that capitalism has used this to reinvigorate itself after its mid-1970s crisis. And, third, although Castells argues that the reflexive use of knowledge can enhance the socialisation of the productive forces 'in a cumulative feedback loop between innovation and uses of innovation' (1996: 32), he also notes that this reflexivity is mostly practised by specific fractions of capital in their own interests (1996: 52, 58–60, 81–90).

Such ambivalence in a noted theorist of contemporary social transformation suggests a tension between the dynamics of informationalism and capitalism. This prompts the question whether the increasing socialisation of the productive forces in a knowledge-driven economy (expressed in dynamic forms of networking and learning) is coming into conflict with capitalist dominance in the social relations of production. This could involve capital hindering the realisation of an information society and/or informationalism eroding private control through its emerging networked forms of governance. These may not, of course, be the only alternatives. But exploring potential contradictions between informationalism and capitalism certainly provides an interesting way to think about the state's role in what is often described as a 'knowledge-driven economy'.

The contradictions of capital accumulation

Marx explored the implications for capitalism's dynamic of treating labour power *as if* it were a commodity. For this shapes the forms of economic exploitation, the nature and stakes of class struggle between capital and labour in production, and the competition among capitals to secure the most effective valorisation of labour-power. It also affects the forms and stakes of politics and the overall nature of societalisation. An analogous argument can be made for knowledge. Knowledge has always been important economically and especially in the major shifts associated with long waves of technological innovation. What is novel in the current period is the growing application of knowledge to the production of knowledge in developing the forces of production; and the increased importance of knowledge as a fictitious commodity in shaping the social relations of production. This does not mean that knowledge is a real commodity or that its exchange-value equals the costs of the commodities consumed in its reproduction. For knowledge is a collectively generated resource and, even where specific forms of intellectual property are produced in capitalist conditions for profit, this depends on a far wider intellectual commons. The exchange-value of commodified knowledge is also hard to measure, of course, owing to the well-known peculiarities of the economics of information. These include the phenomenon that the use-value of knowledge *qua* non-rival good does not diminish when that knowledge is shared – and may even increase thanks to network economies – with corresponding problems for a purely market-led determination of output and price. The complexities of knowledge generation and its different forms of embodiment and embeddedness – especially in a networked economy – also make it hard to establish how knowledge in its various forms contributes to surplus-value and profits. All of this renders implausible a naturalised 'knowledge theory of value' (Bell 1973: 127) but it does still permit a *value theory of knowledge* that would assess the implications of treating knowledge *as if* it were a commodity.

It is worth noting here at least three processes involved in transforming knowledge into a fictitious commodity: the first is its formal transformation from a collective resource ('intellectual commons') into intellectual property (e.g., patent, copyright) as a basis for revenue generation; the second is the formal subsumption of knowledge production under exploitative class relations through the separation of intellectual and manual labour and the transformation of the former into wage labour producing knowledge for the market; and the third is the real subsumption of intellectual labour and its products under capitalist control through their commoditisation and integration into a networked, digitised production–consumption process that is controlled by capital (on the first, see, for example, Aoki 1998; Dawson 1998; on the second, Schiller 1988: 33 and Sohn-Rethel 1978; on the third, see Menzies 1998: 92–3; and Kelly 1998: 77).

These changes transform the contradiction between the socialisation of the productive forces and the private appropriation of profit. This is now reflected in the contradiction between knowledge as intellectual commons and as intellectual property. This is hardly surprising. For this fundamental contradiction has distinctive forms in different times and places. In the case of capitalism, for example, its core contradictions can be analysed in terms of: (a) the general contradictions inherent in the commodity form – as reinforced by (b) the specific contradictions inherent in generalising this form to money, land, and, above all, labour-power and (c) the inevitable dependence of the commodity form not only on fictitious commodities but also on various non-commodity forms of social relations. The very process of commodification rooted in the spread of the market mechanism generates contradictions that cannot be resolved by that mechanism itself. For example, the commodity is both an exchange-value and a use-value; the worker is both an abstract unit of labour power substitutable by other such units (or, indeed, other factors of production) and a concrete individual with specific skills, knowledge, and creativity; the wage is both a cost of production and a source of demand; money functions both as an international currency and as national money; productive capital is both abstract value in motion (notably in the form of realised profits available for re-investment) and a concrete stock of time- and place-specific assets in the course of being valorised; and so forth. These contradictions are linked to strategic dilemmas concerning the relative primacy of their different 'moments'. They are also more or less manageable depending on specific 'spatio-temporal fixes' and the nature of the institutionalised class compromises with which these fixes may be associated.

Spatio-temporal fixes

Reproducing and regularising the capital relation involves imposing a 'spatio-temporal fix' on its extra-economic as well as economic moments and seeking some structural coherence in handling the contradictions inherent in its various structural forms and strategic dilemmas as they present themselves in particular periods. This fix has both strategic and structural dimensions.[2] Strategically, since the contradictions and dilemmas are insoluble in the abstract, they can only be resolved – partially and provisionally at best – through the formulation-realisation of specific accumulation strategies in specific spatio-temporal contexts (Jessop 1983). Such strategies seek to resolve conflicts between the needs of 'capital in general' and particular capitals by constructing an imagined 'general interest' that marginalises at least some capitalist interests. Moreover, since capital accumulation depends on extra-economic as well as economic conditions, accumulation strategies also seek to institutionalise class compromise and address more general problems of social cohesion.[3]

Accumulation strategies and/or hegemonic projects typically displace the costs of such institutionalised compromise beyond the social, territorial, and temporal boundaries of that compromise. This can involve super-exploitation of internal or external spaces outside the compromise, super-exploitation of nature or inherited social resources, deferral of problems into the future, and the exploitation and/or oppression of specific classes or social categories. Success in hegemonic struggles over these strategies always depends on particular spatio-temporal fixes which cannot be guaranteed; and is often secured only through a trial-and-error search that reveals the requirements of 'capital in general' more through repeated failure than sustained success (Jessop 1983; 1999). Different aspects of capital's structural contradictions and/or alternative horns of resulting strategic dilemmas may be handled through a scalar division of labour within or across different institutions, apparatuses, or agents.

There may also be a temporal division of labour with different institutions, apparatuses, or agencies specialising in action over different time horizons. In both cases there may also be scope for 'meta-governance' activities to re-balance the role of these institutions, apparatuses, or agencies in various ways (on meta-governance, see Dunsire 1996; and Jessop 1998).

Regulating Atlantic Fordism

Atlantic Fordism benefited from a spatio-territorial matrix based on the congruence between national economy, national state, national citizenship, and national society; and from institutions relatively well adapted to combining the tasks of securing full employment and economic growth and managing national electoral cycles. Given the terms of the Fordist industrial paradigm and the temporality of its business cycle, the various contradictions of capitalism were dealt with largely in a national context by treating the wage relation as the primary site of contradiction and articulating its regularisation primarily to the money form. This is the core significance of the Keynesian welfare national state (hereafter KWNS). Thus, within relatively closed national economies which had been institutionally-discursively constituted as the primary objects of economic management, national states aimed to achieve full employment by treating wages primarily as a source of (domestic) demand and managed their budgets on the assumption that money circulated primarily as national money. The diffusion of mass production (and its economies of scale) through expanding Fordist firms as well as the development of collective bargaining indexed to productivity and prices were the primary means through which wages as a cost of production were brought under control. And the combination of the Bretton Woods monetary regime and the GATT trade regime ensured that the (still limited) circulation of free-floating international currencies need not seriously disturb Keynesian economic management based on state control over the national money. Welfare rights based on national citizenship helped to generalise norms of mass consumption and thereby contributed to full employment levels of demand; and they were sustained in turn by an institutionalised compromise involving Fordist unions and Fordist firms. Securing full employment and extending welfare rights were in turn important axes of party political competition and were a political reflection of the institutionalised class compromise.

Some costs of the Fordist compromise and the KWNS were borne within Fordist societies themselves by the relative decline of small and medium firms, by workers employed in disadvantaged parts of segmented labour markets, and by women subject to the dual burden of paid and domestic labour. Other costs were borne beyond Fordist societies by economic and political spaces that were integrated into international regimes (such as those for cheap oil or migrant labour) necessary to Atlantic Fordism's continued growth but not included within the Fordist compromise. Atlantic Fordism was also enabled through a Janus-faced temporal fix. On the one hand, it depended on the rapid exploitation of non-renewable resources laid down over millennia (notably the 'subterranean forest' of fossil fuels as well as raw materials); and, on the other hand, it produced environmental pollution and social problems that it did not address within its own temporal horizons – as if working on the principle of *après moi, la déluge* (see, for example, Altvater 1993; Brennan 1995; Stahel 1999).

The crisis of Fordism is inevitably overdetermined. But a major contributing factor was the undermining of the national economy as an object of state management – notably through the internationalisation of trade, investment and finance. This led to a shift in the primary aspects of its two main contradictions. Thus the wage (both individual and social) came increasingly to be seen as an international cost of production rather than as a source of domestic demand; and money came increasingly to circulate as an international currency, thereby weakening Keynesian economic demand management on a national level. This shift in the primary aspect of the contradiction in the money form is related to the tendency for the dynamic of industrial capital to be subordinated to the hypermobile logic of financial capital and the tendency for returns on money capital to exceed those on productive capital. At the same time the relative exhaustion of the Atlantic Fordist growth dynamic posed problems of productivity growth and market saturation (which combined to intensify an emerging fiscal crisis of the state) and problems of how best to manage the transition to the next long wave of economic expansion (which entails changes in the temporal horizons of state economic intervention and thus in the forms and mechanisms of such intervention). The crisis of US hegemony is also reflected in struggles over the shaping of new international regimes and the extent to which they should serve particular American interests rather than the interests of capitalism more generally. New conflicts and/or forms of struggle have also emerged that escape stabilisation within existing structural forms: two major examples are the crisis of corporatism and the rise of new social movements. New problems have also emerged, such as pollution and new categories of risk, which are not easily managed, regularised, or governed within the old forms. Finally we should note that, relative to the growth phase of Atlantic Fordism, some contradictions have increased in importance and/or acquired new forms. Three of these are directly relevant to the knowledge-driven economy.

New contradictions in a globalising, knowledge-driven economy

This section discusses three of these new (or newly important) contradictions together with a major conflict that marks the current transition from Fordism to post-Fordism. The contradictions comprise: first, a dissociation between abstract flows in space and concrete valorisation in place; second, a growing short-termism in economic calculation vs. an increasing dependence of valorisation on extra-economic factors that take a long time to produce; and, third, the contradiction between the information economy and the information society as a specific expression of the fundamental contradiction between private control in the relations of production and socialisation of the forces of production. In addition, although it does not as such constitute a structural contradiction, serious conflicts arise over the appropriate horizons of action for the spatio-temporal fix, if any, within which the old principal contradictions of Atlantic Fordism and the newly important contradictions of the current period might prove manageable.

The first contradiction is an expression of the fact that 'the new economy operates in a "space" rather than a place, and over time more and more economic transactions will migrate to this new space' (Kelly 1998: 94). This is a complex, non-propinquitous, multidimensional, cyberspace with novel spatial dynamics grounded in the possibilities that cyberspace offers for simultaneous co-location of myriad entities and relationships. Nonetheless cyberspace is not a neutral, third space between capital and labour, market and state, public and private: it is a new terrain on which conflicts between these forces, institutions and domains can be fought out. An often-cited expression of this contradiction is the institutional separation of hypermobile financial capital from industrial capital – with the former moving in an abstract space of flows, the latter still needing to be valorised in place. But it also appears in the individual circuits of financial, industrial and commercial capital as well as within their interconnections. For, however much capital migrates into cyberspace, it still depends on territorialisation. In the case of global finance capital, of course, this 'fix' is provided by the grid of global cities (Sassen 1996). In the case of industrial capital, it is innovation milieus, industrial districts, etc., as well as the physical infrastructure described by Harvey (1982). Even e-commerce needs such an infrastructure, even if it involves a 'celestial jukebox' sending digitised music on demand. Thus, the globalising knowledge-driven economy does not signal the final transcendence of spatial barriers but effects 'new and more complex articulations of the dynamics of mobility and fixity' (Robins and Gillespie 1992: 149).

The second contradiction is grounded in the paradox that '(t)he most advanced economies function more and more in terms of the extra-economic' (Veltz 1996: 12). The paradox rests on the increasing interdependence between the economic and extra-economic factors making for structural or systemic competitiveness. This is linked to the growth of new technologies based on more complex transnational, national, and regional systems of innovation, to the paradigm shift from Fordism with its emphasis on productivity growth rooted in economies of scale to post-Fordism with its emphasis on mobilising social as well as economic sources of flexibility and entrepreneurialism, and to the more general attempts to penetrate micro-social relations in the interests of valorisation. It is reflected in the increasing emphasis given to social capital, trust, and communities of learning as well as to the increased importance of competitiveness based on entrepreneurial cities, an enterprise culture and enterprising subjects.

This generates major new contradictions that affect the spatial and temporal organisation of accumulation. Thus, temporally, there is a major contradiction between short-term economic calculation (especially in financial flows) and the long-term dynamic of 'real competition' rooted in resources (skills, trust, heightened reflexivity, collective mastery of techniques, economies of agglomeration and size) that may take years to create, stabilise, and reproduce. Paradoxically, reflexivity enhances this contradiction: it takes time to create collective learning capacities but '(t)hose firms, sectors, regions and nations which can learn faster or better (higher quality or cheaper for a given quality) become competitive because their knowledge is scarce and cannot be immediately imitated by new entrants or transferred, via codified and formal channels, to competitor firms, regions or nations' (Storper 1998: 250). Spatially, there is a fundamental contradiction between the economy considered as a pure space of flows and the economy as a territorially and/or socially embedded system of extra-economic as well as economic resources and competences. The latter moment is reflected in a wide range of emerging concepts to describe the knowledge-driven economy – national, regional, and local systems of innovation, innovative milieus, systemic or structural competitiveness, learning regions, social capital, trust, learning-by-doing, speed-based competition, etc. This poses new dilemmas if the capital relation is to be stabilised over an expanding range of scales and over increasingly compressed as well as extended temporal horizons of action.

A third contradiction that becomes important once again in the after-Fordist (or, at least, the post-industrial) accumulation regime is that between the increasing socialisation of the productive forces and private control in the social relations of production. For the nature of networked knowledge-driven economies – in which economies of agglomeration and 'economies of networks' gain in importance – heightens the contradiction from both sides. Two features enhance the socialisation of productive forces in networked knowledge-driven economies. First, 'economies of networks' are generated in and through multi-actor, polycentric and multiscalar networks rather than by single (or quasi-vertically integrated) organisations, which are better able to realise economies of scale. Second, almost exponentially increasing returns to network size mean that 'each additional member increases the network's value, which in turn attracts more members, initiating a spiral of benefits' (Kelly 1998: 25). These two features pose collective action problems rooted in the conflict between socialisation and private appropriation – problems that are linked to tendencies to market failure noted even in orthodox studies of the 'economics of information'. In particular, the socialisation of knowledge production makes it hard to distinguish legally between different firms' intellectual property (Kundnani 1998–9: 56) and this reinforces the tendency for network economies to be captured by the network – albeit often asymmetrically – rather than by a single firm (Kelly 1998: 26–8). This suggests the need for new forms of enterprise able to capture such network economies without destroying any broader network(s) involved in generating them. 'Virtual' firms and networked firms are said to correspond to this need (e.g., Castells 1996: 151–200). However, unless the 'virtual' firm becomes co-extensive with all those involved in production, the contradiction is still reproduced on the side of the social relations of production. For, whereas every capital wants free access to information, knowledge, and expertise, it also wants to charge for the information, knowledge, and expertise that it itself can supply.

A fourth site of problems concerns the appropriate horizons of action for the spatio-temporal fix, if any, within which the old principal contradictions of Atlantic Fordism and the newly important contradictions of the current period might prove manageable. This is closely related to a new complexity of time-space in informational capitalism due to the interaction of new forms of 'time-space distantiation' and 'time-space compression'. Time-space distantiation stretches social relations over time and space so that they can be controlled or co-ordinated over longer periods of time (including into the ever more distant future) and over longer distances, greater areas, or more scales of activity. In this regard, then, globalisation results from increasing spatial distantiation reflected in the growing spatial reach of divisions of labour in different fields and is enabled by new material and social technologies of transportation, communication, command, control, and intelligence. Conversely, time-space compression involves the intensification of 'discrete' events in real time[4] and/or increased velocity of material and immaterial flows over a given distance. This is linked to changing material and social technologies enabling more precise control over ever shorter periods of action as well as 'the conquest of space by time'. Differential abilities to stretch and/or compress time and space help to shape power and resistance in the emerging global order. Thus the power of hypermobile forms of finance capital depends on their unique capacity to compress their own decision-making time (e.g., through split-second computerised trading) whilst continuing to extend and consolidate their global reach. It is the differential combination of time-space distantiation and time-space compression that was facilitated by new ICTs and enthusiastically embraced by some fractions of capital (and some states) that contributed to the erosion of the spatio-temporal fix of Atlantic Fordism.

This is now reflected in a 'relativisation of scale' (Collinge 1996). The current period involves a proliferation of spatial scales (whether terrestrial, territorial, or telematic, cf. Luke 1994), their relative dissociation in complex tangled hierarchies (rather than a simple nesting of scales), and an increasingly convoluted mix of scale strategies as economic and political forces seek the most favourable conditions for insertion into a changing international order. The national scale has now lost the taken-for-granted primacy it held in the economic and political organisation of Atlantic Fordism; but this does not mean that some other scale of economic and political organisation (whether the 'global' or the 'local', the 'urban' or the 'triadic') has acquired a similar primacy. Indeed there is intense competition between different economic and political spaces to become the new primary anchorage point of accumulation. As yet the new politics of scale are unresolved – although I suspect that 'triads' will eventually replace the nation as the primary scale for managing, displacing, and deferring the contradictions and dilemmas of a globalising, knowledge-driven economy.

Knowledge and its contradictions

I now focus on knowledge and some of the economic issues it raises. If knowledge is becoming fictitiously commodified, one can see it as a site of contradictions analogous to labour-power, land, or money (Polanyi 1957). Five issues can be mentioned here:

- The *primitive accumulation* of capital (in the form of intellectual property) through private expropriation of the collectively produced knowledge of past generations. This enclosure of knowledge takes several forms, including: (a) the appropriation of indigenous, tribal, or peasant 'culture' in the form of undocumented, informal, and collective knowledge, expertise, and other intellectual resources and its transformation without recompense into commodified knowledge (documented, formal, private) by commercial enterprises (Frow 1996: 97–9; Coombe 1998) – bio-piracy is the most notorious example; (b) divorcing intellectual labour from the means of production – embodying it in smart machines and expert systems; and (c) a creeping extension of the limited nature of copyright into broader forms of property right with a consequent erosion of any residual public interest (Frow 1996: 104).
- The role of 'intellectual technology' in the real subsumption of intellectual as well as manual labour. Bell himself notes how this plays a role analogous to that of machino-facture in the subordination of manual labour to capitalist control (Bell 1974: 29; 1979: 167) and Robins and Webster also note how it serves to appropriate the knowledge of the collective labourer (1987: 103).
- The dynamics of *technological rents* generated by new knowledge and their disappearance once the new knowledge (whether as knowledge or as intelligent means of production) becomes generalised and thereby comes to define the socially

necessary labour time embodied in commodities. This problem is intensified by reflexive accumulation. For 'the conditions which a firm, region or production system must now satisfy in order to win are manufactured and remanufactured more thoroughly and more rapidly than ever before, creating a moving target for success and a shifting minefield of risks of failure' (Storper 1998: 249–50). This increases the pressure on firms, regions, or production systems to stay ahead of their competitors so that ever-renewed technological rents and increasing market share can alleviate the normal tendency for super-profits to be competed away. It also encourages attempts to protect vulnerable *monopolies in knowledge or information* by embedding them in technology, standards, tacit knowledge, or legally entrenched intellectual property rights.

- These considerations underline the self-defeating character of the informational revolution from the viewpoint of capital, insofar as each new round of innovation is prone to ever more rapid devalorisation.
- This in turn has implications for social inequality and polarisation within and across national societies. If firms in the information economy are to maintain profit rates despite the tendency for technological rents to be competed away, less technologically advanced sectors must secure below average profits. This is one of the driving forces behind globalisation and the tendencies towards unequal exchange and uneven development with which it is associated. In the longer term, however, this poses problems of demand for the products of the information economy on a global scale.

The contradictions and conflicts considered above are especially clear in the currently dominant neo-liberal form of globalisation. This is reflected in a general tension between neo-liberal demands to accelerate the flow of abstract (money) capital through an increasingly disembedded space and the need for the more concrete forms of capital to be 'fixed' in time and place, as well as embedded in specific social relations, as a condition for their valorisation. The state has a key role in managing this tension.

The nature of the state

Capitalism rests on the institutional separation of the economic and extra-economic. This is needed to secure the appropriate balance between the inherent capitalist drive to commodification and its dependence on non-commodity forms of social relations. This separation is traditionally (but inadequately) understood in terms of the trinity of market–state–civil society. But the changing forms of competitiveness associated with globalising, knowledge-driven economies also involve redefinition of the economic and extra-economic. This is reflected not just in changing forms of state intervention into the operation of market forces but by a more fundamental restructuring, rescaling, and retemporalisation of market–state–civil society relations. This complex process provides the focus of this and the next section.

Let us begin with a brief (and far from exhaustive) list of functions that the state might perform *vis-à-vis* the information economy and/or informational capitalism. They include: (a) managing the fundamental contradiction between the socialisation of productive forces and the relations of production as expressed in the general tension between information society and information economy; (b) re-designing the relationship between the economic and extra-economic in the light of the changing forms of competitiveness associated with the knowledge revolution, reflexivity, and learning; (c) articulating the (de/re-)commodification of knowledge in this context – especially given the fictitious nature of knowledge as a commodity – and dealing with its more general articulation with intellectual and manual labour; (d) articulating the interlinked processes of (de/re-)territorialisation and (de/re-)temporalisation associated with new forms of time-space distantiation and time-space compression in order to create a new spatio-temporal fix for managing the contradictions of the capital relation; and (e) addressing the political and social repercussions of the changing forms of appearance of the structural contradictions and strategic dilemmas of accumulation. Some of these issues were noted above, some are reviewed later. Only the first two are explored in this section.

First, states at all levels help in managing the contradictions rooted in the distinction between the intellectual commons and intellectual property. On the one hand, '(t)he intellectual commons is fundamental to the production of knowledge' (Dawson 1998: 281); and, on the other, intellectual property is a key basis of accumulation in informational capitalism. This contradiction was recognised in Bell's claim that a free circulation of knowledge offers no incentives to firms to produce it so it must be created by some 'social unit, be it university or government' (1979: 174). With hindsight one can see that Bell's proposal is rooted in the earlier logic of the Fordist mixed economy rather than the emerging logic of the networked economy. Nonetheless one can agree with his broad conclusion that states must commit themselves to designing 'a socially optimal policy of investment in knowledge' (1979: 175). Different states are, of course, situated differently in this regard. They tend to polarise, first, around interests in protecting or enclosing the commons (e.g., North–South) and, second, around the most appropriate forms of intellectual property rights and regimes on different scales from global to local. Thus some states are more active than others in promoting the primitive accumulation of intellectual property, in privatising public knowledge, and in commoditising all forms of knowledge; others are more concerned to protect the intellectual commons, to promote the information society, and to develop social capital. Given its competitive advantage in ICT products and the knowledge revolution, the US state is especially important in promoting the neo-liberal form of the knowledge revolution on a global scale.

In all cases states must attempt to resolve various contradictions and dilemmas whilst eschewing any direct, hierarchical control over knowledge production. For example, they 'must balance the need to protect and maintain the intellectual commons against the need to stimulate inventive activity' (Dawson 1998: 278); and, even in the latter context, they need to balance the protection of individual intellectual property (to encourage technological rents) and the general diffusion of its applications 'by creating open systems, by moving key intellectual properties into the public domain, by releasing source code democratically' (Kelly 1998: 28) with the result that individual technological rents are competed away. In many cases this is perhaps best achieved through state promotion of innovation and diffusion systems (including social capital), broad forms of 'technological foresight', co-involvement and/or negotiated 'guidance' of the production of knowledge, and the development of suitable meta-governance structures (see Messner 1997; Willke 1997). This is reflected in the state sponsorship of information infrastructures and social innovation systems on different scales; in the development of intellectual property rights regimes and new forms of governance and/or regulation for activities in cyberspace; and in the movement away from national utility structures with universal supply obligations suited to an era of mass production and mass consumption to more flexible, differential, multi-scalar structures suited to a post-Fordist era.

Second, insofar as 'structural' or 'systemic' competitiveness is held to depend not only on an extensive range of long-acknowledged economic factors but also on a broad range of formally extra-economic institutional forms, relations, resources, and values, etc., there is a fundamental redefinition of the 'economic sphere'. This encourages a transition to a Schumpeterian workfare orientation. This orientation can be described as Schumpeterian because it promotes innovation, competitiveness, and entrepreneurship tied to long waves of growth and to the more recent pressures for perpetual innovation. It can be described as workfare because social policy is increasingly being tied to the demands of flexibility, reskilling, and reflexivity in a 'learning economy'. This is linked to the creation of new capabilities and skills (including, of course, ICT literacy both as producers and consumers) and a growing commitment to life-time learning. States thereby get locked into the pursuit of technological rents on behalf of capital and this leads to the subordination of the totality of socio-economic fields to the accumulation process so that economic functions come to occupy the *dominant place* within the state. Other functions thereby tend to gain direct economic significance for economic growth and competitiveness and this tends in turn to politicise those formerly (or still formally) extra-economic domains that are now direct objects of state intervention. In this context states also get involved in managing the conflicts between time horizons associated with time-space distantiation and compression – especially in regard to protecting the social capital embedded in communities, promoting longer-term economic orientations, and designing institutions that sustain innovation. But this expanding field of intervention means that the state finds it harder to reconcile its responses to ever more insistent economic imperatives with the more general demands of securing general political legitimacy and social cohesion (Poulantzas 1978).

The overall consequence of these changing functions is a transformation in the state. In Atlantic Fordism, this involves a shift from nationally specific versions of the Keynesian welfare national state (KWNS) to nationally specific versions of what I have called the Schumpeterian workfare post-national regime (SWPR). In East Asia, it involves a shift from forms of Listian workfare national state[5] to other versions of the SWPR. Indeed, it was in part the apparent superiority of East Asian economies in catching up with the West and, especially, Japan's record of innovation in some knowledge-based industries that prompted the reorientation of Atlantic Fordist economies in a more Schumpeterian workfare direction. Economically, the SWPR tries to promote flexibility and permanent innovation in open economies by intervening on the supply-side and tries to strengthen as far as possible the competitiveness of the relevant economic spaces and their extra-economic supports. This in turn leads to increased pressure to subsume these factors under the logic of capital. Indeed valorising the extra-economic is a key dimension of current accumulation strategies oriented to so-called 'strong' competition based on flexibility and innovation. The process and pace of the re-articulation of the economic and extra-economic have been reinforced and economic strategies have become more concerned with the social and cultural embeddedness of innovation and competitiveness as well as more reflexive about how to promote accumulation.

Three trends in the state and their counter-trends

These shifts in function are associated with three general trends in the nature of the state. First, there is a general trend towards the de-nationalisation of the state. This is reflected in the 'hollowing out' of the national state apparatus with old and new state capacities being reorganised territorially and functionally on subnational, national, supra-national, and trans-local levels. State power moves upwards, downwards, and sideways as state managers on different territorial scales try to enhance their respective operational autonomies and strategic capacities. This shift is closely related to the state's new economic roles in the globalising, knowledge-driven economy – especially its turn to the supply-side (including the need to penetrate the micro-pores of society), to the management of self-reflexivity and 'connexity' (Mulgan 1997), to the governance of cyberspace, and interscalar articulation intended to improve overall economic policy 'co-ordination' across different states in the face of a 'relativisation of scale'.

Second, there is a trend towards the de-statisation of the political system. This is reflected in a shift from government to governance on various territorial scales and across various functional domains. This trend is closely related to the changing nature of the economy, which is too complex, dynamic, etc., to be managed by top-down government. It is also related to the emergence of a complex, multidimensional, non-propinquitous cyberspace that undermines 'concepts such as sovereignty and the use of distinctions such as public/private, ownership/access, foreign/local, external /internal, and economic/political' (Katsh 1995: 1717). Guiding these activities requires a bottom-up, self-organising approach akin to a giant networked information processing grid. Accordingly there is a movement from the central role of the official state apparatus in securing state-sponsored economic and social projects and political hegemony towards an emphasis on partnerships between governmental, para-governmental, and non-governmental organisations in which the state apparatus is often only first among equals. Governments have always relied on other agencies to aid them in realising state objectives or projecting state power beyond the formal state apparatus. This reliance has been re-ordered and increased. The relative weight of governance has increased on all levels – including not only at the supra-national and local or regional levels but also in the trans-territorial and inter-local fields. This increase in governance need not entail a loss in the power of government, however, as if power was a zero-sum resource rather than a social relation. Indeed, resort to governance could enhance the state's capacity to project its influence and secure its objectives by mobilising knowledge and power resources from influential non-governmental partners or stakeholders.

Third, there is a complex trend towards the internationalisation of policy regimes. The international context of domestic state action has extended to include a widening range of extra-territorial or transnational factors and processes; and it has also become more significant strategically for domestic policy. The key players in policy regimes have also expanded to include foreign agents and institutions as sources of policy ideas, policy design, and implementation. This trend is reflected in economic and social policies as the state becomes more concerned with 'international competitiveness' in the widest sense. Of increasing significance for the globalising, knowledge-driven economy, of course, is the World Trade Organisation with a remit that has extended to intellectual property issues. This trend affects local and regional states below the national level and is linked to the growth of inter-regional and cross-border linkages across different national formations.

All three trends have been presented in a one-sided and undialectical manner. Each is linked to a counter-trend that both qualifies and transforms its significance for political class domination and accumulation. Countering the de-nationalisation of statehood are national states' growing involvement in interscalar articulation. Whilst it might be thought that there is a simple continuity of function in this regard, a major discontinuity has been introduced through relativisation of scale. Nonetheless, without a supranational state with equivalent powers to those of national states, de-nationalisation is linked to the attempts by the latter to re-claim power by managing the relationship among different scales of economic and political organisation. Not all states are equal in this regard, of course; within each regional bloc there is usually one hegemon and, on a global scale, the USA is the key political force in the rescaling of politics.

Countering the shift towards governance is government's increased role in *meta-governance*. Governments on various scales are becoming more involved in organising the self-organisation of partnerships, networks, and governance regimes. They provide the ground rules for governance; ensure the compatibility of different governance mechanisms and regimes; deploy a relative monopoly of organisational intelligence and information with which to shape cognitive expectations; act as a 'court of appeal' for disputes arising within and over governance; seek to re-balance power differentials by strengthening weaker forces or systems in the interests of system integration and/or social cohesion; try to modify the self-understanding of identities, strategic capacities, and interests of individual and collective actors in different strategic contexts and hence alter their implications for preferred strategies and tactics; and also assume political responsibility in the event of governance failure. Such tasks are conducted by the state not only in terms of their contribution to particular state functions but also in terms of their implications for political class domination.

Somewhat ambiguously countering yet reinforcing the internationalisation of policy regimes are national states' efforts to shape international policy regimes in the interests of the capitals most important for their economic growth. This is clear in struggles over international regimes for IPR, WTO, TRIPS, etc. As Aoki notes, for example, '(w)idely divergent concepts of property and ownership, originating in extremely diverse political, economic, and social circumstances, provide the fuel for hotly contested and seemingly unresolvable disputes' (1998: 462–3). The United States is currently most influential in promoting international policy regimes for a globalising, knowledge-driven economy and it promotes thereby 'the sovereignty of domestic U.S. intellectual property owners' (Aoki 1998: 461). There are, of course, comparable conflicts on the triadic level – notably in attempts to harmonise regimes across the European Union. A second, and equally ambiguous countertrend, is the *'interiorisation'* of international constraints as the latter become integrated into the policy paradigms and cognitive models of domestic policy-makers. However, 'interiorisation' is not confined to the level of the national state: it is also evident at the local, regional, cross-border, and inter-regional levels as well as in the activities of so-called 'entrepreneurial cities'. The relativisation of scale makes such 'interiorisation' significant at all levels of economic and political organisation.

Concluding remarks

This chapter has covered much ground in an effort to provide a firm theoretical basis for analysing the changing form and functions of the state in relation to the globalising, knowledge-driven economy. At the same time too little ground has been covered. I have not been able to explore how informationalism and networking are transforming the understanding of warfare (e.g., in the US advocacy of 'network-centric warfare'), the approach of the state to disciplinary normalisation and surveillance, the prospects for a decentralised cyber-democracy, or the cultural field. I have also related informationalism to the logic of capital rather than to other aspects of the system- or life-worlds. Some of my future work will address these issues and correct the one-sidedness of the present contribution. In this sense the arguments set out above should be interpreted as defining a research agenda rather than presenting firm empirical conclusions.

Nonetheless the above analysis is based on four broad, theoretically-informed remarks that are worth restating. First, the globalising, knowledge-driven economy cannot be adequately understood by regarding knowledge as a natural(ised) factor of production. Instead it is based on the contradictions between knowledge as a collective resource and as intellectual property – contradictions which are rooted in its fictitious commodification. Second, the increased importance of knowledge in this contradictory sense is related to changes in the primary modalities of competitiveness that transform the relationship between the economic and the extra-economic and thus the modalities of state intervention. Third, information and communication technologies have played a key role in extending and re-articulating time-space distantiation and time-space compression. This too has implications for the modalities of competitiveness, for re-scaling and re-temporalising of competition, and for the relative dominance of different fractions and sectors of capital. Fourth, as a consequence of the above, the globalising, knowledge-driven economy involves a transformation not only in the primary and secondary aspects of the contradictions of capitalism but also in the relative importance of different contradictions.

Together these changes have had major repercussions on the economic and social mode of regulation of the emerging accumulation regime and thus on the role of the state and politics in helping to secure some of the conditions for profitable accumulation and the reproduction of labour-power as a fictitious commodity. They also have major repercussions on the spatio-temporal fixes that are appropriate to the current stage of capitalism and thus on the role of the state and politics in 'chronotopic governance', i.e., the management of the spatialities and temporalities, the spatial and temporal horizons, of capital accumulation. And, finally, they have major implications for the relationship between government, governance, and meta-governance. It is hardly surprising, then, that these are 'interesting times' for capital, labour, and state managers and that there is, as yet, still no stable accumulation regime that has replaced the Atlantic Fordist model that dominated in the thirty golden years of postwar Western capitalism.

Notes

1   This paper has benefited from discussions with Markus Perkmann, Ngai-Ling Sum, Andrew Sayer, Chao-Ming Tseng, John Urry and members of 'TeamTheory'.

2   The idea of a spatial fix was introduced by Harvey (1982: 149). Harvey referred mainly to the spatial fixity or immobility of a part of total capital as a precondition for the mobility of other parts; he also gave the term a primarily physical, infrastructural character. The concept has since been extended to include different forms of socio-cultural as well as material embeddedness. I use the term in this broader sense.

3   The forms and content of institutionalised compromises and the associated ways of addressing social cohesion depend on the specificities of the state and political systems as well as on accumulation regimes.

4   This can occur either by reducing the time a given 'event' takes to produce within a given spatial frame of action; or by increasing the ability to discriminate more steps in an 'event' and so enhancing opportunities to modify its course or outcome by intervening into the event as it happens.

5   The principal exception among the leading East Asian economies was the Ricardian workfare colonial regime in Hong Kong. On this, see Sum (1998).

6 Just in time?

The prevalence of representational time and space to marketing discourses of consumer buyer behaviour

Pamela Odih and David Knights

Introduction

Although there are a variety of frameworks for understanding what constitutes marketing knowledge, a central feature is its unproblematic allegiance to positivist methods within an empiricist epistemology. This allegiance can in part be understood by marketing's seemingly insatiable urge to achieve parity with 'higher-order' scientific disciplines. To achieve this, marketing phenomena within the social world must be amenable to causal analysis involving their reduction to a set of quantifiable variables similar to those of the natural sciences. The social world is assumed to exist independently of the observer, although made knowable only through sense perception of social action and events. It is axiomatic to this form of epistemology that 'time' and 'space' should be ascribed the characteristics of abstract, quantifiable, singular units. Consumption, as represented in marketing literature, is assumed to occur within a given tempo-spatial context (Davies 1994; Engel *et al.* 1995). 'Time' and 'space' are conceptualised as existing externally to the individual, in fixed rationally co-ordinated spheres. Yet 'time' and 'space' are bound up with the socially constructed meanings that are a reflection of, and in turn reproduced by, the consumption event (Jackson and Thrift 1995). For we consume spaces, places and times the instant they are generated in and through the consumption event (Jackson and Thrift 1995: 213).

However, consumption events rarely have the effect of subverting the linear assumptions about time and space prevailing within the institutions that produce goods and services. Whether subscribing to an 'objective', 'subjective' or 'social' model of time and space, consumer research reflects and reinforces these linear assumptions. Aided by an array of postmodern deconstructive techniques we identify how, to varying degrees, these three tempo-spatial discourses serve to sustain the hegemonic status of quantitative linear time and thus, inadvertently, subjugate and/or marginalise alternative discourses. Deconstruction uncovers the socio-historical grounding and political role of representational knowledges, not with an eye to providing a better foundation for knowledge, but rather in order to dislodge their dominance and create a social space, which is tolerant of subjugated discourses. One such subjugated alternative is the discourse of 'postmodern time(s)' and this chapter will reflect on its constitutive elements and its contribution to the discourse of consumption.

The chapter contains three main sections. Section 1 identifies how the prevalence of representational epistemologies in marketing discourses (Morgan 1992; Knights and Sturdy 1993; Knights and Morgan 1993) encourages a disembodied and highly rational approach to 'time' and 'space'. Based on a technical-rational view of knowledge, the dominant view of marketing assumes a parallelism between the natural and social world. The world, whether natural or social, is conceived as an objective phenomenon which operates according to strict laws. Deterministic in structure, 'reality' lends itself to accurate observation and measurement by appropriate research instruments. It is axiomatic to this form of knowledge that 'time' and 'space' should be conceptualised as abstract entities existing externally to the individual, in fixed immutable units. Within the 'objectivist' time allocation models of Becker (1965), for instance, consumers are depicted as rationally allocating time within fixed spatial contexts – i.e., households. Objectivist studies, with their focus on abstract linear time and space, rarely take account of the polymorphous rhythms of social life let alone comprehend them. For representational linear time is premised on abstraction, the rational ordering and control of space and the denial of difference. But despite its hegemony, representational time is struggling to contain radical challenges to the Sovereignty of the Cartesian dualism between subject and object that is its condition of possibility. The 'demise of the episteme of representation' (Benhabib 1992: 211) leaves 'objectivist' models of linear time stranded.

Emerging through fractures and fissures in the hegemony of representational time, psychological discourses have proposed alternative 'subjectivist' tempo-spatial models. Here, time is presumed to represent a kind of synoptic view of events residing in the peculiar nature of human consciousness, the mind or reason, which therefore precedes all human experience as its condition (Holman 1981). The subjective, person-centred, conceptual orientation, which binds these studies, is presented as a significant departure from 'objective' rationalistic approaches to time and space. However, these subjective tempo-spatial models, like their objectivist counterparts, prioritise linear time. That is, they draw upon a linear conception of time to evaluate subjective temporal perceptions (Hirschman 1987). This is best exemplified by the 'memory construct' in decision-making models. According to this construct, memory arises from 'traces' of previous experience, neuronal traces, stimulated or activated by present consciousness. But this merely attends to representational discourses of linear time, in which events are ordered in a linear chronology, with prior memory having a causal influence upon succeeding occurrences. In Section 1.2 we illustrate decision-making models as failing to recognise how memories are embodied in places, spaces and time(s). As Game (1994: 16) expresses it, 'space transforms time in such a way that memory is made possible'.

Objectivist models assume time and space to be 'objective' facts existing independently of human beings and subjectivist models assume 'time' and 'space' to be a mere 'subjective' notion rooted in human nature. Neither model challenges the 'objectivist' status of linear time-space as a representational 'given'. In Section 2, we identify how this is partially rectified within the third predominant tempo-spatial discourse of consumer research – the notion of 'social' 'time-space'. Discourses of 'social' 'time' and 'space' present a notion of temporality which is closer to the views expressed in this chapter. The concept of 'social' time recognises 'time' and 'space' as both medium and outcome of the social construction of meaning (Hassard 1990; Nowotny 1992). We learn in the various discourses of social time that specific moments come to acquire particular meanings (Sorokin and Merton 1990; Lewis and

Weigart 1990; Gurvitch 1990). Accordingly, meaning comes to associate an event with its tempo-spatial setting.

Clear epistemological distinctions separate discourses of social time from their representational counterparts. In contrast to the 'objective' social world of representational linear time, social realities are recognised here as constituted in and through meaning. Consumption is re-represented here as involving the communication of meaning/signs (Douglas and Isherwood 1978; McCracken 1990). Individuals actively engage with the symbolic meanings attached to goods in an endeavour to both construct and to inform others of their 'life-style' or identity (Featherstone 1991). For the self is not conceptualised here as stable, continuous and consistent between distinct linear time frames as presumed in objectivist models of linear time-space. Rather subjects live time-space contexts as well as being constituted through them. Geographies of consumption have long since revealed modern identity formations as predicated on the 'consumption' of spaces, places and times (Urry 1995). Time and space are site or a medium through which identities are affirmed and contested via specific acts of consumption (Jackson and Thrift 1995: 227).

The concept of social time has also featured within consumer research, in one of two capacities. In the first it represents a structural phenomenon – a medium through which we co-ordinate complex meaning structures and reproduce everyday life. Exemplified by studies highlighting the relativity of time perception to a given socio-cultural context (e.g., Nicosia and Myer 1976; Graham 1981; Gronomo 1989), this discourse of time prioritises structure over action. Although fundamentally social phenomena, the time and space constitutives of consumer subjectivity are assumed, in these studies, to be the outcome of structural relations. A shopping trip, for example, is recognised as constitutive of social time and space. The meanings attributed to these relations of time and space, and thus their form and function, are presumed, in these models, to be the outcome of specific structural relations (e.g., age, ethnicity and social class). Conversely, and by way of a reaction to the determinism of previous models, 'experiential' models (e.g., Bergadaa 1992; Gibbs 1993) recognise time and space as determining and also forming part of self-identity. Bracketed off from the concealment of structural relations, in these essentially phenomenological discourses, the self emerges as embedded in and constituted through time-space events. Returning to the previous example, phenomenological discourses of social time render problematic the fixture of meaning necessary to derive structural explanations for the time-space relations generated through the shopping 'event'. In the absence of social structure, mutually informative measurements can no longer be made between tempo-spatial events. Each event is thus unique to the individual's first order interpretations. But this focus on the existential dimension of the self and its 'being' in the world deflects attention from the social relations of which the self is both medium and outcome (Knights and Willmott 1983).

Moreover, its dualistic oppositions between action/structure, presence/absence reproduce the Cartesian subject/object mind/body dualisms through which linear time as a representational given is constituted (Knights and Odih 1997; Odih 1998). For it is never shopping time or clock time, Easter or spring, New Year or the January sales. Rather, a multiplicity of times impact on our lives simultaneously. Section 2.2. presents Giddens' (1979) theory of structuration as partially resolving the dualistic tendencies of social time-space. Instantiated through social interaction, the structural outcome of social practices are premised upon their time-space characteristics, since these time and space defined practices serve simultaneously as patterns of interaction and through their impacts on human experience as foundations for motivation to future practices. But, this discourse also has a tendency to produce a 'closure on meaning' (Derrida 1978) and thus takes for granted part of that which it seeks to explain. In an attempt to avoid the representational problematics of social time, we turn to postmodern discourses of time and space.

Section 3 identifies the dual effects of the dissolution of the autonomous, centred subject and the collapse of universal historical narratives in postmodern discourse as radically dislocating conventional experiences of temporality. Reflective of modern culture, the rational ordering and control of time and space, it may be argued, is founded on instrumental rationality, a progressive history and the suppression of difference. Conversely, Baudrillard's (1968) postmodern world of simulacra and 'hyper-reality' is a world without 'fixed' references of meaning. The effect of such a breakdown in the signifying chain is to reduce experience to a series of 'pure' unrelated 'presents' in time (Jameson 1984). The corollary of this loss of temporality has been the pervasive 'flattening' of space. Postmodern spatiality dissolves distinctions between inside and outside, surface and depth, front and back. The extraordinary tempo-spatial fragmentation (postmodern time(s)) fractures, disrupts and overflows the fixed and discrete boundaries that are necessary to time and space as a representational object of something in the world. In so doing, postmodern discourse presents a radical challenge to the linearity that theories of consumption in marketing take for granted.

The synthesis of leisure and consumption activities invited by postmodern consumption sites, such as shopping arcades and redeveloped out-of-town commercial areas, extends beyond literal forms of consumption to embrace complex combinations of social behaviours. Shield (1992: 7), for example, identifies how the emergence of postmodern consumption spaces is accompanied by an emphasis on various forms of *flanerie* (loitering, aimless strolling) which present a marked shift from the purposive instrumentality of 'old' consumption spaces. In the concluding section of this chapter we examine the implications of this analysis for studying marketing discourse and its conceptions of consumption.

1   Representing time and space in marketing discourse

Companies world wide are working to develop products and services, process orders, and satisfy customers faster than ever before. These companies are . . . focusing on a potent business idea: the power of TIME to improve customer satisfaction and corporate performance. Looking at your company through the lens of time will show you new and better ways to introduce innovative products and services, manage distribution channels, tailor products and services to specific customer needs, eliminate unnecessary costs, even transform your organisation.

(Harvard Business School 1994: 2)

Marketing is constituted here, as elsewhere in academic marketing literatures, as a neutral observer operating in a given environment. One should not underestimate the intensity of this claim to neutrality, as it is central to marketing's fervour to be recognised as a scientific discipline. As Hunt expresses it:

Theories are systematically related sets of statements, including some law-like generalisations, that are empirically testable. The purpose of marketing theory is to increase scientific understanding through a systematised structure capable of both explaining and predicting phenomena.

(Hunt 1983: 10)

Natural science is above all else, an empirical activity embedded in objective observation devoid of judgement, interpretation and other forms of subjective mental operations. The adherence of scientific practice to objective observation is said to have enabled the production of the much avowed generalisations known as scientific laws. To emulate the apparent success and prestige of scientific practices, marketing phenomena within the social world must be amenable to causal analysis involving their reduction to a set of quantifiable variables similar to the natural sciences. The social world is assumed to exist independently of the observer, although made knowable only through sense perception of social actions and events. It can be thought of as a stable, predictable structure composed of a network of determinate relationships between constituent parts. Reality is to be found in the concrete behaviour and relationships between these parts. In the same way that knowledge and 'control' of the 'natural' world have been made possible through scientific methods and engineering technologies, so it has been thought that society can be measured and 'controlled' once the 'correct' techniques of analysis are devised and developed.

Within this technical-rational view of knowledge, consumers are knowable, limited entities, the characteristics of which can be captured in the same way as can the characteristics of natural phenomena. Nowhere is this assumption more evident than in the prevalence of cognitive decision-making models to discourses of consumer buyer behaviour (see Engel *et al.* 1995). Consumers, within these often a-social models, are represented as problem-solvers systematically progressing through a sequence of cognitively mediated, deliberative stages. Knowledge of the consumer resides in the elicitation of those objective variables, which govern decision making. Social relations and identities, fixed, constrained and subject to invariably predictive forces, are 'external' to marketing activity (Engel *et al.* 1995). Marketing's position, here, embodies a correspondence theory of truth, namely, that the truth of a statement is to be determined by its correspondence with the 'facts'. While it is recognised that statements can only be validated through their correspondence with representations of 'reality', there is no epistemological questioning of the 'reality' of representation. Language, within this 'episteme of representation' (Benhabib 1992), mirrors the social world and scientific vocabularies are ascribed the capacity to uncover the essential structures and dynamics of human behaviour. It is axiomatic to this form of epistemology that time and space should be ascribed the characteristics of abstract, singular units that can be measured. Time is perceived as a homogeneous unity that is nonetheless divisible into discrete units – a flow which, is however, unaffected by the transformation it charts (Adam 1993: 166). Similarly, space is presumed to be an empty, and or, dimensionless vesicle within which events can be recorded and given boundaries. Time and space conceptualised as abstract, quantifiable entities are evident in two predominant approaches to consumer behaviour in marketing discourse. These are the 'objectivist' and 'subjectivist' tempo-spatial models, which we now discuss in turn.

1.1 Objectivist tempo-spatial models: consuming in absolute time and dimensionless space

The first approach, described here as 'objectivist' tempo-spatial studies, conceptualises time and space as unproblematically existing external to the individual, in fixed immutable units (Gibbs 1993). These units in turn yield an implicit utility to the consumer. This implicit utility facilitates the transformation of time and space into other entities such as money or products (Gibbs 1993). This conception of time and space as a quantifiable, compartmentalised, objective entity constituted by uniform parts linked in a linear and sequential fashion to one another, aligns with the positivistic epistemological foundations of marketing knowledges. It is therefore no surprise that the 'clockwork precision' (Gibbs 1993: 8) of this 'objective' notion of time and space, has encouraged its application to a wide variety of buying instances and issues. Time is conceptualised as, for example: a controllable scarce resource (Becker 1965); a 'pressure' in decision making (Engel *et al.* 1990); an exchangeable commodity in 'resource models' (Jacoby *et al.* 1976); a 'risk' in that time might be wasted if the search behaviour results in post-purchase cognitive dissonance (Schiffman and Kanuk 1991). More specifically, in the time allocation models of, for example, Becker (1965), consumers are perceived as inherently concerned with 'time scarcity'. This is the state whereby consumers are willing to use discretionary income to save what they have come to see as non-discretionary or 'fixed' time (Gross 1987: 3). In short they are prepared to pay for convenience. Consumption, in these studies, occurs within a given temporal context in which time can be allocated, distributed and exchanged for money. Similarly space, as a quantifiable phenomenon, is presumed to fix, constrain and/or facilitate consumption in predictive, objectifible ways. It is an 'objective' situational variable in 'time allocation models' (Hornik 1984) or a fixed parameter in which the sequels of consumption unfold in linear succession (Nicosia and Meyer 1976).

Although objectivist studies have both time and space as their focus, neither are problematised. A consequence of failing to problematise time and space beyond their apparent 'objective' existence is that they are constituted in terms of their immediate context. To study the experience of duration, the estimation of an interval, people's orientation within horizons, or the timing, sequencing and co-ordinating of behaviour is to define time and space as duration, interval, passage, horizon, sequencing and parameter (Adam 1990: 94). Neither time nor space is an 'emergent' concept in these studies, but rather both are pre-defined in advance of whatever is being studied (Adam 1990: 94). Consumers are depicted as purchasing and using goods within a pre-given context which is defined prior to the event in accordance with the subject of study. This is achieved through reference to an a-social space and a-social time, divorced from the generative consequences of human existence. Clear parallels can be found between this concept of time-space and the problematics of Newtonian physics. Both assume time to be an absolute quantity, invariant, infinitely divisible into space-like units, measurable in length, quantifiable and reversible (Adam 1990). Steeped in their world of Newtonian motion, objectivist discourses forget, all too easily, that qualitative variations precede the uniform abstract quality of human origin (Adam 1990). This is despite the revolutionary developments in science generated by Einstein's theory of relativity. Einstein demonstrated, through relativity, that:

there is no such thing as a fixed interval of time independent of the system to which it is referred. There is no such thing as simultaneity, there is no such thing as 'now', independent of a system of reference.

(Barnett 1957: 41)

Einstein established that time and space are relative to observers and their frames of reference. While we are used to the simple measures of clock-time, our various subjective experiences of time are clearly different. Consider such phrases as, *tempus fugit*, or 'time flies', often drawn upon when, as a result of being completely absorbed in an activity, time and space pass unnoticed. Recall the sense of awe and spatial disorientation invoked by a new building, district or place. Recollect how, as if by magic, gargantuan spaces are readily transformed into familiar departments, corridors and offices. These experiences and phrases indicate just how perceptions of time and space vary with our social context. But contemporary social science continues to follow a Newtonian conception of time and space although, as we shall see shortly, both science and social life have rendered the simplicity of mechanical linearity suspect. For, despite its hegemony, representational linear time is struggling to contain radical challenges to the sovereignty of the Cartesian subject–object dualism, its rational ordering of space and its subversion of *différance*. As Ermarth expresses it:

> Across a broad range of cultural manifestations a massive re-examination of Western discourse is underway: its obsession with power and knowledge, its constraints of language to primarily symbolic function, its ethic of winning, its categorical and dualistic modes of definition, its belief in the quantitative and objective, its linear time and individual subject, and above all its common media of exchange (time, space, money) which guarantee certain political and social systems . . .
>
> (Ermarth 1992: 6)

Within marketing discourses of consumer behaviour these challenges appear in the form of 'subjectivist' and 'socio-temporal' models of time and space.

1.2 Subjectivist temporal models: I think, therefore I buy

Subjective notions of time, in contrast to objective approaches, focus on the cognitive, perceptual apparatus by which individuals comprehend time (Hirschman 1987; Gibbs 1993). Time is presumed to represent a kind of synoptic view of events residing in the peculiar nature of human consciousness, or the human mind or reason, which therefore precedes all human experience as its condition (Elias 1992). The theoretical and philosophical antecedents of this subjectivist notion include the writings of Descartes. However, it found its most influential expression in the philosophy of Kant, who saw time and space as representing an *a priori* synthesis. Kant (1788/1956) argued that all our representations, whether they be due to the influence of 'outer things' or produced though 'inner causes' belong to the mind's inner experience – to what he calls 'inner sense'. Time, according to Kant, is the formal condition of inner sense, necessarily structuring our experience. If we could not distinguish time in the sequence of one representation upon another, the single representation would be useless for knowledge. For each of them, as far as they are contained in a single moment, could not be anything but a single unit. All our knowledge is thus 'finally subject to time' (1788/1956: 131). Kant deals with space in exactly the same way as time, that is they are *a priori* principles of knowledge. They exist intuitively and without this we could have no knowledge of anything – no representations.

Kant's treatment of time and consciousness reveals an inherent tension at the centre of his treatise. This tension relates to the 'objective' or subjective ontology of time. On the one hand, time is internalised; it is as he says, 'nothing but the form of inner sense, that is, of the intuition of ourselves and of our inner state' (1788/1956: 78). It is 'a purely subjective condition of our (human) intuition'; of itself 'apart from the subject', it is 'nothing' (1788/1956: 78). But the manner in which time is, for Kant, subjective does not make it less objective than the 'outer' world of spatial objects. The mind's awareness of temporal succession must indeed be mediated through its awareness of outer appearances. We can get at succession only through the mediating role of representations in outer sense. Our experience of time, therefore, presupposes an objective, necessarily connected ordering of representations in accordance with the categories of the mind. The Kantian subjective versus objective divide thus collapses, to reveal an 'objective' expression and experience of time. A similar problematic is evident in marketing discourses of subjective temporal experiences.

1.2.1 Subjectivist studies of time within consumer behaviour research

The subjectivist approach to 'time', popularly embodied in psychological temporal models, has been used within consumer research to ascertain subjective perceptions of duration (e.g., Holman 1981; Hornik 1984), time allocation (e.g., Hendrix 1980; Wilson and Holman 1980), time-inconsistent buying preferences (e.g., Hoch and Loewenstein 1991) and individual positioning in relation to past, present and future (e.g., Holman 1981; Morello 1988; Daz 1991). The subjective, person centred, conceptual orientation which binds these studies is presented as a significant departure from objective rationalistic approaches to 'time' (e.g., economic temporal models). However, these subjective temporal models also prioritise linear time. They draw upon a linear conception of time to evaluate subjective temporal perceptions (Hirschman 1987; Gibbs 1993). This is exemplified by the psychophysical research traditions, whereby time is envisaged as an 'external flow of physical units', and used to evaluate individual subjective temporal perceptions (Hirschman 1987: 67).

Time as memory

Within the influential decision-making models of Howard and Sheth (1969), Bettman (1979), Engel *et al.* (1968) and Engel *et al.* (1995), the 'memory construct' plays a pivotal role. In line with their rationalist discourse, it is assumed that memory (represented as an objective observable entity) has the capacity for the storage of semantic information. Bettman has stated that semantic storage is

> organised as a network of nodes and links between nodes, with the nodes representing concepts and the links denoting relationships among concepts . . .

new information is integrated by developing a configuration of links to already existing concepts.

(Bettman 1979: 57)

Memory is presumed, within decision-making models, to be physically represented somewhere in the central nervous system. It is thought that memory can be reduced to an essential psychological component, the Memory Trace, in that memory is assumed to be what is left from previous experience, neuronal traces, stimulated or activated by present consciousness. It is therefore argued that through physical acti-vation – the application of an energising force on the trace – the past can be released from the raw, physical matter where it is believed to reside. In short it is assumed that the past is carried forward in time by the brain's physicality (Arcaya 1992: 303). An everyday example of this cognitive process is repeat purchases through brand loyalty. Recollection of past purchases, stimulated through advertising, is thought to directly influence both present and future consumption. Underlying this is an implicit view of temporality, in which events are ordered, in a linear chronology, with prior memory having a causal influence upon succeeding occurrences. This is, of course, the temporal template of linear time, whereby time exists as a series of presents, displacing themselves from an experienced past to an infinite horizon – the future.

The essentially linear tempo-spatiality of the memory construct is best exemplified by the often cited Engel *et al.* (1968) 'consumer decision-making' model. Here 'felt need/problem recognition' – the start of the 'decision-making process' – is described as a felt disparity between a present and anticipated state. An everyday example here might be realising the 'need' for a new domestic appliance. This disparity, or 'need', is reconciled through 'search' (i.e., product information gathering and its delibera-tion). Once the 'need' for a new domestic appliance has been realised, the 'rationally motivated' discerning consumer then sets about deliberating and choosing between a plethora of options. This search is facilitated by approaching a particular space (e.g., supermarket) and drawing upon a specific past experience – stored information, or memory. Buyer behaviour is thus a product of past experience, present stimulation (e.g., advertising) and future anticipations. It is understood as a chronological event-based process brought about by deliberate, cognitive activity.

Critique

The account of memory that underlies this understanding of buyer behaviour, unsurprisingly, has been heavily criticised (e.g., Strauss 1966; Zimbardo *et al.* 1971; Arcaya 1992), particularly in relation to its disembodied perspective of time and space. For subjectivist models of time and space consistently fail to recognise the transcendent powers of consciousness and the construction of the self through time. Phenomenologists (e.g., Heidegger 1971; Husserl 1928; Merleau-Ponty 1945) understand memory, to be (re)constructed in and through time-space events. Any re-invocation of the past from physical fragments (e.g., having to replace an old domestic appliance) or intangible influences (e.g., an advertisement for a more advanced version of the domestic appliance) in the present, requires the individual to engage in a 'bi-temporal' (Strauss 1966) splitting of consciousness. This is to appre-ciate that basic to all forms of human reading is a conscious organism living in time and capable of uniting the literal with the virtual or linking one temporal order (the present) with others (the past and future). Time is, therefore, inextricably bound up with places, spaces and the body. For people 'do not so much think real time but actually live it sensuously, qualitatively' (Urry 1995: 6). To paraphrase Urry (1995: 24), memories are both symbolically and materially localised and so the temporality of memory is spatially rooted.

Conversely, consumer behaviour models of decision making assume that present stimuli evoke the 'mind' to recapitulate its history devoid of any (re)interpretation. Thus the past is assumed to be remembered outside the reflexive evolving, transforming contemporary self. This is to assume that humans remember in the same mechanistic way as computers or photographic plates, where a past image is simply replicated in the present. Returning to the previous example, when finding our old appliance we are assumed to remember and understand its significance as if it were bought only yesterday. In contrast, it may be suggested that the human memory is tied to the social and psychological context of its use. That is to say, as humans, we selectively interpret (or remember) the past in terms of what is meaningful to us in the present and in relation to particular projects for the future. Consciousness has the capacity to 'order or date the object of perception as a temporally evolving phenomenon, placing it within a set of past occurrences and future possibilities' (Arcaya 1992: 308).

For example, the domestic appliance cited above might have been part of a wedding gift or bought during a significant period of the individual's life. The recognition of its replacement stimulates a (re)interpretation of its value both as a wedding gift and as an appliance which has now become outdated. At the centre of these merging tempo-spatial horizons is found the (re)constituted memory of the domestic appliance. The future is therefore not discrete or apart from memory – as decision-making models imply – but intertwined in the act of remembering. As Arcaya (1992: 311) has argued, 'lacking in phenomenological descriptions of remembering, cognitivists have been more guided by the heuristics and logic suggested by their presuppositions rather than by the experience of their own consciousness'.

If memory is not considered a mechanism but rather as an aspect of personal consciousness, it is necessary to acknowledge the socio-contextual and existential quality of temporality. So far, consumer behaviour models in marketing have tended not to do so. One might suggest that they simply insert this conception of memory into existing models of consumer behaviour. But how? For this would require recognising the self as a social process, situated within social, physical and temporal orders that provide it with substance and meaning. This would involve a dramatic paradigmatic shift away from the existing 'psychodynamic',[1] cognitivist focus that guides much of consumer research. Indications of a paradigmatic shift of this magnitude are evident in the anti-positivistic research of: Arndt 1985; Foxall 1986, 1987, 1990; Anderson 1989; Hirschman and Holbrook 1992. Arndt (1985), for example, argues that marketing has for too long been dominated by a highly reductionist structural-functional approach. He contrasts this with a 'subjective world paradigm' drawn from social constructivism and symbolic interactionism. This 'interpretivist' approach has, in recent years, gained a considerable following in consumer behaviour (see Hirschman and Holbrook 1992; *European Journal of Marketing* 1994). The increasing popularity of 'interpretivist' and 'interactionist' perspectives has resulted in the emergence of radical alternatives to linear discourses of time and consumer behaviour.

2 Social time and marketing discourse

> Understanding consumers' perceptions of time is basic to understanding their behaviour. This basic understanding is thwarted if the researcher assumes the rational model of consumer behaviour, as this model has an implicit perception of time built into it. That is, for perceptions other than linear-separable, the use of rational man [*sic*] assumptions interprets perfectly rational behaviour as irrational or as 'noise'.
>
> (Graham 1981: 342)

The principal manner in which marketing discourses construct their version of consumer behaviour is to assume the existence of an 'objective' concrete reality of which positivist representations are an exact correspondence. An alternative formulation is to postulate social realities as being constituted in and through meaning and to say that social realities cannot be identified in abstraction from the language in which they are embedded. Meaning is profoundly to do with language considered not as a system of grammatical and syntactical rules but as social interaction. Language, considered as modes of social relationship, is constitutive of 'reality', is essential to its being the kind of reality it is. According to this interactionist perspective, language and meaning are not private subjective things, but rather, public and intersubjective. Meanings are not finitely specifiable but receive their sense from a background of context and interpretations employed by speakers of the language.

Consumption, in this epistemological tradition, involves the communication of cultural meaning (Douglas and Isherwood 1978). At the core of this perspective lies the idea that material objects embody a system of meanings, through which we express ourselves and communicate with each other. Goods not only communicate social categories and hierarchies (e.g., superior/subordinate, avant-garde/conservative) but also a highly varied, specific and symbolically charged range of meanings (Douglas and Isherwood 1978). Meaning is constantly flowing to and from these objects, aided by the collective and individual efforts of designers, producers, advertisers and consumers (McCracken 1990).

Consumption as the communication of cultural meaning represents a radical alternative to the a-social, 'psychodynamic', individualistic models of consumer behaviour which dominate marketing discourse. In the writings of, for instance, Douglas and Isherwood (1978) we come to understand consumption as situated within a social cultural context which shapes consumer trends, the products we consume, and the 'needs' and identities that are constructed, sustained or changed through processes of consumption. Whilst the linkage between consumption and identity is not a new thing, it is argued that a transition has occurred from an ideology of consumption to consumer subjectivity (Knights and Sturdy 1993). Whereas in the former situation, individuals had to be manipulated to consume to avoid the economic consequences of over-production, in the latter, individuals are continually seeking out commodities with which both to identify almost in conformist fashion (e.g., designer label clothes) and to differentiate themselves from others (e.g., the particular design) (Knights and Sturdy 1993).

Consumption in Western societies is, therefore, characterised by a diversified or pluralistic range of consumption patterns, where the search for a degree of distinctiveness through the consumption of highly differentiated products and services is both the condition and consequence of a multiplicity of social identities. Individuals in their capacity as consumers engage with a proliferating multitude of goods for the purposes of 'fashion[ing]' (Bauman 1988: 808) a distinct and differentiated subjectivity.

Identity, in this formulation, does not relate to fixed attributes of personality or self, still less to certain fixed forms of behaviour. Instead, as Giddens (1990: 54) states, identity lies now 'in the capacity to keep a particular narrative going'. It has acquired the status of an ongoing 'project'. Identity construction in our late modern era has become a story

> which a person writes and rewrites about him/herself, never reaching the end until he/she dies and always rewriting the earlier parts, so that the activity of writing becomes itself part of the story.
>
> (Gabriel and Lang 1995: 86)

This conception of identity, self and subjectivity as an almost arbitrary intervention in the flow and flux of social experience through time and space has significant implications for linear conceptions of objective time and absolute space. For the self is not conceptualised here as stable, continuous and consistent between distinct linear tempo-spatial frames as presumed in discourses of linear time. Rather, the self is embedded in and through time-space 'events'. The literature of contemporary geography is particularly helpful here in conceptualising identity as a fluid reflexive process instantiated through the time-space events, which it also serves to constitute (Giddens 1979, 1991; Mort 1988; Lefebvre 1991). In the urban geography of Mort (1988: 218), for example, we come to recognise time and space as mediums of social meaning which shape identities and the senses we have of ourselves. Similar examples exist within consumer research, where consumption has become part of the place-based constitution of social identities.

2.1 Consuming places, spaces and time

The significance of social time and space for identity formation has featured in consumer research in two distinct, although inextricably linked capacities. In the first, time and space represents structural phenomena – a medium through which we co-ordinate complex meaning structures and reproduce everyday affairs. Exemplified by studies highlighting the relativity of time perception to a given socio-cultural context (e.g., Nicosia and Myer 1976; Graham 1981; Gronomo 1989), this discourse of time-space prioritises structure over action. Consumer subjectivity is constituted within these discourses as an emergent property of specific structurally determinant tempo-spatial processes. Graham (1981), for example, in his model of cultural variations in time perception, contrasts 'anglo' time which he describes as being 'linear separable', with the 'circular-traditional' time of most Latin American countries. According to this model, consumer subjectivity arises through the assimilation and or socialisation into these spatially distinct forms of time. Consumption, whilst embedded in a social context, occurs in accordance with responses to specific dimensions of social stratification (age, ethnicity, sex) which then combine in particular places and at specific times (Jackson and Thrift 1995: 229). An example of this perspective might be that of defining temporal perceptions of service duration in terms achieving synchrony between the customer's timetable and that of the organisation. Customer satisfaction is thus assumed to be the outcome of two structural determinants – i.e., harmony between organisational and personal timetables. Time is represented here as a static relationship between specific parts and the whole (Giddens 1979). In spite of some emphasis on individual consciousness (see Venkatesan and Anderson 1985), structuralist discourses of social time tend not to recognise individuals as active subjects, reflectively monitoring their action through the time-space events which constitute them.

However, self-referential, or reflexive monitoring of behaviour is evident in the second of the two predominant approaches to social time. In the 'experiential' temporal models of Hirschman and Holbrook (1992), and Bergadaa (1992) and the 'relativity' temporal model of Gibbs (1993), time and space are conceptualised as determining and also forming part of self-identity. Bracketed off from the conceal-ment of structural relations, in these phenomenological discourses, the self emerges as embedded in, and constituted through, time-space 'events'. Rather than assume the customer's perception of duration to be the outcome of negotiated timetables, phenomenological discourses render problematic the fixture of meaning needed to make mutually informative measurements between diverse events.

Whilst overcoming the determinism of structural temporal models, the phenom-enological focus on the existential dimension of the self and its 'being' in the world deflects attention from the social relations of which the self is both medium and outcome (Knights and Willmott 1983). Moreover, its dualistic oppositions between action/structure and presence/absence reproduces the Cartesian subject/object and mind/body dualisms through which linear time as a representational given is constituted (Knights and Odih 1997; Odih 1998). An alternative procedure is to recognise that 'agency and structure are temporally and spatially specific, and that societies and individuals are embedded in a particular historical configuration of time and space which itself is the creation of history, society and individual action' (Gregson 1997: 61). In Giddens' (1979) theory of structuration we find recognition of the continuous dialectic of individual and society as social systems unfold through time and space. Instantiated through social interaction, the structural outcome of social practices is premised upon their time-space characteristics, since these time and space defined practices serve simultaneously as patterns of interaction and, through their impacts on human experience, as foundations for the motivation to future practices. The production and (re)production of systems is, according to Giddens (1979), accomplished by 'knowledgeable' actors; their choices to act or not to act mean that structures are reproduced only in the instant in action. As Giddens (1979: 71) says: 'structure is thus the mode in which the relation between moment and totality expresses itself in social reproduction'.

In summary, the writings of Giddens' (1979, 1981, 1984) provide for a means of reconciling the time-space contextuality of agency and structure; whilst temporal and spatial organisation limit individual action they are at the same time the creation of history, society and individual action. But despite his efforts to demon-strate fluidity and spontaneity in structuration, Giddens also attends to a notion of concreteness. He proposes a fixture of meaning in the humanly created and representationally generated meanings that are subsequently embedded in routines (Storper 1997b).

Whilst seemingly challenging the hegemony of linear time through proposing that individuals structure time and space socially, Giddens (1979) also seems committed to the very episteme of representation through which linear time has achieved its hegemony (Odih 1998). What is needed is a radical discursive shift away from representational suppositions and the dismantling of not only absolutes but also single perspectives and scientific objectivity. Postmodernism and postmodern discourses of time represent a discursive shift of this nature.

3 Marketing postmodern time and space

> Diverse brands follow one another, are juxtaposed and substituted for one another with an articulation or transition. It is an erratic lexicon where one brand devours the other, each living for its own endless repetition. This is undoubtedly the most impoverished of languages; full of signification and empty of meaning. It is a language of signals. And 'loyalty' to a brand name is nothing more than the conditioned reflex of a controlled affect.
>
> (Baudrillard 1968/1996: 17)

The postmodern condition suggests that we are experiencing an intense phase of time-space compression and their fragmentation both globally and locally. Moreover, the postmodern condition entails the treatment of time and space as finite or tied to the social context of their use. They are, from a postmodern perspective, no longer unproblematic media whose neutrality permit 'grand narrative' comparison and communications across temporal and spatial distances. There are no longer common times and spaces 'in' which we all live in more or less mutual relevance. On the contrary, time and space are constituted as a local definition, a dimension of an event, a unique and unrepeatable location or period (Adam 1990; Ermarth 1992).

The Enlightenment project involved a rational ordering and delineation of time and space where no such conceptualisations had existed under classical forms of life. Innovations in transport and communication radically developed our relationship to time and space resulting in their becoming compressed and relational (Harvey 1991). The 'instantaneity' and 'simultaneity' characteristic of the postmodern condition mitigates against the delineation of time and space into rationally co-ordinated spheres. One might suggest this intense time-space compression is simply an intensification of the dilemmas that have, from time to time, beset capitalist procedures and their demand for a constant revolution of the forces of production. Whilst the economic, cultural and political responses may not be exactly new, the range of these responses differs in certain important respects from those which have occurred before. The intensity of time-space compression in Western capitalism since the 1960s with all of its emergent features of excessive ephemerality and fragmentation in the political and private, as well as in the social realm, does seem to indicate an experiential context that makes the condition of postmodernity somewhat distinctive (Harvey 1991). This is not to argue that all that went before is displaced but only to suggest that tenses, in the sense of past, present and future, and distance, or what is remote, comparatively close and intimate, are blurred. There is clearly intensity and tension (Cooper and Law 1995) around postmodern temporal/spatial relations that was perhaps not so evident within representational regimes where the boundaries between one time and another or different spaces were more clear-cut.

Adam (1992) identifies how, with global communication and air travel, day and night have lost their fixed social meaning, and have become thoroughly relativised. Instantaneous global communications reflect and reproduce a relativisation of the dimensions of space and time. Through multi-tasking operating systems we can simultaneously fax, telephone, video-phone and email an individual. Space, it seems, is no longer an obstacle to communication in such instances, or at least it has been rendered less relevant. Adam has stated that

> When virtual simultaneity of communication and instantaneity of feedback are not only the accepted but an expected norm, then the speed of light appears to be the only absolute limiting factor to the transmission of information.
>
> (Adam 1992: 177)

Advances in information technologies make possible a new kind of relationship between place and space, 'through their capacity to transgress frontiers and subvert territories. They are implicated in a complex interplay of deterritorialisation and reterritorialisation' (Morley and Robins 1995: 75). Places are no longer rationally defined and ordered by their boundaries, borders and frontiers. Rather, technologies and geographical locations have become continuous (Morley and Robins 1995: 75). Instead of demarcating places, the boundary becomes 'permeable, an "osmotic membrane", through which information and communication flow' (Morley and Robins 1995: 75). Consequently, previously distant time and space relativities now form part of the daily interaction of global communities, financial markets, business corporations, academic institutions and so on.

The postmodern condition radically disrupts linear representations of events. The almost instantaneous communications via satellites negate any simple delineation of the past, present and future and their linear succession. It provides instant feedback and simultaneous expression across vast spaces. Its communication extends beyond the present to real and fictitious events of the past as well as projected happenings of the immediate and long-term future (Adam 1992). The innumerable channels provided by satellite television give simultaneous coverage of past, present and future events from all over the globe. The war in Bosnia, the plight of Apollo 13 and the Jurassic dinosaurs simultaneously bombard our senses. The past, present and future lose their absolute distinction when the universe, our history and current events unfold indiscriminately as present on our television screens. Baudrillard (1983) identifies consumer culture and television to have produced a surfeit of images and signs that have given rise to a simulational world which has effaced the distinction between the real and the imaginary, a depth-less aestheticised hallucination of reality.

Central to Baudrillard's (1983) world of simulacra and hyper-reality is a loss of a sense of history and the fragmentation of time into a series of perpetual 'presents' in which there is the experience of multi-phrenic intensities. Without a coherent or unified sense of the subject, it becomes increasingly difficult to speak of temporality in terms of memory, narrative and history. We are condemned to a perpetual present, the immediacy of seemingly random unconnected signifiers. Similarly, Jameson (1984) describes postmodern culture as a culture of stylistic diversity and heterogeneity, of an overload of imagery and simulations that have lost any sense of a point of reference, or stable object of representation. Time and space are fragmented into a series of 'present' moments or positions. The inability to chain signs and images into narrative sequences leads to a schizophrenic emphasis or vivid, immediate, isolated, affect-charged experience of presentness of the world – of intensities (Jameson 1984). Here the channel-hopping MTV television viewer's fragmented perception of the world is presented as the paradigm form.

The spatial corollary of this loss of temporality has been the pervasive 'flattening' of space. Postmodernism discredits any sense of a fixed signifier. For the postmodernist, meaning is perpetually deferred, constantly slipping beyond our reach. Whilst modernism conceptualised space as static and inert – an emptiness which can be filled with objects – postmodernism dissolves distinctions between inside and outside, surface and depth, front and back. The ability of the postmodern sign to disrupt fixed spatial distributions is, according to Jameson (1984), evident in the design of the Westin Bonaventure Hotel. Jameson provides a detailed description of the tempo-spatial fragmentation he experienced during a visit to this example of postmodern architecture. The following excerpt barely captures the profound sense of de-centring and disorientation which he describes.

> We may conclude all this by returning to the central space of the lobby itself. . . . The descent is dramatic enough, plummeting back down through the roof to splash down in the lake. What happens when you get there is something else, which can only be characterised as milling confusion, something like the vengeance this space takes on those who still seek to walk through it. Given the absolute symmetry of the four towers, it is quite impossible to get your bearing in this lobby; recently colour coding and directional signals have been added in a pitiful and revealingly, rather desperate attempt to restore the co-ordinates of an older space. I will take as the most dramatic practical result of this spatial mutation the notorious dilemma of the shop-keepers on the various balconies; it has been obvious since the opening of the hotel in 1977 that nobody could ever find any of these stores, and even if you once located the appropriate boutique, you would be most unlikely to be as fortunate a second time; as a consequence, the commercial tenants are in despair and all the merchandise is marked to bargain prices . . .
>
> (Jameson 1984: 43–4)

Jameson proceeds to identify the Bonaventure Hotel as a profound example of postmodern hyperspace's success in 'transcending the capacities of the human body to locate itself, to organise its immediate surroundings perceptually, and cognitively to map its position in a mappable external world'. The human subject, unprepared for the spatial fragmentation and mutation in which it finds itself, experiences an alarming disjunction point, a schizophrenic fragmentation between, self, the body and its built environment (Jameson 1984: 43–4). Similar narratives exist in recent geographies of consumption. Urry (1995) describes how the demise and blurring of spatial boundaries intensify competition between places to present themselves as attractive to potential investors, tourists, consumers and so on. Within these struggles over spatial definition have emerged spectacular monuments to postmodern consumption. Examples include the multifarious shopping arcades that engulf urban and increasingly suburban landscapes. Modernism involved the segmentation of culture and the separation of life into separate value spheres: culture differentiated from economy, work from leisure, public from private and so on. Combinations of leisure and consumption that characterise postmodern sites of consumer culture subvert these modernist demarcations. In so doing, they invite elaborate combinations of social behaviour which exceed the rational economic activities of representational consumption studies (e.g., Engel *et al.* 1995). Shield describes how

> everyday shopping activities are foregrounded as if on a theatre stage, to be observed by passers-by who may vicariously participate in the bustle and lively activity of consumption without necessarily spending money.
>
> (Shield 1992: 6)

The diaphanous spaces of the arcade, or precinct shopping centre, invite forms of *flanerie* (loitering, aimless strolling) which significantly challenge the notion of purposeful, situated, buyer behaviour evident in objectivist, subjectivist and social tempo-spatial models. Neither of these models accounts for consumption spaces as anything but sites for consumption. They fail to recognise non-consumption, ideally browsing and gazing aimlessly, as also involving the 'consumption of spaces' (Shield 1992). For postmodern consumer culture displaces the coherent reflexive and responsible modernist self and in its place presents us with a fragmented, multivocal plethora of tempo-spatial 'event' based subjectivities. In the following section we examine marketing's 'response' to postmodern consumer culture.

3.1 Postmodern consumer culture; advertising and the floating signifier

The modernist separation of economy and culture invites consumers to constitute their identities through the purchase of products whose stories and images echo historically specific grand narratives (Firat 1994). By contrast:

> the consumers of postmodernity seem to be transcending these narratives, no longer seeking centred, unified characters, but increasingly seeking a 'feel good' in separate, different moments by acquiring self images that make them marketable, likeable, and/or desirable in each moment. . . . Thus occurs the fragmentation of the self. In postmodern culture, the self is not consistent, authentic or centred.
>
> (Firat 1994: 91)

Firat (1994) is here beginning to infer an inter-subjective movement away from regarding goods merely as utilities having a use and exchange value. Rather than related to some fixed system of human needs, exchange involves the consumption of signs (Baudrillard 1975, 1981) that are 'free floating' – not tied to an object of signi-fication but simply circulating in a space of signifiers. For Baudrillard (1975) a central feature of the movement towards mass production of commodities is that the negation of the 'essential' use-value of goods by the dominance of exchange-value under capitalism has resulted in the commodity becoming a sign. In line with Saussurian semiotics, this understanding of representation goes beyond the reductive notion of the sign as vehicle of meaning and signification. Rather the relation between sign and referent are completely arbitrary 'determined by its [their] position in a self-referential system of [floating] signifiers' (Featherstone 1991: 85). Signification means simply difference and nothing else. For Baudrillard (1968/1996) the only meaning that signs retain is their difference from other signs; and this is the end of meaning. Signs are entirely self-referential, making no attempt at signification or classification, their only point being to make a temporary impact on our consciousness. This detached status of the code enables a by-passing of the 'real' and opens up what Baudrillard (1975) has famously designated as 'hyper-reality', i.e., the becoming 'real' of what was originally hype or simulation, as well as the inclination of the postmodern consumer to experience or live the simulation rather than the 'real'.

Advertisers are constantly involved in complex processes of meaning transfer, whereby commodities come to be imbued with cultural meanings only arbitrarily linked to the referent that they originally signified. Advertising in the current era lives on the playful and self-reflexive nature of postmodern culture. Advertisers 'attach signifiers to disparate objects and just as quickly detach them, all in the name of novelty and play' (Knights and Sturdy 1993: 227). In the hyper-real world of postmodern consumption everything mutates into everything else, all is image, appearance and simulation. Time as a function of position, as a dimension of particular self-referential events is fractured, multiple and discontinuous. The search for an authentic, integrated self through consumption is displaced, fractured, fragmented by the 'pure' sign that resides within a self-referential tempo-spatial context and almost coincidentally collides with 'products' (which are themselves mere signs). The 'I', ego, Cogito of postmodern consumer subjectivity exists situationally; specific duration then disappears or undergoes transformation into some new state of being (Ermarth 1992).

Consumption spaces, places and times in postmodernity are host to unique multifaceted forms of consumer subjectivity. With the negation of universals, linear time and space can no longer defend their hegemony; they can no longer be central in a pastiche of fractured times. The presence of postmodern, multivocal consumer subjectivities suggests the need for methods of representation which radically depart from all that went before. But should the challenge to representational time-space, and its prevalence in consumer research, merely involve its negation and replacement with the fleeting, floating, referential times and spaces of postmodernity?

Whilst embodied and embedded in local difference, fragmentation and contingency, we are simultaneously exposed to the universal scheduling, timetables and 'constraints' of linear time and space (Adam 1995). Clock time is a socially constituted, mechanistic means of structuring activity and therefore a discursively constituted representation but this socially constituted artefact permeates every aspect of our daily existence. The project should not be that of valorising contingency at the expense of our lived perceptions of, and relations to, linear time and space. Rather our emphasis should be that of rendering the given status of representational linear time and space problematic, and decentring the power/knowledge configurations that reinforce and reproduce its hegemonic status. This involves, amongst other things, identifying the partiality of this man[sic]-made construct for ordering human existence and drawing attention to the co-existence of multiple dimensions of time and space. Postmodern discourses of time and space represent resistance to the discursive configurations through, and in which, linear time and space are constituted. In calling into question universals and decentring absolutes, our attention is drawn to the problematic nature of linear time and space. In their identification of pluri-tempo-spatiality we come to recognise our tempo-spatial world as constituted by a plurality of times and spaces which exist in more than a simple opposition to linear time. However, to dissolve the existence of linear time and space into a pastiche of situated fragmented times and spaces involves a denial of our lived experiences of these representational constructs.

An alternative procedure is that of deconstructing linear time and spaces, thus revealing these constructs as grounded in specific power–knowledge relations. Linear time and space are revealed as existing in, and through, a plurality of discourses which are informed, and inform, particular types of knowledges that come into being through relations of power. Representational time and space are thus recognised as emerging through discursive formations and discursive practices that privilege specific regimes of truth whilst suppressing or subjugating discourses that do not conform to and concur with the regime of truth they establish (Foucault 1982).

Conclusion

Time and space tend to be taken for granted and, even when a focus of social attention, remain largely unquestioned. All too often analysis of them is guided by operational definitions that relate to their measurement. Time and space are presupposed as objective, quantifiable entities divisible into discrete units. They are parameters for the location of organised and structured actions.

It has been our intention in this chapter to identify marketing approaches to time and space as reinforcing, to varying degrees, the hegemonic status of linear time. The models deployed by marketing explicitly presuppose time and space as fixed, invariant and measurable entities. Although the subjectivist models are presented as diametrically opposed to objectivist understandings of time and space, a number of basic assumptions are common to both traditions. In both, time and space are represented as facts of nature, in one case an 'objective' fact existing independently of human beings and in the other as a merely subjective notion rooted in human nature. Moreover, a predominant feature of these respective expressions of time and space is their reliance on an understanding of linearity and fixedness. The focus of the subjectivist model is the cognitive perceptual apparatus by which individuals comprehend time and space. Variations in cognitive characteristics will produce differences in perceptions of time and space. However, they draw upon linearity and solidity when evaluating subjective temporal and spatial perceptions. This was exemplified by the psychophysical research traditions and the 'memory construct' of decision-making models.

Within marketing, the centrality of linearity and solidity to the objectivist and subjectivist models can, in part, be explained in terms of the dominant psychodynamic paradigm (Gibbs 1993). A factor contributing to the resilience of this paradigm is the failure of critics to do more than substitute one positivist framework for another when objecting to psychodynamic research. While appearing to displace the unobservable mental processes of psychodynamics, for example, Radical Behaviourism (Foxall 1990) relies implicitly upon the 'mind' as a repository of sensory experiences, recollection of which can be readily brought to memory for purposes of classifying stimuli as positive (pleasurable) or negative (painful). This memory, in turn, is dependent upon a linear conception of time where stimulus–response experiences can be stored sequentially in a history bank, as it were, to produce for subjects instantaneous guides to behaviour. The epistemological principle of sensory experience as the foundation of scientific knowledge is thus preserved, to the exclusion and/or subjugation of alternative discourses of behaviour.

Discourses of social time, with their focus on social constructions and inter-subjective perceptions, appear to represent a position that stands in contradistinction to dualistic conceptions of time and space, where the present has a linear relationship to the past or the future and spaces are perceived as if they were independent of the objects that fill them. However, an unintended consequence of their continued reliance on Enlightenment epistemological traditions is that of inadvertently reinforcing the hegemonic status of linear time and spatial solidarity. By failing to challenge this hegemony, alternative discourses of social time and space inadvertently take for granted precisely what they need to question.

Postmodern discourses of time and space represent a direct challenge to the Enlightenment episteme. These postmodern critiques subvert the metaphysics that posits essences such as stable self-identities, non-discursive laws and other universals that reflect and reproduce modern conceptions of time and space. They present a direct challenge to these stabilities, fracturing their hegemony and fragmenting time(s) and space(s) into a plethora of situated, embedded finitudes. Their redefin-ition of time and space as a function of position, as a dimension of particular events where both position and event are described in terms of language, provides a unique insight into the constitutive capacities of language. Our world is thus (re)represented as constituted by a plethora of times and spaces that come to exist through a multiplicity of competing discursive positions. But should we valorise the fragmented time-spatial conditions of postmodernity over our everyday lived experience of time and space?

Power and knowledge are inextricably linked to our perceptions of, and relations to, time and space. However, we need not elevate embedded and situationally contin-gent, fragmented times and spaces at the expense of our lived experience. We need only recognise that postmodernism deconstructs the power/knowledge configurations that privilege dualistic conceptions of time and space. In doing so, we have to engage theoretically with the discursively constituted existence of linear time and bounded space. We can then see how alternative discourses (e.g., social and social construc-tivist) inadvertently reinforce the hegemonic status of linear time and bounded space thereby contributing to their own marginalisation. In an attempt to avoid these problems, our research has involved various deconstructions to reveal how linear time and bounded space are the condition and consequence of specific marketing discourses of consumption. In addressing time and space in this chapter, we have sought to draw attention to the inextricable link between dominant conceptions of time, space, power and knowledge.

Note

1 The 'psychodynamic' paradigm, in broad terms, relies upon a model of consumer behaviour drawn from cognitive psychology which presumes individuals possess a set of 'needs' that through a process of rational cognition they seek to satisfy. All that is required, then, is for the producer to identify the needs and design products and services that are efficient in satisfying them. (For a critique see Knights *et al.* 1994.)

Part II

Knowledge at work in space and place

7 Creating and sustaining competitiveness

Local knowledge and economic geography

Edward J. Malecki

All places are not alike, and these differences are most pronounced beneath the surface, in the things that cannot be seen by the tourist or casual observer. Just as customs vary and reflect local cultures, so too do firms and institutions in particular places engage in long-standing patterns and behaviours that reflect local culture. This chapter addresses the interplay between the knowledge that firms control and the knowledge that is 'in the air' and spills over to be shared by other, even all, firms in any particular place (Marshall 1909). Places – localities and regions – are better off when they promote shared or public knowledge from which many firms can benefit. Indeed, such knowledge forms the basis of successful high-tech regions. To create sufficient knowledge that it forms a basis for local competitiveness is difficult, and to sustain it is even more difficult as competitor regions continue to emerge.

Large firms and local entrepreneurs alike try to utilize local knowledge to serve customers in untapped market niches. However, a great deal of knowledge is local, and to have access to it requires that firms have a presence in a place. The variations in knowledge, and the mobility of some knowledge and immobility of other knowledge, are important parts of what comprises the economic geography of our world. Earlier chapters in this volume have addressed the general role of knowledge in capitalist economies. This chapter examines in greater detail the culture of technology as a knowledge system that creates and sustains technology-based economic development.

The chapter begins by identifying the various forms that knowledge takes in the economic geography of places. Next, knowledge as an input to economic growth and the transfer of knowledge are discussed, including a consideration of how local knowledge has become central to understanding firms. The focus of the chapter then shifts to 'sticky' knowledge, which is more difficult to exchange and move from place to place, and the sticky places where it is found. The Boston region is proposed as one region that has met the standard of the 'learning region' and regional innovation system where renewed and sustained competitiveness is attained as a result of accumulated and ongoing knowledge acquisition by regional firms and institutions.

What (and where) is knowledge?

Knowledge, as used in this chapter, includes the skills of workers, the experience of managers and owners, and the 'pulse' of customers' needs and demands. The accumulation of skills and knowledge in particular places has been recognized as a phenomenon in the location of economic activity for decades, and arguably has been increasing in importance in recent years. This phenomenon has taken two principal forms: first, the location of some industries depends on localized knowledge; and, second, the innovative or knowledge activities of most sectors can be called knowledge-based.

The location of knowledge-based industries

The importance of skilled workers was identified decades ago. Marshall (1909) saw that both employers and workers benefit from the localization or concentration of industry (Krugman, 1991: 36–7). Hartshorne (1927/1968: 26) observed that 'workers skilled in a particular industry are to be found in large numbers only in districts where that industry or allied industries are well established, is developed in new areas only at great expense, and is extremely difficult to transfer from other districts'. The skills represented in such districts result in labour constituting a high proportion of the cost of finished product, including 'the manufacture of jewelry, silverware, and other articles of exceptional value' (Hartshorne 1927/1968: 26). The *high value-added industries*, as we would call them today, are concentrated geographically. The importance of skilled workers in high-technology industries, which comprise the majority of sectors that are high value-added, is well-established (Oakey 1981). Knowledge-based manufacturing sectors today include those with high research and development (R&D) intensity: aerospace, pharmaceuticals, computers and office equipment, communications equipment and semiconductors, scientific instruments, and electrical machinery (OECD 1995). These 'high-technology' industries are remarkably concentrated in all countries where they are located (Breheny and McQuaid 1987).

In addition to these manufacturing industries, a large number of producer services – those services for whom firms are the principal customers – are knowledge- and information-intensive, such as financial and legal services, marketing, and R&D (Lee and Has 1997; Marshall *et al.* 1988). Many services, both in firms and in the public and non-profit sectors, require the work of *symbolic analysts* (Reich 1991a). Their work includes problem-solving, problem-identifying, and strategic-brokering activities common in information-related jobs such as lawyers, software engineers, and management consultants. Face-to-face contact and collaboration are essential for these activities because of their need to tailor or customize their service to specific clients (Reich 1991a; Von Glinow 1988). Table 7.1 includes three categories of knowledge-intensive economic activities: units of large corporations, firms specialized in technical or business services, and a number of public and non-profit organizations. All of these are located disproportionately in the largest cities (Daniels 1982, 1993; Knox and Taylor 1995; Moulaert and Tödtling 1995).

Perhaps more useful than an over-arching conception of knowledge is a continuum on which routine information, such as that processed in 'back offices', is the lowest level (Wilson 1995). The next level is structurally ordered information, which Törnqvist (1983) and Andersson (1985) call knowledge. Above these in importance are competence (embodied knowledge) and, at the highest level, creativity. 'Creative regions', which also contain high levels of information, knowledge, and competence, are creative because they are centres of communications, with links to other places and sources of knowledge (Andersson 1985).

In the context of services, Törnqvist (1983) and Illeris (1994) suggest a distinction among three types: 1) back offices, which use telecommunications, rendering location near customers as unimportant; 2) unsophisticated, customized services (also called interpersonal services), which must be near customers; and 3) highly specialized services, including most producer services, which rely on face-to-face contact so large cities are the best locations. The third type includes the managerial and financial activities that gravitate towards the 'capitals of capital' for the various types of agglomeration economies that are found there (Edwards 1998; Moulaert *et al.* 1997).

The location of innovative activities

A second effect of skilled workers is that 'the presence of labor skilled in such industries is the most important element in the concept of "advantage of an early start"' (Hartshorne 1927/1968: 26). To understand the effect of the 'early start' and its impact on regional economic change and dynamics requires going well beyond the location of industries to examine the effect of R&D and innovation (Richardson 1973: 113–32). The framework of the product life cycle and its counterpart, the profit life cycle, clarifies the early start. In general, the first firm to sell a product or service is able to charge a high (monopolist's) price, since there are no competitors with the knowledge to compete. As imitations or equivalent products come on to the market from other firms, the prevailing price normally falls rapidly, as price becomes a more important, and eventually the primary, basis of competition. The only way out of this dilemma for a company is to invest in R&D on new products (or marked enhancements of existing products), from which the monopolist's 'super profits' can be made (Markusen 1985).

In many sectors, such as electronics and scientific instruments, product life cycles are short, both in the duration of the product's life and in the overall sales level that can be attained. In these industries, and in others as product differentiation and the need for flexibility become the norm, a greater proportion of knowledge workers are needed to sustain the flow of new products necessary to compete. Indeed, the

Table 7.1 Knowledge-intensive economic activities

Industrial corporations	Specialized technical and business service firms	Public and not-for-profit organizations
Corporate headquarters	Law	Federal agencies
Research and development	Engineering	State agencies
Regional offices	Accounting	Universities
Divisional offices	Finance	Music and the arts
Computer centres	Advertising	Hospitals and clinics
Training centres	Public relations	Cancer centres
	Insurance	Professional associations
	Seminars and conventions	Federal Reserve banks
	Communications	Foundations
	Airlines	Museums
	Consultants	Consulates
	Business information services	

Source: Knight 1982: 56.

proportion of a firm's products that are new within the past three years has become a common measure of a firm's competitiveness. In fact, however, many 'new' products are refinements – major or minor – of existing products. This is to be expected: knowledge within firms is cumulative and follows established trajectories (Nelson and Winter 1982).

Like the industry-specific skills that influence value-added manufacturing to concentrate geographically, non-routine information-intensive or knowledge-intensive office activities and R&D are concentrated geographically. Ullman (1958) demonstrated the concentration within the USA of science, invention, innovation and, generally, of 'brains' and 'great minds'. This concentration remains in force today, seen in the urban location of patents (O'hUallachain 1998). This agglomeration of knowledge occurs for several reasons, as seen in other chapters in this volume. Labour pooling and large numbers of specialized producers help firms in concentrations or clusters to minimize transaction costs. Beyond transaction costs, proximity facilitates a social division of labour – which allows firms to make choices between what they do internally and what they do externally – enhancing specialization (Storper 1998). Specialization, not scale, is the critical strategic choice in a com-petitive environment of differentiated demand (Porter 1985). More importantly, the localization of transactions, that is, of traded interdependencies, is inadequate. Untraded interdependencies help to explain dynamic agglomeration economies, which enhance the possibility of technological learning, as opposed to simply reductions in the unit costs of production with a given technology (Martin and Sunley 1996; Storper 1995, 1998).

Large urban regions generally have shown considerable resilience as high-tech locations over long periods of time (Castells and Hall 1994). These regions provide the concentration of brains, money, research and education, and quality-of-life attrac-tions that are essential both to firms and to skilled workers (Malecki and Bradbury 1992). R&D, the principal corporate innovative activity, is, like producer services, disproportionately located in urban regions, and shows little tendency to disperse (Malecki 1987). It is mainly more routine production and some engineering that has dispersed to 'technical branch plants'; R&D remains centralized (Glasmeier 1988). For high-tech industries and R&D and other knowledge-based activities, there is 'a new entry to the list of currently important location factors influencing the geographic location of industry: the knowledge assets of particular local, regional or national milieus' (Maskell *et al.* 1998: 24).

Knowledge in the firm

Knowledge production is different from the production of goods in three ways. First, it 'entails a deeper element of uncertainty' (Howitt 1997: 12). Indeed, uncertainty is central to the entire process of technological change. Second, knowledge is embodied not only in capital goods, but also in people, a phenomenon addressed conceptually, but inadequately, by the concept of human capital. Third, knowledge also is embodied in organizations, taking the form of organizational routines. The resultant knowledge is greater than the sum of the individual knowledge possessed by the firm's employees (Nahapiet and Ghoshal 1998). None of these aspects is well-understood and none has been modelled adequately in endogenous growth theory (Howitt 1997: 12).

Knowledge and competence are not merely inputs to production; they are 'strategic assets' (Winter 1987). Firms make decisions based on knowledge accumulated from the outcomes of previous decisions. The trajectories follow different paths for each firm that are not random, but are akin to the biological development of an organism (Nelson 1995). In the case of firms and industries, however, the result is not necessarily a larger organism (or organization), but new technologies, new products, and new services. This evolutionary or neo-Schumpeterian economics provides a framework that need not result in equilibrium (Nelson and Winter 1982).

Building upon these ideas is the *competence theory of the firm*, which differs markedly from both neoclassical economics and transaction-cost economics. Competence is 'a typically idiosyncratic knowledge capital that allows its holders to perform activities – in particular, to solve problems – in certain ways, and typically do this more efficiently than others'. In short, 'firms as seen *essentially* as repositories of competence' (Foss 1996c: 1). This perspective – actually a distinct theory of the firm – has grown from several roots: Nelson and Winter's evolutionary theory of the firm, a series of contributions to the 'resource-based view' (for a review, see Oinas 1995), Hamel and Prahalad's (1994) work on the 'core competence' of the firm, and related research in technology management focused on learning (Teece and Pisano 1994). The origins of the competence perspective is reviewed masterfully by Knudsen (1996); a review focused more on the economics literature is found in Carlsson and Eliasson (1994).

In the competence perspective, the firm is essentially a repository of skill, experience and knowledge, rather than merely a set of responses to information or transaction costs (Hodgson 1998b; Langlois and Robertson 1995). These *capabilities* of a firm comprise 'the ability to identify, expand, and exploit the business opportunities' that arise (Carlsson and Eliasson 1994: 694). There are four types of capabilities:

- technical (functional) ability relating to the various functions within the firm, such as production, marketing, engineering, R&D, and product-specific capabilities;
- organizational capability;
- selective, or strategic capability: the ability to make innovative choices of markets, products, technologies and organizational structure; to engage in entrepreneurial activity; and especially to select key personnel and acquire key resources, including new competence; and, at the highest level,
- learning capability.

(Carlsson and Eliasson 1994: 694–5)

Furthermore,

> what is involved with managerial and entrepreneurial skills is not mere information or knowledge but sophisticated but essentially idiosyncratic judgements and conjectures in the context of uncertainty . . . This is a key difference between contractual and competence-based theories of the firm.
>
> (Hodgson 1998b: 183)

Learning is seen as a key process – the highest of the four levels of capabilities above – that allows a firm to adapt to changed circumstances in its competitive environment. Learning is a process that saw little attention until Nelson and Winter's (1982) work, and it was given a new significance by the work of Cohen and Levinthal (1989), who expanded the conventional concept of R&D to include its 'two faces': innovation and learning. More influential was their work (Cohen and Levinthal 1990) that stressed a firm's 'absorptive capacity' for learning: the ability to evaluate potential knowledge, assimilate it, and apply it. The 'prepared firm' is one that does R&D as much to accumulate related knowledge as to accomplish a specific technological objective (Cohen and Levinthal 1994). However, 'by its nature, learning means creativity and the potential disruption of equilibrium. In short, the phenomenon of learning is antagonistic to the concepts of equilibrium and rational optimisation' (Hodgson 1998b: 185). The potential disruption of equilibrium is why small firms, those least prepared to learn, are the least likely to gather and, especially, to act upon new information (Glasmeier *et al.* 1998).

The transfer of knowledge

Although it is fairly clear in which industries, occupations, and economic activities knowledge is prominent, it is less clear how – and how effectively – that knowledge is transmitted to other settings. A key characteristic of knowledge is the fact that it can be exchanged (i.e. sold or given away) without any loss of knowledge on the part of either party. Not all knowledge is equally transferable, however. It is easiest to transfer *codified* knowledge – that which is tangible in some way, usually in printed form, as in books, patent applications, and scientific papers. Privately-held knowledge and shared expertise, on the other hand, are *tacit* in nature. Tacit knowledge may well also be unteachable or unarticulable, not observable in use, or complex and part of a larger system (Winter 1987: 170). Emergent technologies are likely to be tacit or uncodified, but information about them will diffuse to others outside the firm as they become key technologies used by that firm or by others in the industry (Boisot 1995b). Generally, tacit knowledge is embodied in people, rather than in written form or in objects. Thus, it typically is acquired through hiring, R&D, and interpersonal networking (Faulkner *et al.* 1995; Nonaka and Takeuchi 1995). Firms that do basic research can signal, through papers published by their researchers and patent applications (both codified forms), that they also possess other knowledge that cannot be (or has not yet been) published or otherwise made public.

The key to economic competence, and to competent firms, is that tacit knowledge and information asymmetry render much knowledge untradable. The integration of specialized functional capabilities in the firm and specialized, non-tradable information in the form of tacit knowledge give rise to synergy or scale effects. These synergy effects cannot occur in a neoclassical world (Carlsson and Eliasson 1994: 696). Furthermore, 'compared with goods and other services, information and knowledge cannot be so readily "bought as required" . . . we do not know the value of information until after it is purchased' (Hodgson 1998b: 183). One firm's knowledge can be acquired by another through interpersonal contacts, which are most easily accomplished in geographic proximity.

The degree to which one can appropriate, or profit from, technical knowledge is an important issue. Patents, copyrights, and trade secrets (such as recipes and chemical formulas) are legal means of keeping a technology out of competitors' hands, but they do not guarantee that imitation cannot take place (Teece 1986; Teece and Pisano

1994). Intellectual property law dilemmas related to legal copying *versus* piracy have exploded since digital technologies have emerged (*The Economist* 1996; Lamberton 1994).

It is rarely easy to transfer complex knowledge from one person to another. On-the-job training, on-site engineering, and other means of learning technologies have been central to the process of technology transfer, but few attempts have been made to translate these mechanisms to more general situations. To some degree, 'learning by doing' and 'learning by using' were early ways of capturing these kinds of learning that take place outside of formal R&D. We now recognize that user-producer interaction is a key mechanism for how outside knowledge and technologies are obtained, understood, and incorporated. But, in addition, it is essential that the recipient have sufficient absorptive capacity for knowledge transfer to occur, even between units within a firm (Szulanski 1996).

Technology transfer requires technical competence in *both* the provider and the recipient. If firms are 'learning organizations' then the firm's employees must be able to learn to 'gain knowledge' from many sources (Leonard-Barton 1995; Nevis *et al.* 1995). The simplest form of learning is 'learning by doing', suggested over 30 years ago by Arrow (1962) as a form of technology acquisition outside of formal R&D. The list of types of learning has expanded greatly over the years, now encompassing learning by using, operating, training, hiring, searching, trying, interacting, selling, borrowing, failing, and from inter-industry spillovers (Malecki 1997: 59).

Transfers of knowledge within and among firms depend critically on the absorptive capacity to learn as well as on the nature (tacit or codified) of the knowledge being communicated. Complex knowledge, such as that used to set up new machinery or an entire new production plant, is 'sticky' because it can only be transferred as tacit knowledge or as learning by doing or by using, thereby requiring iterative consultation and travel (Gertler 1995; von Hippel 1994). In general, key activities dependent on sticky, tacit knowledge should be kept inside the firm rather than trusted to outsiders (Chesbrough and Teece 1996). In effect, however, as the next section details, the local culture of some regions operates as 'internal' and facilitates knowledge creation and widespread learning. In addition, the spectrum of complementary assets, which encompasses a range of capabilities which support and sustain the development and enhancement of technology, also makes some places better in many ways for knowledge-based efforts. In short, 'many transactions are highly sensitive to geographical distance by virtue of their substantive complexity, uncertainty and recurrence over time' (Storper and Scott 1995: 507–8).

In general, then, 'the knowledge required to make and sell any firm's products resides in the structure of direct and indirect capabilities within that firm, supplemented by the structure of indirect capabilities that connect it with other firms' (Loasby 1998: 154). 'Some routines and competences seem to be attributable to local or regional forces that shape firms' capabilities at early stages in their lives' (Teece and Pisano 1994: 550). These local and regional forces, the topic of the next section, in fact continue to shape firms' capabilities over time.

The local nature of knowledge

The sensitivity to distance of not only transactions, but an array of 'untraded interdependencies', makes it possible for a region to be able to build and keep its distinctive competence. Because tacit and idiosyncratic knowledge are decentralized, co-location is required at several locations, and the necessary knowledge cannot be centralized in a single point (Grant 1996). It has been emphasized thus far that knowledge (and the information flows or contacts needed to obtain it) is far from evenly distributed. Just as knowledge is not ubiquitous, it also is not possible to claim that it is uniformly and perfectly available, as neo-classical economic models would have us assume. If knowledge is not found everywhere, then *where* it is located becomes a particularly significant issue. While codified knowledge is easily replicated, assembled and aggregated – witness the spread of nuclear weaponry – other knowledge is dependent on context and is difficult to communicate to others. Tacit knowledge is localized in particular places and contexts. (To economists, 'localized knowledge' typically means localized within a firm or pertaining to a specific technology (e.g. Stiglitz 1987).)

Learned skills become partially embedded in habits. Habits grow into routines or customs – or conventions – when habits become a common part of a social culture (Storper 1998). Institutions are formed as durable and integrated complexes of routines and customs. Thus, habits, conventions, and routines preserve knowledge, particularly tacit knowledge in relation to skills, and institutions 'act through time as their transmission belt' (Hodgson 1998a: 180). This is especially true of collective knowledge, which is embedded in a social setting and comprises largely tacit (or implicit) knowledge (Spender 1996).

> The local business environment functions as a social context which not only produces scale economies through external relations and offers an efficient division of labour . . . but also makes it possible to economise on transaction costs, and promotes entrepreneurship and innovations and the development of dynamic learning externalities and technological spill-over.
>
> (Maskell et al. 1998: 183)

In other words, firms become – and remain – competitive by conceiving and implementing strategies that utilize – directly or indirectly – a number of valuable traits of their place of location. These traits enable those firms to earn a profit when faced with otherwise similar competitors located in other places (Maskell et al. 1998: 51).

The British motor sport industry and the Swiss watch industry are two such environments that have benefited from, in the former case, a great deal of untraded interdependencies rather than transactions (Henry et al. 1996), and in the latter case, from institutional strengths that allowed the regional industry to respond to the threat from electronic technologies (Glasmeier 1991; Maillat et al. 1995). Perhaps the best example of localized competitiveness is the Danish wooden furniture industry. Wooden furniture is a low-tech, price-competitive commodity, made in virtually every country, yet Danish firms have managed to create country-based, local competitive advantages (Maskell et al. 1998).

The local milieu is not independent from the production processes centred there. The local context provides 'the values, the knowledge, the institutions and the physical environment necessary for its continuance' (Becattini and Rullani 1996: 161). Going further, an *innovative milieu* combines learning (from both local and non-local – even global – sources of knowledge) and interaction, or co-operation with respect to innovation (Camagni 1995). The collective learning dynamic is 'characterized by the ability of the milieu's players to adapt their behaviour, in the course of time, to the changes in their environment (through innovation, company formation, production of specific know-how, etc.)' (Maillat 1995: 162). Maillat sees innovative milieus as a special case of industrial districts; some districts are innovative, while others are merely highly interactive. Conversely, technopoles tend to be innovative without sufficient interaction to develop a milieu.

Empirical evidence tends to support these generalizations. Johannisson *et al.* (1994) and Massey *et al.* (1992) have found that science parks do not necessarily constitute innovative milieus because interaction among firms in the parks often does not take place. Similary, Hardill *et al.* (1995), comparing small clothing firms in France and the UK, found little use of local networks and resources in the East Midlands and, not coincidently, lower levels of innovativeness. While firms in such environments can be competitive by relying on non-local networks, the stronger local environment for firms is one in which both local links are abundant and flows of knowledge to and from other places are common.

Ubiquitification

All knowledge begins as local knowledge, known only to those who have gone through the learning steps needed. This new knowledge also exists only in tacit form, known only to those with whom this knowledge is shared. While not all such partners need be local, some (even most) are likely to be. When formerly tacit knowledge is converted into a codified form, it rapidly becomes a ubiquity, accessible on the global market at a scale not possible when the knowledge was internal to the firm. For the possessor of formerly tacit knowledge, the immediate effect of codification is the same as with all other former competencies from which the firm might have benefited, but which later become available to everyone: by becoming ubiquitous – the process of ubiquitification – the knowledge loses its potential to contribute to the competitiveness of the firm (Maskell 1999). 'No firm can build competitiveness on ubiquities alone, and little economic progress would be made in any local milieu, region or country if everyone was able to do exactly the same in all other places at once . . . As ubiquities are created, localised capabilities are destroyed' (Maskell *et al.* 1998: 22–3).

Globalization has accelerated the rate at which heterogeneous resources – localized capabilities – are turned into ubiquities. No firm exposed to international competition and located in a high-cost area can depend solely on already fully codified knowledge. The more easily codified knowledge is accessed by everyone, the more crucial will tacit knowledge therefore become in sustaining or enhancing the competitive position of the firm (Maskell 1999). Both codified and tacit knowledge are 'translated' by local cultures into their ultimate use as products and services. Some places can make the translation to profitable activities; others cannot.

Sticky local knowledge

The tacit nature of knowledge and the localness of tacit knowledge contribute to its 'stickiness' and allow interfirm communication and co-operation to take place smoothly and cheaply. The strategic nature of this local knowledge derives from its tacit nature, and the fact that it is difficult to transfer. More importantly, perhaps, is the role of trust. In order to communicate even partially tacit knowledge, a high degree of mutual trust and understanding are normally necessary. This involves a shared language, values, and culture. Because it takes time to build trust relations, there are 'time compression diseconomies' that slow – but do not permanently prevent – competitors from imitating successful business behaviour of established firms (Jevnaker 1997; Maskell *et al.* 1998: 110).

Some of the knowledge created through local interfirm co-operation is embedded in the area's business culture, making it even more difficult for outsiders to imitate. 'Localised capabilities thereby translate into sustainable competencies, enabling firms to survive and thrive in spite of an unfavourable local cost-structure' (Maskell 1999: 50). The development of specialized skill depends on accumulated experience and a variety of experience found in a local area, but that variety can only be encompassed within a network of connections (Loasby 1998).

Likewise, the purchase of new machinery – a seemingly minor change in a firm's routine and one that allows it to take advantage of the new technology embodied in the machinery – also can entail great incompatibilities. The difficulties in implementing German-made machines, for example, within Canadian firms is a complex product of cultural distance, organizational distance, and the difficulty of overcoming long-distance physical distance for face-to-face interaction (Gertler 1995, 1997).

Learning takes place within large firms as they attempt to 'transplant' their factories to new locations. Japanese auto firms, for example, have been unable to replicate precisely the conditions found in Japan. The result, in both Europe and North America, are 'hybrid factories' that blend management structures and practices proven successful in Japan with circumstances regarding labour and suppliers necessary for the local culture and setting (Abo 1994; Thompson and Rehder 1995). These hybrids represent an ongoing learning process as Japanese companies learn what can – and what cannot – be transferred directly from Japan. Mair (1997) suggests that Honda may be the most successful Japanese firm at 'strategic localization'. All of these gaps between producers and users are a result of the 'stickiness' of tacit knowledge.

What is it that happens in some places that keeps them creative, and enables them to remain innovative milieus? Region-specific competence 'implies a connection with the resources of certain localities and regions, and that stretching them means trouble of some kind' (Törnqvist 1983: 98). The avoidance of 'trouble' is what makes some places 'sticky'. Industrial districts, the localized clusters of small, specialized firms found in many settings around the world, are among the 'sticky' places. A great deal of attention has focused on the districts in Italy, where co-operation and trust and social interaction provide external economies or spillovers that are greater than merely the 'Marshallian' kind based on the fact that 'the secrets of industry are in the air' (Harrison 1992; Malecki 1997).

Markusen (1996) identifies 'sticky' places as those that 'are able to sustain their attractiveness to both labor and capital' (p. 293). She says that flexibly specialized or new industrial districts owe their stickiness to the role of small, innovative firms, embedded within a regional cooperative system of industrial governance which enables them to adapt and flourish despite globalizing tendencies . . . Stickiness connotes both ability to attract as well as to keep, like fly tape.

(Markusen 1996: 294)

The stickiness of industrial districts derives from the collective nature of the territory or milieu, whereby the district itself can be seen as a *collective entrepreneur*, with not only firms, but inter-firm associations, worker organizations, financial institutions, and governmental agencies also playing important roles (Best 1990; Lorenz 1992). This is an example of the social capital that firms are able to build through their ties to their (local or regional) institutional environment (Levinthal 1994). The *territory* itself plays a role, as a place of co-ordination and learning, that goes beyond transaction cost-based agglomeration economies (Belussi 1996; Courlet and Pecqueur 1991; Courlet and Soulage 1995).

Some of the local character of an industrial district is a cultural and social tradition that fosters tight-knit networks and the passing down of industrial traditions and skills. An example from outside Italy, Gnosjö, Sweden, is such a place, where its 'entrepreneurial spirit' and strong local traditions have been cited (Tornqvist 1983; Karlsson and Larsson 1993). However, Gnosjö has shown signs of having difficulty keeping up with the knowledge-intensive demands of contemporary competition. The firms in the region are less oriented toward external or global knowledge, thus threatening its long-term development sustainability (Karlsson and Larsson 1993).

To the classic Marshallian district, and its Italianate and California variants (including both Silicon Valley and Orange County), Markusen adds three other types of industrial districts, all of which she identifies as 'sticky places':

- hub-and-spoke districts, centered on one or more large, vertically integrated firms, as in Detroit and Seattle in the USA, Pohang in Korea, and Toyota City in Japan;
- satellite industrial platforms, based on attraction of investment by firms from elsewhere; and
- state-anchored districts, generally based in military, research, or other state activity.

Satellite industrial platforms are sticky in the sense of 'sunk costs' (Clark and Wrigley 1997). If the plants and facilities concentrate on 'higher-end' activities, they may be able to become sticky by means of the skilled labour force. Some places, such as Silicon Valley and other large metropolitan areas, exhibit elements of all four models.

Markusen's four types of districts bear resemblance to the two ideal types of local productive systems proposed by Becattini and Rullani (1996): (1) the industrial pole, a local system organized around one or more large companies (i.e. hub firms), and the (2) industrial district, which 'is integrated not by dominant companies but by a local market, a common cultural hallmark and by a strong feeling of belonging' (Becattini and Rullani 1996: 166). These local productive systems would seem to be among the sticky places that Markusen asserts are complex products of corporate strategies, industrial structures, profit cycles, state priorities, and local and national politics. Local institutions alone do not explain them, because their companies, workers, and other institutions (e.g. universities and government installations) are embedded in external relationships. Despite these external links, hub firms play an important role through their proactive diffusion of new forms of production organization through their supplier networks, thus helping to upgrade the skills of localities to 'higher end' activities (Ettlinger and Patton 1996; Florida 1996).

How do some places create, attract, and keep economic activity? In other words, how do they maintain their competitiveness in a world of increasingly global competition? They do so because people there 'make connections' with other places, retaining close connections with other local systems and with global knowledge (Amin and Thrift 1992; Becattini and Rullani 1996; Malecki and Oinas 1999). The innovative milieu is the seat of permanent processes of adjustments and transformations to external changes, such as competition and technological discontinuities. These adjustments are able to take place when interaction and co-operation are the social norm, and where learning and innovation are able to respond to and incorporate new knowledge. These are externalities of proximity, variety and accessibility that are usually found in larger cities (Maillat 1998: 127).

The external connections do not 'just happen'; they rely on intermediaries, especially wholesalers, who buy and sell outside the region, bringing new information and competing products back to the region (Meyer 1998). Other key individuals act as bridges that cross industry lines, serving as community entrepreneurs (Cromie *et al.* 1993). If there are widespread technical contacts, wholesaling, producer services, and financial services, its long-term vibrancy is more secure (Meyer 1998).

The 'Learning Region' and the role of policy

Perhaps the most difficult task is to institutionalize collective learning processes in a locale for the benefit of firms and residents (Knight 1995). It is this collective learning that is the key to development. The current 'best practice' in regional and local development is to be a *learning region* (Asheim 1996; Florida 1995a; Morgan 1997). The external economies generated by strong firms, a labour and supplier base, and knowledge base are more often local or regional than national (Porter 1990). The archetype of a learning region is the industrial district or innovative milieu. The vitality of these places revolves around two phenomena: (1) the learning efficiency of these areas, based on cultural norms that are not the same everywhere nor easily transplanted or created (Saxenian 1994; Sweeney 1996), and (2) interaction, which enhances innovation and learning. Significantly, agglomeration, while often related to learning dynamics, is no guarantee of them (Asheim and Isaksen 1997; Johannisson *et al.* 1994).

When found in combination with a technical culture and strong institutions, learning proceeds efficiently based both on local knowledge and assets and on links to networks elsewhere (Maillat 1995; Oinas and Malecki 1999; Storper 1995). Learning is territorialized (localized or endogenous) in three ways: agglomeration, indirect externalities, and untraded interdependencies (Storper 1995). The strength of the local learning system depends greatly on competent governments and firms working to understand and support one another for the benefit of the region and its population (Maskell *et al.* 1998). This mutual support system hinges on trust and social capital, and is exhibited in a variety of ways in different places, as the various regional innovation systems compiled by Braczyk *et al.* (1998) illustrate.

Despite the ubiquitous example of Silicon Valley, it does not take high technology or frontier industries to sustain an economy or to create the benefits of economic development – local linkages and continuing entrepreneurship. Indeed, high-technology firms in a sparse environment may well provide far fewer benefits than low-tech firms (Chandra and MacPherson 1994). Furniture, toys, and food products from Denmark demonstrate that low-tech industries also can be innovative and competitive on a global scale (Hansen 1993; Maskell *et al.* 1998). In order to be competitive, firms need a web of interactions and information flows within and into the region, including high-tech knowledge translated to be applicable to the industry. A focus on learning removes the artificial distinction between high-tech and low-tech sectors and between innovative and non-innovative firms (Oinas and Virkkala 1997).

The policy answers based on the linear model of innovation failed to generate local (or indigenous) growth because they focused on R&D, when they should be more broadly based on incremental learning and technological change (Malecki 1997; Oinas and Virkkala 1997). However, R&D – especially that by firms – is important in other ways, because it represents an 'active' outlook, an absorptive capacity, a level of technological progressiveness, and an 'open mind', reaching out for new information and being receptive to it. Technically progressive firms take part in information exchange, they continually search for information, and they maintain internal communication (Sweeney 1987). From a regional perspective, these *vertical links* and the presence of gatekeepers have been significant (Cromie *et al.* 1993; Flora and Flora 1993; Flora *et al.* 1997). One-way flows and linkages to the outside (e.g. via branch plants) drain a region of its ideas, talent, and control, unless they are balanced by a receptivity to new ideas coming from the outside.

These vertical links as well as the horizontal ones within the region are the essence of the 'munificent environments' identified by Dubini (1989). Put briefly, access to other entrepreneurs, to consultants, and to sources of information are far more readily available in munificent settings than in 'sparse environments'. The characteristics of a successful place depend on the ability of firms to assemble critical sets of factors (such as talent, technology, capital, and know-how) – mainly from the local environment (Malecki 1997). The environment for a firm or a community involves a host of industrial, technological, and economic linkages, many of which are public resource endowments in the locale or region. As with almost everything in regional development, the ideal environment is a setting where firms take advantage of agglomeration and proximity to utilize nearby sources of information, skilled labour, technology, and capital. Peripheral and rural areas generally stand in sharp contrast: networks have not developed, innovation and technology are not native to the local culture and economy, and firms struggle to remain competitive. But they do so by turning to external sources, in effect, substituting non-local resources for the sparse set of (or non-existent) local resources (Vaessen and Keeble 1995). Such locales are among Dubini's sparse regions, where little networking or innovative entrepreneurship takes place (Davidsson 1995). Unable to maintain an R&D effort and 'a systematic technology watch', small firms use their networks as 'antennae' and 'filters' of information (Estimé, Drilhon and Julien 1993: 56).

Sustaining competitiveness: can places remain competitive?

Competitiveness can diminish, despite policy efforts. Florida (1995b) identifies the Great Lakes region of the USA as a learning region, and Sabel (1992) lauds the efforts of the state of Pennsylvania. Deitrich and Beauregard (1995) examine the same evidence and conclude that, in the midst of a massive shift from manufacturing to services, little positive has taken place in the Pennsylvania economy. Thus, it is less than certain that we know a learning region when we see one (Oinas and Virkkala 1997). Cooke (1995) believes that learning regions show institutional learning, institutional reflexivity, and regional associationalism, and that these are found only in Germany (Baden-Württemberg and North-Rhine-Westphalia) and in Italy's Emilia-Romagna. The issue remains open, although Cooke's (1996a) research suggests that a regional industrial policy such as that in Emilia-Romagna serves as current regional best practice. Such a policy includes technology centres, telecommunications, expert governance systems, science and technology (S&T) foresight, networked information and services provision, and gateway services to international markets.

   Public policies for a learning- and knowledge-based economy are different from those that were favoured in the past. Subsidies are unlikely to work, and a strong element of unbalanced regional development is likely to result. A primary need for knowledge-based development for 'learning regions' is simply knowledge and information. To a great extent, this means employing competent people and rewarding them (Maskell *et al.* 1998; Pfeffer 1998). Information about state-of-the-art machinery and processes and of other market trends is critical. Just how much these kinds of learning activities are *regional* as opposed to organizational or firm-specific is an open question, and one that makes policy actions difficult (Oinas and Virkkala 1997).

The case of the Boston area

Boston is a city whose firms have been criticized for their lack of flexibility in this age of flexibility, resulting in a less entrepreneurial culture, especially in comparison to Silicon Valley (Saxenian 1994). Can Boston, which has gone through three reindustrialization processes in the last 50 years, based on a 'commitment to excellence in education and research' (Castells and Hall 1994: 38), do so a fourth time? Documenting the state's commitment to the innovation process and human, technology, capital, and infrastructure resources, Massachusetts shows signs of doing so again (Massachusetts Technology Collaborative 1997). Indeed, the resilience of the area has led many to cite its successful revitalization through a sequence of industrial waves during the past 50 years: the first was the defence-oriented complex of the 1940s and 1950s. The second was less successful, as semiconductor technology evolved in Silicon Valley; the third wave was the computer boom of the late 1970s and 1980s. All of these were dependent to a significant degree on defence spending, and the current reindustrialization will be the first that is organizationally and technologically not tied to defence (Castells and Hall 1994). The overall performance of the Massachusetts economy, and its Boston core, remains strong if not on top nationally on nearly all indicators (Massachusetts Technology Collaborative 1997). A closer examination of the local culture of the Boston area sheds light on the ability to sustain regional competitiveness through knowledge.

The high-tech cluster of computer and electronics firms along Route 128 surrounding Boston attracted attention as a model, inspiring imitators elsewhere. It is one of a handful of regions that has attained the status of a 'self-sustaining cluster', able to muster an infrastructure and agglomeration economies to facilitate business formation (Miller and Coté 1987). (Silicon Valley, Minneapolis, Los Angeles, and West London are the others named by Miller and Coté; Herbig and Golden (1993) list Silicon Valley, Route 128, Austin, the Research Triangle, and Pittsburgh). Within the United States, Boston and northern California's Silicon Valley have been the twin towers of high-tech growth for several decades, until recently. Saxenian's (1994) comparison of the computer industries in the Boston area and Silicon Valley concluded that the Boston area's conservative, large-firm culture is unable to respond to change in high-technology industries. This critique has had tremendous influence. Since its publication, Silicon Valley has been held up as the model of resilience and entrepreneurial drive, partly on the strength of Saxenian's critique (*Business Week* 1997; Mickelthwaite 1997; Postrel 1997). But is the Boston area in a state of decline? Or is it resilient and self-sustaining as an innovative region – a learning region (Florida 1995a)?

Boston is 'a place where knowledge industries are key and economic success derives from knowledge-based innovation' (Kanter 1995: 202). Boston is the most diversified area for advanced technology and the area 'has the greatest intrusiveness of high technology into other, more traditional forms of business' (Bolland and Hofer 1998: 276), creating a business culture that facilitates the entrepreneurial process: successful role models and private venture capital networks play large roles in creating an environment for entrepreneurship (Roberts 1991). Despite the decline of the minicomputer industry and in defence-related sectors in the 1980s, Boston has sustained its competitiveness by shifting to new industries. Although the networks of computer firms were weak in comparison to those in Silicon Valley, 'the Boston area also has networks and institutions of a "generic" nature, allowing the region to diversify and develop new industries . . . and high-tech firms in general' (Tödtling 1994: 337). The formation of the biotech industry in the 1980s, for example, drew on the support of the 'inherited' network, namely the same institutions (universities and venture capitalists) that had helped to start and develop the computer-related firms of Route 128 (Tödtling 1994). In addition to biotechnology, the software, telecommunications, and health care sectors are well-established, rendering the decline of the computer sector and defence spending merely 'bumps' (albeit large bumps) for the local economy to travel over (Kanter 1995).

The educational strengths – the agglomeration of 'brains' – in the Boston area is unmatched anywhere else. In addition to Harvard and MIT (Massachusetts Institute of Technology), there are ten world-class graduate schools, and 65 colleges and universities in the greater Boston area (Rosegrant and Lampe 1992). There is no area of the country where university-based expertise, investors, and entrepreneurs have more successfully interacted to form new companies. Because Boston's long-standing history of high technology, its educational and transportation infrastructure, and the availability of highly educated (and relatively low-cost) technical labour continue to attract high-technology businesses, the city is still a magnet (Bolland and Hofer 1998: 276–7; Castells and Hall 1994).

The significance of education in the local culture has helped institutions in Boston to craft educational and training systems outside the university context as well. Educational and training institutions in the area have adapted to changing technologies and changing needs of local employers (Flynn 1993). Far more than merely a spin-off model from elite research universities, the Boston area has evolved diffusion-oriented models of academic outreach at all levels of local education (Moussouris 1998). Consequently, in Boston high technology is found in service and supplier firms, software shops, consultants, and others that subcontract with the largest companies (Bolland and Hofer 1998: 276–7).

The area's defenders point to Route 128's current success in software and networking. The region has developed a 'porous structure, unthinkable at the old minicomputer and defence electronics firms' that 'fosters more sharing of information . . . the key to Silicon Valley's success' (Campbell 1995: 9–10).

The resilience of the region also is seen in its various cultures – regional culture, regional industrial (or technical) culture, and organizational culture – which together comprise a key aspect of regional economic development (Oinas 1998). A change took place 'in the region's collective culture or a parallel culture evolved as new high-tech industries grew'. In addition, a generational change occurred as young software and networking firms 'cherry-picked the most compatible people' for the new culture (Campbell 1995: 10). Despite this change in culture, Campbell believes that 'Boston's cluster of talent and its modes of interaction, while stronger than in most regions, remain thinner than Silicon Valley's' (Campbell 1995: 11). In other respects, the two regions have marked similarities, such as a declining quality of life, which prompted Herbig (1994) to look to the 'inevitable decline' of Silicon Valley.

The culture of Boston, however, is rooted in the excellence of its educational institutions. As the home of a number of world-class institutions, Boston is 'intimately tied to the international network of scholars' (Rosegrant and Lampe 1992: 20). A substantial part of this network derives from the elite schools, whose graduates are found world-wide. In the case of MIT, graduates are likely to have founded a company; 4,000 MIT-related companies were identified in a recent study (BankBoston 1997). This number includes firms whose founders include an MIT graduate or a former member of faculty or staff, or companies spun off from a major MIT lab or which were founded based on licensed MIT technologies. MIT connections are both local within Massachusetts and within Silicon Valley and the rest of California: these two states account for 70 per cent of all MIT-related electronics firms, 68 per cent of software firms, and 63 per cent of drug and medical firms (BankBoston 1997).

Boston has changed since the 'Massachusetts miracle' days of the 1980s, and shows an ability on the part of firms, institutions and other local actors to adapt to new conditions. The outstanding universities, and a history and culture of entrepreneurship that continue to nurture new start-up firms in emerging technologies, combine to suggest that the Boston region has grown beyond its decades-long dependence on military spending. Has Boston become a 'learning region', if there can be such a concept? The region's 'productive culture' is centred around firms and other institutions, and has never been government-led, suggesting that the cultural basis has the systemic qualities of a *regional innovation system* (Cooke et al. 1997).

Conclusions

Knowledge has become a central organizing concept for those concerned with regional economic development. This chapter has traced the role of knowledge, from high-tech and high-value-added industries, to capabilities and competence of firms, to technology transfer, and locally 'sticky' knowledge. The case of the Boston area provided a test for ideas concerning knowledge as a basis for regional economic development, on which the region fares rather favourably.

Going well beyond the concern for high technology of the 1970s and for flexibility of the 1980s, *learning* has become the best way to understand regional economic change. The importance of tacit knowledge, rooted in regional and local cultures and its intrinsically human or soft characteristics that resist economic modelling, is fundamental to understanding both agglomeration or clustering of economic activity and the ability of places to develop 'sticky' competitiveness. Policy efforts have turned to the nurturing of 'learning regions' and the complex set of public and private institutions and actions needed to sustain competitiveness as the twenty-first century begins.

Can all places become 'learning regions' and develop regional innovation systems? Not likely. It is essential to form networks of interaction, both within a region and to other regions, and this is less likely to occur where 'brain drains' are the norm. Outmigration saps a region of its young people – potential entrepreneurs and in most cases among the most productive workers. Outmigration occurs for at least two reasons: first, the quality of life is not competitive with that in other places, making it a less frequent choice of mobile professional and technical people and, second, the economy and productive culture is weak, failing to provide opportunities that must be sought elsewhere. Learning regions must be 'magnets' as Boston is; great universities help to attract young minds from far outside the local region. In addition, Cooke (1996b) suggests that four I's are needed: identification, intelligence, institutions, and integration – all linked to regional stakeholders, strategy, standards, sectors, and skills. Business service firms play a key role here, serving as knowledge brokers for manufacturers and other service firms (Bryson 1997). It is evident that these conditions for learning do not exist – and probably cannot be created – in all places. Where it does happen is where knowledge takes centre-stage in the culture of the region. This is the challenge for other places in our competitive economy.

8 (The) industrial agglomeration (of Motor Sport Valley)

A knowledge, space, economy approach

Nick Henry and Steven Pinch

Introduction[1]

With the advent of economic globalisation has come a renewed recognition of the salience of agglomerations or clusters of production in the contemporary economy. Interestingly, this recognition and the explanation of geographical specialisation has become central to two competing strands of what has been termed the 'new economic geography'. On the one hand there is the new 'geographical' economics associated with leading names such as Paul Krugman, Michael Porter and W. Brian Arthur. Focused on the spatial agglomeration of industry and long-run convergence of regional incomes, this work has attracted sympathetic but ultimately critical reviews from economic geographers (Dymski 1996; Martin and Sunley 1996) with, for example, Martin (1999a) concluding most recently that '"the new geographical economics" represents a case of mistaken identity: it is not that new, and it most certainly is not geography', (p. 3).

On the other hand, there is the ever-growing corpus of work of economic geographers 'refiguring the economic in economic geography' (Thift and Olds 1996; Lee and Wills 1997) in the face of the postmodern challenge (Martin 1994b; Bryson *et al.* 1999), and coalescing around a 'new heterodoxy' (Storper 1995) of industrial and economic geography in which geographical specialisation is viewed as the construction of a socially embedded and 'relational' economic system.

Using a case study of the agglomeration of the British motor sport industry in 'Motor Sport Valley', this chapter will contrast these opposing explanations of geographical specialisation put forward by the two schools of thought of 'new economic geography'. Expanding on the latter school of thought, the chapter will demonstrate a 'knowledge, space, economy' approach to understanding (the) agglomeration (of Motor Sport Valley). It will be argued that Motor Sport Valley is best conceptualised as a 'knowledge community': a socially *and spatially* embedded economic system which has developed the capacity to generate and rapidly disseminate knowledge about superior ways of designing and manufacturing racing cars. The chapter concludes with the implications of the 'knowledge, space, economy approach' for understanding contemporary agglomerations of economic activity.

Motor Sport Valley: what is it?

The British motor sport industry is a classic example of a world-leading regional agglomeration of small-and medium-sized firms (see Figures 8.1 and 8.2). Conservatively estimated to employ 30,000 in engineering, with a collective annual turnover of

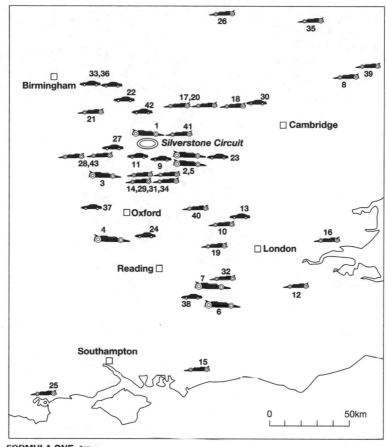

FORMULA ONE

1. Benson and Hedges Total Jordan Peugeot
2. Danka Arrows Yamaha
3. Mild Seven Benetton Renault
4. Rothmans Williams Renault
5. Stewart Ford
6. Tyrrell Racing Organization
7. West McLaren Mercedes

OTHER FORMULAE and TOURING/RALLY CARS

8. Argo Cars
9. Audi Sport
10. Bowman Cars
11. BMW Team Schnitzer
12. Elden Racing Cars
13. Ford Motorsport
14. Galmer Engineering
15. G Force Precision Engineering
16. Hawke Racing Cars
17. Jedi
18. Lola Cars Ltd.
19. Lyncar
20. Magnum
21. Marrow-Jon Morris Designs
22. Mitsubishi Ralliart

23. Motor Sport Developments
24. Nissan Motorsport Europe
25. Penske Cars Ltd.
26. Pilbeam Racing Design Ltd.
27. Prodrive
28. Pro Sport Engineering Ltd.
29. Ralt Engineering
30. Ray Mallock
31. Reynard Racing Cars Ltd.
32. Ronta
33. Rouse Sport
34. Spice Racing Cars
35. Spider
36. Total Team Peugeot
37. TWR Racing

38. Valvoline Team Mondeo
39. Van Diemen
International
40. Vector Racing Car
Constructors
41. Vision
42. Volkswagen-SBG sport
43. Zeus Motorsport
Engineering

Figure 8.1 Motor Sport Valley, circa 1998
Source: Pinch and Henry (1999b).

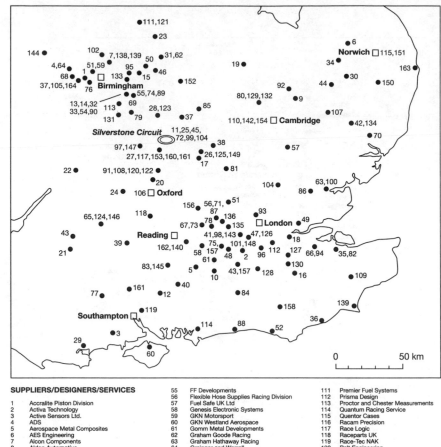

Norwich □ 115,151

Birmingham

Silverstone Circuit

Cambridge

Oxford

Reading □

London

Southampton □

0   50 km

SUPPLIERS/DESIGNERS/SERVICES

1	Accralite Piston Division	55	FF Developments	111	Premier Fuel Systems
2	Activa Technology	56	Flexible Hose Supplies Racing Division	112	Prisma Design
3	Active Sensors Ltd.	57	Fuel Safe UK Ltd	113	Proctor and Chester Measurements
4	ADS	58	Genesis Electronic Systems	114	Quantum Racing Service
5	Aerospace Metal Composites	59	GKN Motorsport	115	Quentor Cases
6	AES Engineering	60	GKN Westland Aerospace	116	Racam Precision
7	Alcon Components	61	Gomm Metal Developments	117	Race Logic
8	Aldon Automotive	62	Graham Goode Racing	118	Raceparts UK
9	Aleybars	63	Graham Hathaway Racing	119	Race-Tec NAK
10	Alfred Bull	64	Grainger and Worrall	120	Ralt Engineering
11	Aluminium Radiators and Cooling Systems	65	Griffin Motorsport	121	Ramair Filters UK Ltd
12	Alresford Tectonics	66	Guy Croft Tuning	122	Rapid International
13	Andy Rouse Engineering Ltd.	67	Hewland Engineering	123	Ravic Engineering
14	AP Racing	68	Hi-Tech Motorsport	124	Rayfast
15	Arrow Race Engine Components	69	HJS Motorsport Catalysts	125	Ray Mallock
16	Astratech	70	Holbay Racing Engines Ltd.	126	RESB International Ltd
17	ATL Competition Equipment	71	Icore International	127	Ricardo Consulting Engineers
18	Auriga	72	Ilmor Engineering Ltd.	128	Risbridger
19	Aurora	73	Impact Finishers	129	Rollcentre
20	Automotive Developments	74	Induction Technology Group	130	RT Quaife
21	Avon Racing Division	75	International Spares	131	SAS Motorsport
22	AWF	76	James Lister and Sons Ltd	132	Safety Devices Ltd
23	Aztek	77	Janspeed	133	Samco Sport
24	Berkeley Prowse	78	Janus Technology	134	Scholar Engines
25	BM Motorsport Ltd.	79	John Morris Designs	135	Serck Marston
26	Bob Sparshott Engineering	80	Kartronix	136	Serdi (UK)
27	Brembo	81	KDM Motorsport	137	Servo and Electrical Sales
28	BTB Exhausts	82	Kent Aerospace Castings	138	Spa Aerofoils Ltd
29	C & B Consultants	83	Kulite Sensors	139	Spider Engineering
30	Champion Manufacturing	84	Lane Electronics	140	Spot On Control Cables
31	Clarendon Eng. & Motorsport Supplies	85	Langford and Peck Ltd	141	Stack Ltd.
32	Colinton Engineering	86	Leda Suspensions	142	Steve Bunkhall
33	Colledge and Morley	87	Lee Products	143	Stone Foundries
34	Composite Wings	88	Lemo UK	144	Staubli Unimation
35	Connaught Competition Engines	89	Lifeline	145	Strain Gauging Co
36	Corbeau Seats	90	Lifeline Fire and Safety Systems	146	Swindon Racing Engines Ltd.
37	Cosworth Engineering	91	MagThink AutomotivenecoTWlr UK	147	The Monogram Co
38	Cranfield Impact Centre	92	Mark Dunham Racing	148	Think Automotive
39	DCL Components	93	McDonald Race Engineering	149	Tickford
40	Dove Composite	94	Minister Motorsport	150	Tony James Component Wiring
41	DPS Composites	95	Mira	151	Trick Machining Ltd.
42	Dunnell Engines	96	Monk Aeroelectronic Design Ltd.	152	Trident Racing Supplies
43	Dymag Racing UK	97	Mo Tec (Europe)	153	Turbo Technics
44	Dynamic Suspensions	98	MM Competition Systems Ltd	154	TWI
45	Earl's UK	99	Motorsport Distribution Ltd	155	Uniclip Automotive
46	Eibach Suspension Technology Ltd	100	Mountline Racing Ltd	156	Universal Air Tool
47	Elaborazione Colasuna	101	MSAS Motorsport Division	157	Universal Grinding Services
48	Electron Beam Processes	102	MTEC	158	Warrior Automotive Research Ltd.
49	Engine and Dynomometer	103	Mugen (UK Base)	159	Willans Racing Harness
50	Engineering and Motorsport Supplies	104	Nimbus Motorsport	160	WP Competition Systems
51	En Tran	105	Omega Pistons	161	WP Suspension (UK) Ltd
52	ERL	106	Oselli Engineering	162	X Trac
53	Eurotech Motorsport	107	Pace Products	163	Zephyr Carns Ltd.
54	Farndon Engineering	108	Pallas Connections	164	Zeus Ltd
		109	Piper Cams		
		110	PI Research		

Figure 8.2 Motor Sport Valley suppliers, circa 1998
Source: Pinch and Henry (1999b).

around £1.3 billion, the industry is clustered within a 50 mile radius around Oxford in southern England in what has been termed the 'Silicon Valley of Motor Sport' or 'Motor Sport Valley' (Henry *et al.* 1996; Pinch *et al.* 1997). Although its roots are that of a network of small enterprises, the industry includes investment from some of the largest car companies in the world including, for example, Ford, Nissan, Subaru, Mercedes-Benz, Volvo and General Motors as well as a series of multi-national sponsors from other sectors (principally, but not solely, tobacco). Thus, approximately three-quarters of the world's single-seater racing cars are designed and assembled in this region including the vast majority of the most competitive Formula One, Championship Auto Racing Teams and Indy Racing League cars. Combined with the region being the base for a large number of rallying teams also, Motor Sport Valley dominates the world 's racing car industry (Henry 1999).

Explaining Motor Sport Valley: the new geographical economics

It might initially seem surprising that geographers have not been impressed by the resurgence of interest by economists in issues of location, regionalisation and industrial agglomeration. Indeed, geographers' reticence about such work might seem especially strange given that it is concerned with issues such as 'path dependency', 'lock-in' and 'knowledge spillovers' that find echoes in much recent geographical work. However, a closer inspection of the new geographical economics shows that there is no great mystery here. As Martin (1999a) argues, the work of writers such as Krugman and Porter is essentially a reworking of regional science. As such, it is based on a model-building approach which attempts to distil reality into a few basic principles (albeit expressed in a complex mathematical form) which can be applied universally. Differences between the model and reality are then explained by missing variables. The problems of this methodology are well known by now. Not only does this approach tend to produce a highly simplified view of the world, but it is often difficult – if not impossible – to test such models adequately. The inevitable gap between the predictions and reality can always be explained away by numerous tacit assumptions or missing variables which means that the tenets of the model are rarely questioned. Furthermore, these models tend to be based upon a simple ahistorical calculus which ignores the complex space and time dependent contingencies that are crucial in economic systems. It is precisely these contingencies and their socially embedded character that geographers have been attempting to grapple with in recent years through their engagements with fields including regulation theory, the sociology of scientific knowledge and cultural studies. Consequently, the concept of space in the new work by economists is an archaic absolute space on to which processes are 'mapped', rather than a constituent part of those social processes. It is therefore hardly surprising that Martin (1999a) comes to the conclusion quoted above in the introduction.

Nevertheless, what is also clear is that the work of economists on location is highly influential both in academia and in policy circles. It therefore needs greater attention than it has hitherto received from geographers. The crucial issue is how do such rival concepts used by economists and geographers work in explaining a particular empirical example?

Drawing on the new 'geographical' economics, Aston and Williams (1996) have provided an explanation of the dominance and continued supremacy of Motor Sport Valley. They start from Krugman's (1991) dictum that nations don't specialise in what they are good at so much as they are good at what they specialise in. Thereafter, they apply his two basic explanatory concepts of 'accidents of history' and 'external scale economies' to explain the remarkable rise of the British motor sport industry (see Pinch and Henry 1999b).

A *chapter of accidents?*

Investigating the history of the motor sport industry, Aston and Williams produce a list of accidental factors that helped the British-based industry to emerge. Amongst others these include: literally an accident in 1953 which killed or injured 183 spectators and led the powerful Mercedes team to withdraw from motor sport; the switch from racing on public roads to designated circuits (so providing a new use for Britain's surplus postwar airfields); a set of vertically-integrated continental European manufacturers such as Ferrari and Porsche who built cars and components only for themselves, as against small British-based constructors willing to sell products to anybody who wished to race them; a 1965 ban on cigarette advertising on British television triggering circumvention of the rules through sponsorship of British-based racing teams (beginning with Lotus in 1968); and the availability of a lightweight, high-powered aluminium firepump engine (the Coventry Climax) capable of being adapted to motor racing.

Illuminating though the game of 'spot the accident' might be, this approach is misleading as an explanation for the British motor sport industry for two reasons. First, the approach gives far too much attention to accidental factors. These accidents might have played some small part but, by themselves, they are not sufficient to account for the growth of the British constructors. For example, the withdrawal of Mercedes might have provided a 'window of opportunity' for other manufacturers to enter the field of motor sport, but this raises the question of why the challenge was taken up in Britain rather than, say, in Italy, a country whose interest in motor sport is second to none. Or it should be noted also that the British-based constructors were quite well established *before* the cigarette manufacturers entered the fray in the late 1960s and early 1970s. Their entry into British motor sport was therefore as much a consequence as a cause of the strength of the British motor sport industry. A second, and even more important, flaw in the 'accidents of history' explanation is that it is questionable whether some of the developments listed above were truly accidental. For example, the fact that there 'happened' to be the Coventry Climax engine available, and later the most succesful racing engine ever (the DFV engine produced by Cosworth), seems to be far from accidental. Instead these developments may be interpreted as a reflection of a British tradition of expertise in engines and lubricants (Hebb 1993). To paraphrase Martin's critique of this new economic geography:

> Path dependence [history] thus has a place-dependent character. Path dependence [history] does not just 'produce' geography as in the 'new economic geography' models; places produce path dependence [history].
>
> (Martin 1999a: 16)

External scale economies?

Figures 8.1 and 8.2 suggest that external scale economies have played a part in the development of Motor Sport Valley; there are numerous component suppliers, such as AP Racing of Coventry and Xtrac of Wokingham who supply key components (clutches and gearboxes respectively) to virtually all the racing car constructors of Motor Sport Valley. In such a context, local external economies of scale can be derived from market size effects reducing the average cost of components and transport costs (Pinch and Henry 1999b). Yet this explanation is not a comprehensive one for the success of the British motor sport industry. During interviews with over 50 senior individuals within the industry, our research indicated that for many of the leading specialist suppliers *cost reduction* is not the main factor affecting their competitive position. Winning races is the main objective in motor sport and the 'best' component (fastest, most reliable, strongest, etc.) is the crucial requirement. Furthermore, given their generous sponsorship deals (Grant-Braham 1997), many of the constructors are able to pay handsomely for the privilege of acquiring such components. External economies of scale may enable some of the specialist suppliers to reduce the average cost of their components but this is not the main reason for their survival. Similarly, constructors are not clustering together primarily to minimise transport costs to component suppliers (although most argue that *proximity* does help).

Thus, the new 'geographical' economics version of the new economic geography is by no means convincing as an explanation for the origins and growth of Motor Sport Valley. Indeed, the degree of clustering is expecially remarkable given that firms in the region, as in similar complexes throughout the world, run a great risk of having their key personnel poached by rival companies. This raises the issue of a further element of external scale economies – what Krugman (1986) would term *technological knowledge spillovers*. Krugman dismisses these spillovers as difficult to measure, yet our analysis indicates that these are crucial in understanding the evolution and dynamics of the British motor sport industry. However, in order to appreciate the true importance of this knowledge-based explanation, it is to the second version of the 'new economic geography' that one needs to move.

Explaining Motor Sport Valley: social embeddedness and the new economic geography

> it is precisely the social, institutional, cultural and political *embeddedness* of local and regional economies that can play a key role in determining the possibilities for or constraints on development; and thus why spatial agglomeration of economic activity occurs in particular places and not others.
>
> (Martin 1999a: 11, emphasis in original)

Numerous influences have coalesced to form the 'new economic geography' including heterodox variants of economic theory (neo-Schumpeterian, institutional, the sociology of economics), 'technology and innovation studies' such as the sociology of scientific knowledge and actor network theory, a focus on regulation and 'regulationist' theory, and a recognition of 'discursivity' in the construction of theoretical/real worlds. Storper (1995) argues that these developments have led to a profound shift in the conceptual approach used in economic geography. The 'old orthodoxy' based around the metaphor of economic systems as machines with quantifiable inputs and outputs – as embodied in classic Marshallian notions of external economies – has been replaced by a 'new heterodoxy' based around the metaphor of regional economies as sets of relations. Economic systems are now seen as comprising sets of conventions used by highly reflexive human agents. Hence, regions are not merely bound together by input–ouput interactions but are also integrated by a much wider set of less tangible but crucially important linkages which Storper terms 'untraded interdependencies'. These comprise coventions, rules, practices and institutions which combine to produce possible and real 'worlds of production' (Storper and Salais 1997).

This notion of untraded interdependencies provides a way of unravelling Marshall's tantalising and much quoted speculations about the presence of 'industry in the air' (Marshall 1890: 261). For example, it is frequently observed that industries develop technological trajectories that are difficult, if not impossible, to reverse. It may be argued that these evolutionary paths arise because of uncertainty over the actions of other companies together with a common dependence upon pools of knowledge, suppliers of components and the tastes of consumers of products. Researchers have therefore begun linking the notion of technological trajectories with work on the social construction of knowledge and the social embeddedness of local economies. Attention is therefore now placed upon institutional and cultural factors that help to generate a strong sense of trust, common purpose, skill development and technological innovation (Cooke and Morgan 1998; Grabher 1993a). This has led to the concepts of the 'learning economy' (Lundvall and Johnson 1994) and the 'learning region' (Morgan 1997), in both of which knowledge is the most important resource and learning the most important process.

Technological trajectories and shifts in 'Knowledge Systems'

If we envisage knowledge as a crucial 'glue' which holds regional economies together in the contemporary economy, then a radical change in the technological regime holds out the prospect that economies can become 'unstuck' and reconfigured. The reason for this is that radically new technologies require different input chains in terms of knowledge and material inputs (Storper 1995). The crucial impetus to the dynamic growth of Motor Sport Valley – the element that glued the contingent factors of club racing, airfields, reduced market competition, tobacco sponsorship, etc. together – was a radical shift in technological regime (knowledge system). Just as the shift from valves to transistors and semi-conductors was reflected in a spatial shift in the electronics industry from East Coast USA to Southern California (eventually to be labelled Silicon Valley), so the shift to an aerospace-inspired technological trajectory (see Figure 8.3) ultimately saw the locus of the world's motor sport industry shift from northern Italy to southern England (eventually labelled Motor Sport Valley) (Pinch and Henry 1999a). As Hebb (1993) has noted, Britain had a relatively large aerospace industry after the Second World War. The industry can be traced back to governmental support for aircraft manufacture in the 1920s and 1930s in the face of military rivalry with continental powers. Hence, the British Air Ministry created research facilities at Farnborough and also supported aeronautical departments in universities such as Imperial College, London. Thus, by the 1950s, over 16 per cent of Britain's qualified scientists and engineers were engaged in research and devel-

Innovation	Aviation use	Formula One use
Aviation inspired		
Active suspension	Computer-based control	Improved handling (banned)
Aerodynamics	Efficiency in the air fuel economy, speed, etc.	Speed but also downforce
Aeroengines	Combination of light weight with high power output	Combination of light weight with high power output
Aluminium	Aircraft structures	Widely used in early chassis/cockpit design
Carbon composites	Aircraft structures	Monocoque construction
Carbon disc brakes	High efficiency and lightweight braking	High efficiency and lightweight braking
Carbon fibre clutch	Not directly used but aviation inspired	Weight saving and strength
Computer-based telemetry	Rapid communication of technical data	Rapid communication of technical data
Fly-by-wire	Weight-saving electronic control	Weight-saving throttle-by-wire control
Turbo charging	Extra performance at high altitude	High power output (banned)
Wind tunnels	Testing aerodynamics	Testing aerodynamics
Wings/winglets	Support in the air, drag reduction	Downforce for grip
Automotive inspired (non-aviation)		
Four wheel drive		Improved handling (unsuccessful – wings did the same job)
Mid engine layout		Improved handling
Semiautomatic transmissions		Rapid gear changes
Six wheels		Improved handling (unsuccessful)

Figure 8.3 Key innovations in Formula One
Source: Pinch and Henry (1999b).

opment in the aviation sector (Hebb 1993). Britain's presence in aerospace was no accident of history and was critical in providing knowledge inputs to a motor sport industry reorientating around a new aerospace-inspired technological trajectory.

The Knowledge Community: knowledge generation and circulation in action

A knowledge-based approach is also important in accounting for the continuing growth of the British motor sport industry as well as its origins. Knowledge generation and dissemination have been crucial to maintaining the dominance of the region and have taken place, we argue, through the medium, of the 'knowledge community' (Henry and Pinch 1997). Here we define a knowledge community

> as a group of people (principally designers, managers, and engineers in this case) often in separate organisations but united by a common set of norms, values and understandings, who help to define the knowledge and production trajectories of the economic sector to which they belong.

In the case of Motor Sport Valley, the region has become synonymous with the knowledge community of motor sport; the core of the community's spatial organisation is that of the Valley. A number of key (regionalised) processes can be identified which construct and maintain this community (and thus the Valley) and occur on both a traded and untraded basis.

Staff turnover

One of the most important ways in which knowledge is spread within the motor sport industry is by the rapid and continual transfer of staff between the companies within the industry. Especially at the end of every racing season, there is an intense period of negotiation as designers, engineers, managers and, of course, drivers move between teams. As part of a research project on the industry, the career histories of 100 leading designer/engineers within the industry were mapped (see Figure 8.4 for examples). This mapping revealed a move, on average, once every 3.7 years and an average total of 8 moves in a career in the industry. Similarly, a study of advertisements for *technical* posts in Formula One between 1996 and 1997 revealed vacancies for 93 posts, more than 10 per cent of this marque's total employment at that time.

Career trajectories

Whilst the above evidence highlights the mobility encapsulated within the labour market, its significance also stems from the geographical expression of this labour market. As the sample career histories of Figure 8.4 show, Motor Sport Valley acts as the epicentre of a global, yet highly spatially defined, labour market. In particular, whilst moves occur on an international scale into, and out of, the Valley, overall, the largest number of moves occur between sites *within* the Valley. Moreover, international moves out are invariably followed in quick succession by a return to the Valley as can be seen in the Figure 8.5 career history. Indeed, these particular career histories introduce a constant and overriding theme of the world of production of Motor Sport Valley. As an individual in motor sport, it is rare that you do not, at some point in your career, spend some time within Motor Sport Valley (and most likely the majority of your career). Essentially, this is about joining *the* knowledge community of motor sport production. Yet opportunities exist in this highly mobile industry to sell your knowledge elsewhere and many do so. However, the knowledge of the Valley is continually being reconfigured and advanced (see below): it is in effect, a knowledge pool. A move out of the Valley is also to stretch your relationship within the community, to risk your position within the 'knowledge loop'. Thus, few stay away from the Valley for long, returning once more to refigure their position within the community of knowledge. An analogy for the labour market of Motor Sport Valley is that of (musical) 'industrial chairs' – the music starts the movement, the music stops and the rush for chairs leads to continual shifting bottoms on seats, with occupants often returning after one or two musical interludes to a seat previously occupied. Furthermore, such musical chairs do not take place only at the level of the individual: joint career paths may also be mapped with individuals often moving together, or continually recombining, throughout their careers (see Figures 8.6 and 8.7).

Rory Byrne – R & D and design – South African

Witwatersrand University 1960s
(Degree in Industrial Chemistry –
South Africa)

↓

Polymer manufacturing plant – late 1960s
(Chief Chemist – Johannesburg,
South Africa)

↓

Set up own business – 1967–73
(Importing and selling motor spares and then
engine development and build – South
Africa)

↓

Royale (Formula Ford) 1973–77
(' . . . I was so involved in racing that
there was only one place to go –
the UK.' – Designer – UK)

↓

Toleman/Benetton – 1977–90
(Co-designed with John Gentry first Toleman
car, an F2, in 1980 and then designed, or
co-designed with Gentry, F1 cars from 1981
to 1990 – Witney, Oxfordshire)

↓

Reynard – October 1990–92
(Involved in design of aborted X project –
Bicester, Oxfordshire)

↓

Benetton F1 – 1992–December 1996
(Responsible for R & D –
Enstone, Oxfordshire)

↓

Ferrari F1 – January 1997–present
(Chief designer – Maranello, Italy)

**Frank Dernie – Aerodynamicist/Designer
(D.o.b. 3.4.50)**

Imperial College late 60s/early 70s
(Mechanical Engineering Degree – London)

↓

David Brown Industries 1970s
(Working with gears)

↓

Garrods 1970s
(Record players – Swindon, Wilts)

↓

Hesketh/Williams 1977–79
(Consultancy work – co-designed 1978
Hesketh with N. Stroud – Towcester,
Northants/Didcot, Oxfordshire)

↓

Williams 1979–88
(Aerodynamicist and R & D – co-designed
1986–88 cars with P. Head – Didcot,
Oxfordshire)

↓

Lotus 1988–90
(Chief Engineer – co-designed 1989 car
with M. Coughlan – Norfolk)

↓

Ligier 1990–92
(Technical Director – France)

↓

Benetton 1992–July 1994
(Race engineer and then chief engineer
in 1994 – Enstone, Oxfordshire)

↓

Ligier July 1994–April 96
(Went back to Ligier after team purchased
by Briatore – Technical Director –
France then Oxfordshire)

↓

Arrows April 1996–97
(Technical Director – moved with
T. Walkinshaw's take-over – Oxfordshire)

Figure 8.4 Career histories: Rory Byrne and Frank Dernie
Source: Henry and Pinch (2000).

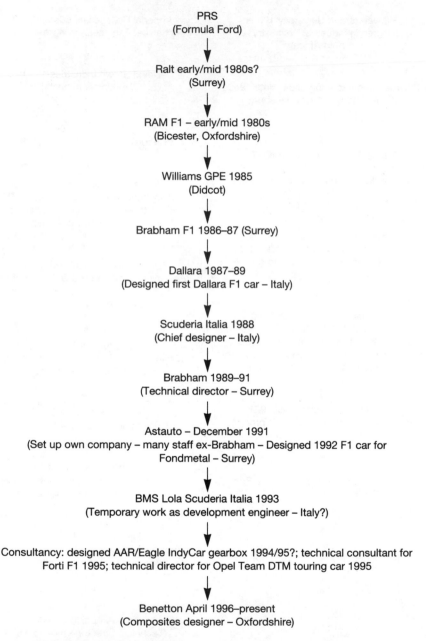

Sergio Rinland – design (from Argentina)

PRS
(Formula Ford)

↓

Ralt early/mid 1980s?
(Surrey)

↓

RAM F1 – early/mid 1980s
(Bicester, Oxfordshire)

↓

Williams GPE 1985
(Didcot)

↓

Brabham F1 1986–87 (Surrey)

↓

Dallara 1987–89
(Designed first Dallara F1 car – Italy)

↓

Scuderia Italia 1988
(Chief designer – Italy)

↓

Brabham 1989–91
(Technical director – Surrey)

↓

Astauto – December 1991
(Set up own company – many staff ex-Brabham – Designed 1992 F1 car for
Fondmetal – Surrey)

↓

BMS Lola Scuderia Italia 1993
(Temporary work as development engineer – Italy?)

↓

Consultancy: designed AAR/Eagle IndyCar gearbox 1994/95?; technical consultant for
Forti F1 1995; technical director for Opel Team DTM touring car 1995

↓

Benetton April 1996–present
(Composites designer – Oxfordshire)

Figure 8.5 Career history: Sergio Rinland
Source: Henry and Pinch (2000).

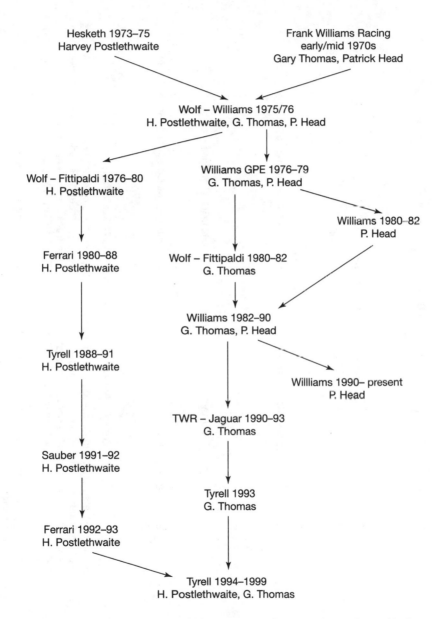

Harvey Postlethwaite, Gary Thomas and Patrick Head link

Hesketh 1973–75
Harvey Postlethwaite

Frank Williams Racing
early/mid 1970s
Gary Thomas, Patrick Head

Wolf – Williams 1975/76
H. Postlethwaite, G. Thomas, P. Head

Wolf – Fittipaldi 1976–80
H. Postlethwaite

Williams GPE 1976–79
G. Thomas, P. Head

Williams 1980–82
P. Head

Ferrari 1980–88
H. Postlethwaite

Wolf – Fittipaldi 1980–82
G. Thomas

Tyrell 1988–91
H. Postlethwaite

Williams 1982–90
G. Thomas, P. Head

Willliams 1990– present
P. Head

TWR – Jaguar 1990–93
G. Thomas

Sauber 1991–92
H. Postlethwaite

Tyrell 1993
G. Thomas

Ferrari 1992–93
H. Postlethwaite

Tyrell 1994–1999
H. Postlethwaite, G. Thomas

Figure 8.6 Joint careers: the Harvey Postlethwaite, Gary Thomas and Patrick Head link
Source: Henry and Pinch (2000).

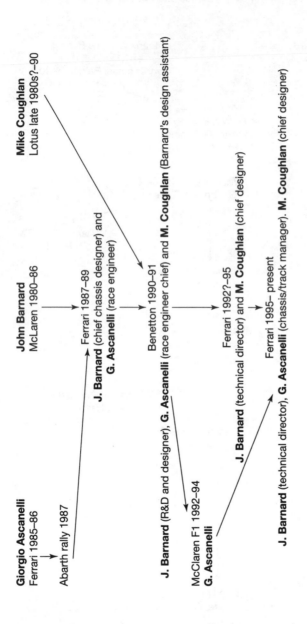

Giorgio Ascanelli
Ferrari 1985–86

Abarth rally 1987

John Barnard
McLaren 1980–86

Ferrari 1987–89
J. Barnard (chief chassis designer) and
G. Ascanelli (race engineer)

Mike Coughlan
Lotus late 1980s?–90

Benetton 1990–91
J. Barnard (R&D and designer), **G. Ascanelli** (race engineer chief and **M. Coughlan** (Barnard's design assistant)

McClaren F1 1992–94
G. Ascanelli

Ferrari 1992?–95
J. Barnard (technical director) and **M. Coughlan** (chief designer)

Ferrari 1995– present
J. Barnard (technical director), **G. Ascanelli** (chassis/track manager). **M. Coughlan** (chief designer)

Giorgio Ascanelli, John Barnard and Mike Coughlan

Barnard: 'I also try to instil my design philosophy in the key development engineers at Ferrari, like Giorgio Ascanelli, with whom I have worked in the past at Benetton and Ferrari. I get lots of feedback from him on the practical aspects of operating the cars at the track and on set-up issues.'
('The Designers – John Barnard', Peter Wright, *Racecar Engineering* 5(6): 41–2)

Figure 8.7 Joint careers: Giorgio Ascanelli, John Barnard and Mike Coughlan
Source: Henry and Pinch (2000).

Firm births and deaths

It is important to note, however, that this mobility of the labour market is both vol-untary and non-voluntary and it is driven not only by movements of individuals but also the continual 'movement' of firms. Another significant feature of Motor Sport Valley is the high firm turnover involving both high rates of new firm formation and high firm death rates. Within an industry of several hundred firms, only a handful employ more than 200 people, and there is a widespread feeling within the industry that even this size is dangerous to the competitive edge of the company. Indeed, in a process of growth reminiscent of Silicon Valley (Saxenian 1994) and the Cambridge Phenomenon (SQW 1985), the hundreds of firms in Motor Sport Valley today can be traced to a few pioneers. For example, Figure 8.8 provides an incomplete family tree for March, a spin-off from Cosworth (Northampton) in 1969. March located in nearby Bicester and began a process which has led this small market town to become the world centre for motor sport production.

Bringing the tree up-to-date, one of March's spin-offs, Reynard Racing Cars, has itself sponsored a number of internal spin-offs in the town and the company has recently opened another large site just outside Bicester. This site is the production centre for the latest entrant to Formula One, British American Racing, a combination of Reynard Racing Cars and the Tyrell Formula One team (also within the Valley) which ended the latter team's 20 year presence in Formula One racing. In fact, this is a classic example of the process of reconfiguration so characteristic of Motor Sport Valley. One of the longest running teams in Formula One, Tyrell will not so much die as be reborn in a modifed and reinvigorated version. Thus Tyrell is in a long line of Formula One constructors who have come and gone; in the last three years this includes Pacific, Forti, Simtek and Lola in a field which rarely exceeds a dozen con-structors a year. Lola, a major Indycar producer, did not survive a season in Formula One with many commentators suggesting that the company had failed to learn from its previous foray into Formula One a decade ago. This time its entry bankrupted the company, including its Indycar production, only for it to be resurrected once again by a new Irish buyer. Its production of IndyCars continues unabated and it still holds ambitions to return to Formula One once again. And it does so from its continued location within Motor Sport Valley, a location held throughout its chequered history.

In similar vein, the recent Formula One bankrupt, Simtek Grand Prix, provides clear evidence of the entwining of labour movement with firm 'churn' (high rates of firm birth and death). On its demise, of the top eight individuals employed by Simtek, six were traced to other motor sport jobs within the Valley within six months.

Supplier linkages

A further important process of knowledge transfer, not involving the movement of bodies *per se*, is through the links constructors and manufacturers have with the numerous component suppliers. Many component suppliers work for more than one constructor making bespoke items of the clients' design (but usually with a substantial contribution of their own expertise). Clearly, it is in the interest of these component suppliers not to disclose secrets to other constructors for whom they also undertake work or they would soon be out of business. Nevertheless, as many of our respondents acknowledged, over time, there is a gradual assimilation throughout the industry of

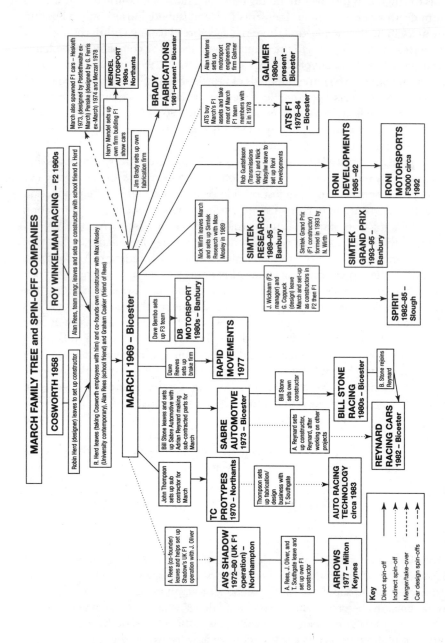

Figure 8.8 March family tree and spin-off companies
Source: Henry and Pinch (2000).

the advantages of particular approaches. This knowledge exchange might occur, for example, because a component supplier subtly steers a constructor away from a sub-optimal way of doing something; other examples of the fine lines of this process of knowledge transfer are evident in the quotations below:

'There is a difference between confidentiality and the knowledge that one builds up within an environment.'

(16, Supplier)[2]

'We did have an innovative idea which would solve a problem or improve performance of a particular component, and we did tell the supplier that, and it ended up being on their standard products and sold to everyone else.'

(14, Constructor)[3]

'It is much easier if we have problems to phone somebody and they say, "yes, we do that for somebody else".'

(11, Constructor)

This means that there will inevitably be some 'leakage' of knowledge throughout the industry as a whole. But perhaps more importantly, any individual team probably has more to gain than lose through the skills which components suppliers learn by servicing a number of different racing car companies. The majority of the leading suppliers are based within Motor Sport Valley and these firms and personnel are part of the knowledge community.

Gossip, rumour and observation

As economic geographers have come to recognise the importance of 'discursivity' in both the philosophical underpinnings of their work (Martin 1994; Barnes 1996; Gibson-Graham 1996; Thrift 1996a; Bryson, Henry et al. 1999) and its empirical application, the importance of gossip, rumour and observation in defining productive knowledge has been highlighted. In the case of the motor sport industry, particular sites and mediums of this process include the pit lane, test track and race meetings and the frenzy of speculation encouraged by the large number of specialist magazines and extensive press and television coverage. The role of 'gossip and rumour' in shaping individual careers and, indeed, the uptake of innovations should not be under-estimated (see Thrift 1994 for similar arguments on the City of London). As Hamilton says

'F1 mechanics are generally chosen by reputation. Few are picked cold from a business outside racing. They need to know what F1 is about, not simply because of the proficiency that will bring but because it means the mechanic will be aware of precisely what he [sic] is letting himself in for. This is not a nine to five job [sic] and the motor racing grapevine will let a team owner know whether or not his [sic] prospective employer can cope.'

(Hamilton 1994: 162)

'I get several calls a week from companies in the same industry asking us if somebody is good for the money or what's happening here or what's happening there.'

(ibid.: 41)

In terms of observation, teams keep a close eye on their rivals' cars and monitor their performance in test periods. Moreover, all the senses are involved in such monitoring with, for example, engine noises, the sounds of gear changes, and the cornering behaviour of cars all providing important clues (for some) regarding the design and construction of a car and its components.

'Spying I do a lot of, I think that's all fair in love and war. So if I get a chance to photograph someone's wing profile or something like that I'm there, and I walk with a camera permanently attached to me. That's so important.'

(7)

'If you call spying trying to get a look under the bonnet of a car or when the wheels are off quickly ducking your head down, if you call that spying, yes there is, but we all do that.'

(56)

The discursive aspect of technological innovation

Once analysis moves into the realms of gossip, rumour and observation – understandings between groups of individuals of the meaning of certain things – the understanding of discourse becomes significant. Discourse highlights shared beliefs and understandings about what is, or should happen and can be related to Storper and Salais' (1997) arguments on the discursive conventions which comprise worlds of production. In an industry dominated by technical innovation, the discursive aspect of technological innovation is prominent.

Elsewhere, we have discussed the insights the sociology of scientific knowledge (SSK) literature provides in understanding the social construction of technological innovation (Pinch and Henry 1999a). Thus, for example, concerning the role of discourse, SSK argues that crucial to the generation and dissemination of scientific and technological advances are shared sets of understandings – discourses that help to interpret the world and whose uncertain construction are encapsulated within SSK through the concepts of *interpretative flexibility* and *social closure*. In effect, a version of truth is 'winnowed' from the various interpretations that are promulgated (Bijker *et al.* 1990). Ambiguities are resolved, usually only partially for a limited period, and certain development trajectories are put in place.

Typically in motor sport, considerable uncertainty abounds about the value of any new approach. Such uncertainty leads to a great deal of blind copying. As one senior engineer commented:

'a car might have a particular style or component which sets it out from the rest, but that may not be the reason why it is doing well, but people just blindly copy it because they think . . . they must have found something, some way of measuring a gain in performance that we can't find, so we'll do it anyway.'

(33)

Another designer graphically explained:

'if painting a race car purple gives it a second a lap, I mean next week they will all be purple . . .'

(23)

The outcome of the above is designs led by strong beliefs and discourses about how cars should look and behave as well as the results of testing. For example, many cars use an aerodynamic device called the Gurney Flap but it is widely accepted that exactly why it works is still unknown.

Furthermore, SSK highlights the spatiality of expert knowledges; how these knowledges reside in knowledge production centres such as scientific laboratories, research centres and factories and how their transportation is a fraught process. Scientific experiments and technical innovations are often dependent upon knowledges of accepted ways of doing things that are seldom written down (Polanyi 1967; Gertler 1995). These forms of knowledge often emerge from sets of people who have undertaken long periods of apprenticeship and are integrated into networks of contacts. For this reason, early scientific experiments are often difficult to replicate in other laboratories. There are so many detailed sets of understandings involved in the formulation of the experiment or in the construction of technological artefacts that not all of these can be incorporated by people attempting replication. For our argument, the point is, of course, that both beliefs and discourses about knowledges are debated and determined within the knowledge community, just as they are learnt, and the knowledge community is based in Motor Sport Valley. One example was a claim by a designer that although some of the important machine tools of the industry were produced in another country:

'They manufacture the tool but they don't know how to use it for that particular use.'

(35)

Or as a foreign national working in the Valley put it:

'To make racing cars in southern California was so difficult because the industry is not prepared. So you may find engineers but their mentality is different. They are aerospace so they think different, they organise different, they are not used to our [sic] racing mentality . . .'

(43)

These examples highlight knowledge as a (community) process which was summed up by this statement by an English engineer who, whilst based in Motor Sport Valley, was working for a foreign-based constructor whose site overseas he visited regularly:

'We are trying to educate them about the English [sic] way of doing things.'

(11)

Indeed, Ferrari have run courses for their engineers in Maranello, Italy, on the English way of doing things.

Thus, SSK helps us to understand how agreements are forged upon understanding the signs, symbols and conventions of racing car production; the theory underlines how technological trajectories may be (discursively) put in place as part of worlds of production. Yet a striking element of racing car production is how some of these trajectories end abruptly, or more precisely, are banned through regulation. Once again, drawing on SSK, one of its fundamentals is its argument concerning the seamless web of society and technology and whilst the above has highlighted the processes of technological discursivity of the designers and engineers of Motor Sport Valley, they are only one of the set of actors in the process of creating a 'fast' car. They are not the only members of the knowledge community.

The message of the seamless web is that technologies are political in that they can be designed, either explicitly or implicitly, to open some options and close others. Motor sport illustrates these points very well, for the design of racing cars has been shaped over the years by many competing interest groups: not least the teams, their sponsors, the television companies and the regulators of motor sport. The major sponsors are keen to ensure that racing is exciting but it must also be relatively safe or the publicity resulting from death or injury to drivers and/or spectators will generally have a negative impact upon the sponsor's image and commercial prospects. Sponsors are all too aware of the historic withdrawal of Mercedes Benz in 1955 and there has been an increased emphasis upon enhanced safety features in motor racing in recent years due to untimely (and unexplained) deaths. For example, following the death of Formula One driver Ayrton Senna in 1994, it was suggested that:

'We are much more bothered about all this than we used to be, even as little as quarter of a century ago. When Jim Clark, the British driving hero and two times World Champion was killed in a Formula Two race at Hockenheim in 1968, the event was front page news, to be sure. But if you look at the *Daily Telegraph* of those times, you do not find the prolonged analysis of the crises in motor-racing or leader page articles agonising over the future of the sport, or enthusiasts pleading for its continued autonomy.'

(*Sunday Telegraph*, 8 May 1994, quoted in Twitchen 2000)

'Clark's death was largely regarded as simply a legitimate consequence of competing in a hazardous sport. The post-mortems were limited and, to a large extent, shrugged aside. A generation later, we live in a very different environment where the freedom of the individual to make a personal choice as to how, or indeed whether he risks his life, is very much hemmed in by wider social constraints.'

(Henry 1994: 6, quoted in Twitchen 2000)

These pressures get worked out also through the technical panels which regulate the various racing formulae. Within motor sport, the technical panels achieve a consensus about the types of cars the various teams can produce. Yet, importantly, the technical panels include the key designers in the industry, so reinforcing their presence within the knowledge community (and the presence of Motor Sport Valley within the regulatory framework of racing car production). This means that there is a very strong cooperative, as well as fiercely competitive, element in racing car design and within the technical panels.

> 'I know some of the other designers and, you know, obviously we are all in competition with each other, but you still might well do that [ring them up] and say, "do you know a good way to . . .?" or "do you know a source of this or that information?"'

(27)

Taking a knowledge-based approach to explanation which draws on the social embeddedness approach of 'new economic geography', provides a highly fruitful route to understanding contemporary processes of agglomeration. As you begin to track shared understanding and meaning – the basis of productive knowledge – so the processes of knowledge production and transfer start to become apparent. In tracking and representing such processes, so do their geographies become apparent. In the world of motor sport production the sites, the people, and the knowledge are first, and foremost, the knowledge community which is Motor Sport Valley.

Conclusions

Conclusions can be drawn from the above in at least three different, but interconnected, realms: thoughts on Motor Sport Valley, thoughts on the competing versions of 'new economic geography', and thoughts on 'knowledge, space, economy'.

Motor Sport Valley: learning region?

In summary, the evidence above highlights that a key characteristic of the industrial organisation and labour market of motor sport are a set of processes leading to the continual 'churning' of people and ideas. Fundamental to the argument of this chapter is the fact that this process of 'churning' is centred on, and within, Motor Sport Valley. In addition, this churning is a process of circulating and producing knowledge within the knowledge community and regional production centre of Motor Sport Valley (see also Henry and Pinch 2000). For example, as personnel move they bring with them knowledge and ideas about how things are done in other firms helping to raise knowledge throughout the industry:

> 'Whoever comes, it doesn't matter from where, he has to change his ways and adapt, and then once he changes his ways and adapts and he gets part of the system, then someone else comes and learns from him. Then he moves to another team and he takes ways of one team into another, you see. And you go from [leading F1 constructor] to McLaren and say "At [leading F1 constructor] we did it this way," and they say "Oh, really. Let's try it." And then vice versa, someone comes from McLaren to here and someone goes to Williams and you get all these exchanges of people and then we learn from each other all the time. Not trade secrets, obviously, but ways of doing things. If someone used to draw wings at McLaren or Williams or whatever and he comes to draw wings here, he is not going to bring the drawings of McLaren under his arm, but he is going to bring

the ways of doing it in here, so he is allowed to explain. He may improve because he is going to see what we have got here, he is going to see the way he did it before and he will say "Ah, I can mix these two and get something even better", and you improve that little bit, and it keeps happening' (43).

Whilst this process *may* not change the pecking order within the industry, this 'churning' of personnel raises the knowledge base of the industry as a whole *within the region*. The knowledge community is continually reinvigorated and, synonymous with this, so is production within Motor Sport Valley. Motor Sport Valley is a knowledge pool (a centre of world class excellence) which is on a constant learning trajectory. To be outside of the Valley is to risk your position within the knowledge community.

It is sometimes claimed that if the dominance of the region is based largely upon knowledge, then it will be easy to shift that expertise to some other area. For example, there has been much speculation about the possible transfer of the British racing car industry to California by the capital-rich US automanufacturers (Coleman 1995). Indeed, precisely just such a scenario of the geographical mobility of motor sport production exploded on to the political scene recently (*Financial Times* 1997; *Guardian* 1997). Yet this chapter suggests that this 'disembodied' and 'spatially ubiquitous' view (that knowledge consists of easily transferable non-material ideas) is highly simplistic. The knowledge of the British motor sport industry is encapsulated in particular people, objects and ways of doing things which are themselves constructed *in a particular place*. Similarly, there has been talk in some circles recently about transferring the production of racing cars to Malaysia, to take account of cheaper labour costs, '100 per cent exemption' on investments and the booming local market for racing cars. Significantly, such proposals envisage maintaining design capacity in the UK but, in addition, this may well be underestimating the (knowledge-laden) production process in this fast-moving industry.

The 'new economic geography'

In some senses, this discussion about the possible mobility of the motor sport industry can be seen to reflect the differing implications of the 'new economic geography' on offer. As Martin (1999) suggests, the new geographical economics could be applauded for wanting to persuade the discipline of economics that 'space matters' but its current version (and vision) of economic geography is a case of mistaken identity (1999: 19–20). Postulating too much mathematics and too little region,

> it is not merely a case of recognising that the mechanisms of economic development, growth and welfare operate unevenly across space, but that those mechanisms are themselves spatially differentiated and in part geographically constituted; that is, determined by locally varying, scale-dependent social, cultural and institutional conditions. This is the stuff of economic geography proper.
>
> (Martin 1999: 19)

One could add its advancement is very much the stuff of the new heterodoxy of 'new economic geography' which is the vision of economic geographers and which recognises the 'sticky places in slippery slopes' of economic agglomeration (Markusen 1996).

Knowledge, space, economy

Finally, assuming an acceptance of the contemporary world of 'the knowledge economy', it is to the delineation and explanation of the spatial organisation of knowledge production that this 'new economic geography' must turn; towards an understanding of how the spatial basis of production is being transformed under the imperatives of the knowledge economy. If Motor Sport Valley exemplifies *one* such mode of organisation, its analysis within this chapter through the lens of the 'knowledge community' must still acknowledge the crudeness of both this concept and the conception of 'knowledge' it draws upon. Whilst we would argue that this chapter's empirical unpacking of 'the knowledge community' provides some insight into the actual *workings* of the nuanced concepts of the new economic geography and agglomeration such as 'industry in the air', 'milieu', 'untraded interdependencies', 'local industrial systems', 'learning regions' and 'institutional thickness', there is much work to be done beyond the mere naming of these concepts as facets of agglomeration if, indeed, agglomeration itself can be accepted as the spatial leitmotif of the knowledge economy.

Notes

1 This research was made possible by a grant from the UK Economic and Social Research Council (Project R0006236144). We would like to thank David Turner who collected much of the data upon which this chapter was based and the editors of this compilation for their thoughts and comments on the chapter.
2 The numbers refer to the transcript code. All respondents were assured of anonymity to enable them to speak freely.
3 A constructor is the company which assembles the car using in-house and external supplies. A constructor is, normally, the racing team (for example, Williams Grand Prix Engineering Ltd run the Winfield Williams BMW Formula One team) but constructors may also run more than one team (for example, Prodrive has run a Subaru team and a Nissan team in the World Rally Championship).

9 Worlds in motion?

'Worlds of production', evolutionary economic change and contemporary retail banking

Jane Pollard and Andrew Leyshon

> Economists have, of course, always recognised the dominant role that increasing knowledge plays in economic process but have, for the most part, found the whole subject of knowledge too slippery to handle.
>
> (Penrose 1959: 77)

Introduction

The category 'knowledge' is increasingly capturing the geographical-economic imagination in discussions of competitiveness. In management literatures, 'information' and 'knowledge' are important buzz words. Early analyses of the 'information revolution' focused on the likely significance of the declining real costs of acquiring, processing and transmitting information and the accelerating pace of technical change (see Porter and Millar 1985). More recently, however, commentators are assigning greater import to different forms of knowledge and emphasising the links between knowledge creation and circulation and innovation, learning and competitiveness.

Examples of this higher profile afforded to knowledge abound. For Lundvall and Johnson (1994), contemporary capitalism is a 'learning economy'. In similar vein, management guru Peter Drucker (1993), regards knowledge – and not land, labour or capital – as *the* essential resource in a 'knowledge society'. Industrial policy literatures in Britain, parts of Western Europe and North America now routinely stress the need for competitive strategy to be based upon continuous innovation and learning rather than cost cutting (Best 1990). At the core of this upsurge of interest in knowledge is the belief that processes of knowledge creation, management and transfer are becoming increasingly significant in the construction of competitive advantage in contemporary capitalism. Nonaka and Takeuchi (1995), for example, argue that the Japanese postwar success story is based not solely on manufacturing prowess, the nature of supplier/customer relationships or access to cheap capital, but rather Japanese firms' ability to create knowledges and disseminate them through people, technologies, products and services in an organisation.

In recent geographical literature, the importance of knowledge has tended to enter the scene via concepts of 'untraded interdependencies', 'milieu', 'learning region', 'embeddedness', 'institutional thickness' and so forth (see Storper 1995; Cooke and Morgan 1993; Amin and Thrift 1994; Lundvall and Johnson 1994). These concepts hint at some of the spatialities of knowledge production and circulation and are often employed to explain the existence of agglomerations or clusters of innovative economic activity, including variety-based manufacturing industries, like the movie industry (Storper and Christopherson 1987), business and financial services (Amin and Thrift 1992) and, most especially, high technology industries, in North America and parts of Western Europe (see Storper and Salais 1997; Henry and Pinch this volume).

This chapter explores the increasing significance of knowledge(s) in a rather different context, that of retail banking in the UK. The knowledge revolution in retail banking presents something of a contrast with that in agglomerated, high technology industries in at least two senses. First, retail banking is a widely diffused activity which is, nevertheless, being restructured as its ability to access, create and circulate different knowledges is enhanced. Second, the shifting knowledges of and in retail banking are rather unusual when compared with those in high technology industries in that they are generating a rescaling of banking operations which is centralising power within organisations and, *contra* much recent debate in economic geography, causing banks to abandon precisely those parts of their organisation that once operated to gather local, tacit economic knowledge (Leyshon and Pollard 2000).

The chapter examines the restructuring of the UK retail banking industry through an analysis of the shifting knowledge bases of the industry. To do this, we draw on, and then sympathetically critique, Storper and Salais' (1997) 'Worlds of Production' framework. Their framework is used because it provides a template for thinking about economic development as an evolutionary process wherein localised knowledges, in the form of cultures, social practices and institutions, are organised and mobilised to develop new products and to respond to competitive pressures. We use this framework to argue that the knowledges required to undertake retail banking in the UK have broadened considerably in the 1980s and 1990s. Banking knowledges in the UK have become, in essence, more porous in that they now comprise a host of technical, regulatory and strategic knowledges generated in other industries, and often from outside the UK. These changing knowledge webs that constitute the practise of retail banking, in turn, generate different retail banking geographies.

The chapter is organised in three parts. The first section briefly examines the growing significance attached to knowledges in contemporary economic geography before focusing on the work of Storper and Salais (1997) and how it can be used to interrogate changes in retail banking. The second section uses the Storper and Salais framework to examine the changing knowledges bases – customer, technical, product, regulatory and so forth – deployed in retail banking in the UK and links these to changing geographies of retail banking in the UK. The final section evaluates Storper and Salais' framework in light of the reality of retail banking and presents some conclusions.

Thinking through knowledge frameworks

There are a number of reasons behind the upsurge of interest in knowledge, and the general assertion that knowledge is becoming more central to economic endeavour. First, and most abstract, is the notion that late industrialism is producing increased volatility and new forms of risk and uncertainty. One variant of this argument is developed by Ulrich Beck (1992) who argues that industrialisation is producing risks – such as atomic accidents, ozone depletion and all manner of pollution – that are global, long term threats that simply defy the 'rational' rules of accounting, insurance and compensation currently employed by institutions which manage risk. Another version of this argument comes from Susan Strange (1999) who suggests that, since the 1970s, the global economy has been characterised by greater volatility, uncertainty and anxiety. Strange acknowledges the seriousness of environmental concerns but also identifies a second 'threat to civilisation', namely 'the casino nature of the financial system' (1999: 3), which is responsible for numerous economic crises and their attendant job losses, bankruptcies and so forth. With volatility and risk comes the attempt to generate knowledges about such risk and volatility, to identify, measure, quantify and manage risk, or in cases where this is impossible, to recognise uncertainty and to manage relationships accordingly.

Beyond these very abstract arguments about the increasing risks and uncertainties of contemporary capitalism are other arguments about the increasing significance of knowledge. Besides the simple observation that there has been a massive increase in the volume of information we can process and transmit, many social scientists are increasingly interested in how, and in what circumstances information is used and, most especially, how knowledge is produced and then transformed from an invention into an innovation. Related to this is Robert Reich's (1992) suggestion that knowledge is of increasing importance because the division of labour is shifting in favour of occupations and skills involved with problem-identifying and solving, what he terms 'symbolic analytic services'. Given that the transmission of information and knowledge is a messy, partial business, industrial policy and management literatures have become interested in organisational forms which allow greater flexibility for the evaluation of decision making. Finally, Thrift (1996b) argues there is a greater interest, in both business and academe, in how organisations can maximise their potential by learning, from employees, from other firms in the same sector, from other industries and so forth. This growing interest in knowledge is reflected in the proliferation of a variety of theoretical perspectives all grappling in some way, shape or form, with flows of knowledge and the knowledge infrastructures that facilitate interaction and learning within and between firms.

Much recent work in geography is sensitised to the growing importance of knowledge. Thrift (1996b) suggests that we are in an era of 'soft' or knowledgeable capitalism. In soft capitalism, the prevalent (academic and business) discourse about the global economy is one which stresses: (i) that knowledge is partial and differentiated; (ii) that the world is complex and qualitatively and quantitatively differentiated; (iii) that the material, imagined and symbolic are bound together and cannot be separated; and (iv) that individuals are socially constructed entities. Thrift argues that the growth of such ideas and beliefs has gradually come to challenge an earlier dominant discourse founded on the notion of transcendent rationality and an ordered, bounded world which could be known by individuals where there was no sense of the world, or individuals, being interactively constituted, as is the case in soft capitalism.

These tenets of soft capitalism inform, to differing degrees, numerous studies of industry change and competitive advantage which examine the changing distribution of knowledge in economic networks. Many analyses have hinted at the importance of different knowledges by highlighting the so-called 'non economic' practices and institutions integral to the co-ordination of economic action (for example see Lee and Wills 1997). One prominent recent example of this is provided by Storper and

Salais (1997). This work deserves particular attention as many economic geographers are engaging with, and following, some of Michael Storper's theoretical work in thinking through the micro-foundations of successful territorialised production systems (see Henry 1999; Lagendijk and Cornford 2000; Leyshon and Pollard 2000; MacLeod 2000).

Worlds of production

Storper and Salais (1997: 12) argue that in order to understand the diversity, heterogeneity and unevenness of production systems in different regions and nations, we must think of production systems as 'humanly constructed orders of routines, cognitive frameworks, institutions, practices, and objects'. Storper and Salais regard so-called 'non-economic' forces (institutions, cultures and social practices),

> as central to the economic process . . . They take the form of conventions – largely implicit rules of action and coordination, generated by humans and routinized – which come together in what we call frameworks of economic action . . . [S]uch frameworks underpin the mobilization of economic resources, the organization of production systems and factor markets, patterns of decision making and forms of profitability.
>
> (Storper and Salais 1997: v)

Storper and Salais' starting point is to identify products as their units of analysis, because 'it is the product that embodies and thus realizes the potentialities of the resources of action' (1997: 14–15), and products are a 'concrete outcome of a complex network of relations between persons' (p. 38). Having privileged products in this way, Storper and Salais then argue that it is possible to classify them along two definitional axes.

- The first axis is based upon the outputted form of the products, which can be either *generic* (that is, designed for mass markets in which risk is measurable and producers classify similar demands) or *dedicated* (that is, where 'demand is unique for the producer' (p. 29).
- The second axis is predicated upon the inputs that go into making of the product, which can be either *standardised* (in that inputs to production are identical and interchangeable) or *specialised* (in that human and material inputs are idiosyncratic).

Combining these two binary distinctions enables Storper and Salais to create four possible 'worlds of production': an 'Industrial world', a 'Market world', a 'World of intellectual resources' and an 'Interpersonal world' (see Figure 9.1).

The nature of these four worlds can be briefly summarised as follows. First, the Industrial world produces generic, standardised products, drawing on economies of scale and such products are sold in markets in which risks are relatively predictable. This is the familiar world of mass production whose difficulties since the 1960s have formed the basis of the Fordist–Post-Fordist debate in Geography and other social science disciplines. The Market world produces dedicated, standardised products, in

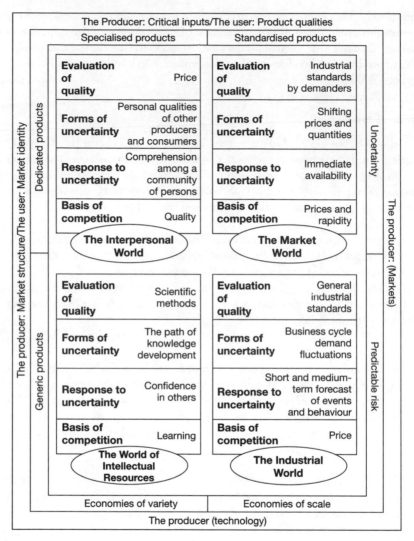

Figure 9.1 Storper and Salais' Four Worlds of Production
Source: Storper and Salais 1997: 33

which production is standardised, but each production run is customised or dedicated to a particular client demand. Storper and Salais cite the example of a nineteenth century industrial merchant using a network of home workers. The World of intellectual resources produces generic, specialised products (or changes qualities of existing products) and actors work in a context where it is not known whether the product will meet demand. Examples here might include R&D activities in corporations wherein workers seek to develop knowledges with general applicability. Finally, there is the extreme uncertainty of the Interpersonal world, a world of dedicated, specialised products in which specialised skills, and their meaning, are constructed and reconstructed within a limited knowledge community. Examples here include cutting edge high technology products, or perhaps business services requiring ' a high degree of attention to particular customer needs' (Storper and Salais 1997: 35). In essence, different forms of economic co-ordination and convention produce different 'worlds of action' in which actors manage uncertainty and produce absolute advantages (in terms of profitability and market share) for firms. Storper and Salais then employ this theoretical template to comprehend how successful real worlds of production are territorialised in regions in France, Italy and the USA.

Worlds of production in retail banking?

This framework generates a number of issues that are pertinent to an examination of the changing knowledge networks of retail banking. First, flows of knowledge and information are central to any analysis. Although Storper and Salais suggest that,

> [T]he development of asymmetric information (whether formalized as science or in the form of traditional skills and uncodified practices) and the reaction to it by actors (firms, individuals, social groups, governments) is one of the central mysteries of social life
>
> (Storper and Salais 1997: 11)

it has long been acknowledged that overcoming information asymmetries has been a central problem within financial markets. As lenders, banks have constantly to manage information asymmetries in dealing with their customers, in that customers always have greater knowledge than the bank as to whether they will be able and willing to repay their debts (Stiglitz and Weiss 1981). Managing information asymmetries and identifying 'good' and 'bad' customers before deciding whether to do business with them is a trait which distinguishes retail banking from many other services, and, until recently, has formed a barrier to entry for other retailing firms interested in selling financial products (Leyshon and Thrift 1999; Alexander and Pollard 2000).

Second, Storper and Salais develop an analytical framework that accepts the existence of diverse, complex and uneven forms of economic co-ordination at intra- and inter-sectoral levels. Shunning the parsimony of unified models of economic growth, Storper and Salais want to appreciate 'the diversity, multiplicity and richness of action frameworks in modern economics' (1997: 306). They argue,

> certain forms of widely available abstract, codified knowledge are taken up more rapidly by some firms, regions, or industries than others, even when the potential direct economic payoffs for adopting such innovations are equalized. In other words, the application of knowledge to product development . . . seems to depend on different routines of actors in relationship to signals. The capacities of actors to 'see' and to 'process' information are highly variable.
>
> (1997: 11)

The retail banking industry exemplifies this point, in that in the 1990s, as we argue below, it is drawing on a range of knowledges that have been circulating for some time in other industries. To give just one example, UK banks have been relatively slow, compared with their counterparts in consumer and industrial goods sectors, to embrace knowledges of marketing strategy and research (Clarke *et al.* 1988; Hooley and Mann 1988; Speed and Smith 1992; Kerfoot and Knights 1993). They are now more open to such knowledges and are, gradually, adopting the languages and practices of what Du Gay and Salaman (1992) term 'the discourse of enterprise' complete with its neoliberal commitment to consumer sovereignty and customer service.

Third, this framework attempts to be dynamic by emphasising the evolutionary nature of worlds of production. Worlds of production are not immune to strong competition, technological changes and conflicts around conventions. Moreover, the positive externalities or untraded interdependencies between firms, technologies and indeed sectors, have a cumulative character and give rise to the notion that technologies, for example, are products of interdependent choices and thus cumulative and path dependent (see Storper 1995; Dosi 1988). In retail banking, there are many examples of how earlier rounds of innovation have shaped the trajectory of contemporary organisational change. The 'industrialisation' of retail banking (Marshall and Richardson 1996: 1854) in the UK (and the United States) – which involves the development of specialist regional centres for data processing and the downgrading of bank branches and their staff – has been facilitated by a series of innovations in the 1950s and 1960s which transformed the way in which banks read, sorted and processed paper. The widespread adoption of Magnetic Ink Character Recognition (MICR) technology in the 1960s was to provide the banking industry with the means to produce faster and more accurate financial information and, ultimately, to eradicate many grades of clerical jobs and to downgrade the role of branches in information processing (Leyshon and Pollard 2000).

Finally, Storper and Salais (1997) also recognise the need to break down any tendency to draw a conceptual divide between the material and the social within economic life. Developments in information technologies have been critically important in facilitating a shift in the types of knowledge privileged by retail banks in determining whether they are to do business with different groups of customers. Moreover, in understanding how banks have introduced new technologies, there is clear evidence of the path dependency of their choices and, indeed, of their learning from other organisations within and outside the UK (Leyshon and Pollard 2000).

For all these reasons, Storper and Salais' *Worlds of Production* represents a rich, ambitious framework with which to understand the diversity of economic development. To what extent is this framework useful in forging an understanding of the organisational logics of the contemporary retail banking industry? To determine the utility of Storper and Salais' approach we now analyse retail banking in terms of its products and attempt to locate their production within one of their four idealised 'worlds of production'.

'Worlds' in motion

Although the number of products available within the retail financial services industry has proliferated in recent years, the industry is still based upon four broad product groups: products for cash handling and transfers (e.g. current or chequing accounts); savings or investment products; debt or loan products; and insurance products. If we attempt to locate these products within Storper and Salais' two-by-two matrix of possible industrial worlds then, on the basis of its current organisation, the retail banking industry would appear to be located firmly within the Industrial world. What are the characteristics of this world of production? It is a world which relies upon standardised inputs, where products are generic in form, and sold in markets of predictable risk and which are amenable to economies of scale. It is, in short, a world of relative stability, within which large organisations possess several competitive advantages.

The retail banking industry displays many of these features. Its products, like cheque and savings accounts, are for the most part reliant upon standardised inputs, they are generic in their outputted form, and they are sold into a mass market for which there is a generally constant, if fluctuating, demand. The long-term stability of the market for the industry's products is assured by the pivotal role of financial services within the social and economic fabric of contemporary capitalist societies. Governments and employers rely on banking infrastructure to pay wages, salaries and make transfer payments, while there exists a general understanding and recognition of the basic retail financial products among the population (Leyshon and Thrift 1999; Leyshon et al. 1998). Therefore, while the demand for particular types of retail financial service product may ebb and flow, financial service products in general have become all but essential for the survival and reproduction of households and individuals.

According to Storper and Salais, each world of production is further defined by the following four characteristics: the way in which quality is evaluated; the forms of uncertainty which pervade production; the response to such uncertainty; and the basis of competition. Within the Industrial world, quality is evaluated by the existence of general industry standards; second, the forms of uncertainty prevailing within an industry revolve around fairly predictable and foreseeable fluctuations, such as the business cycle and variations in demand; third, the response to uncertainty in this world involves attempts to forecast short- and medium-term uncertainty, and to make appropriate adjustments to these; fourth, and finally, competition within this world is predicated upon price.

The retail financial services industry also conforms to the additional characteristics Storper and Salais attribute to products originating within the Industrial world. For example, the basis of competition within the industry in Britain is increasingly based upon price, as a host of new firms like Virgin and Prudential have entered the market in recent years and sought to buy market share with products offering lower costs and higher returns than the industry standard. Further, the quality of retail banking products is evaluated in large part by general industrial standards, which are a product of at least two processes. The first is the existence of a system of self- and statutory prudential regulation, which is a pervasive feature of the financial services industry, given the political importance of probity in personal financial matters. The second is the importance within the industry of recognised practices or conventions for managing access to the financial system. As indicated earlier, overcoming information asymmetries is a core competency of the industry, and the means in which this is achieved is generalised across the industry at any one time. Currently, information about the characteristics of customers is processed through credit scoring systems, which are used to assess the likely profitability of customers and the risk attached to granting them access to cash handling and debt products (Leyshon and Thrift 1999). Computer-based systems are similarly used to both calculate and price insurance risk (Leyshon et al. 1998). By screening out those customers that are likely to be less profitable or even worse, loss generating, the 'quality' of products offered to the remaining customers will increase, both in terms of the range and level of services attached to products and in terms of price. Such systems are also used to both define and respond to the nature of uncertainty within the industry, as the credit scoring systems can be adjusted at different points of the business cycle to either relax or tighten the creditworthiness requirements of customers before they are provided with debt-related or insurance products.

While the industrial model successfully captures many of the features of the contemporary retail banking industry, it is not an exhaustive characterisation. Important parts of the industry do not conform to this typology and may be found located within one or other of the remainder of Storper and Salais' worlds of production. Moreover, the relative importance of these worlds to the industry have changed over time, so that it is possible to trace the temporal movement of the industry through each world of production (Figure 9.2). Indeed, it is possible to use this framework, and the changing combination of inputs and outputs which produce retail banking, to interpret and explain much of the recent historical geography of the industry.

This historical geography begins in the top left hand corner of the diagram, in the Interpersonal world. According to Storper and Salais, this is a world made up of dedicated and specialised products; that is, they are produced through interpersonal action, wherein the skills are highly specialised and produced in often idiosyncratic contexts, so that the outputs are also highly idiosyncratic, dedicated for a narrow band of customers, often on demand, and in response to particular customer requirements. This is, for the most part, a world in which information about quality is exchanged in a non-codified form, usually through spoken and body language via co-presence. It is also through such face-to-face contact that firms understand and respond to uncertainty, which is manifested in the actions of other actors. Through interpersonal contact there develops 'something like a "common language" that can be pressed into service to define products and describe the nature of . . . others' actions' (Storper and Salais 1997: 36). It is this common language, and the close understandings forged through face-to-face contact and co-presence, which ensures that competition within this world revolves around the quality of the product produced.

These characteristics go a long way to describing the organisation of the retail banking industry in the immediate postwar period. Retail banking then was more a class – than mass-market industry, which catered only for the more affluent members of society. In an era of private banking accounts and services and 'relationship banking', the industry relied heavily on face-to-face contact for the gathering of information about both customer requirements and the suitability of customers as patrons of the bank. This information was gathered through contact with customers both within the bank – at interviews and other meetings – and within the community, through social contacts. Therefore, at this time the retail banking industry was located firmly within the Interpersonal world, providing specialised and dedicated products for an affluent minority of the population.

As the postwar settlement delivered the stabilisation of employment conditions and a general increase in living conditions, however, this class market was gradually transformed into a mass market. The spatial imprint of this shift was reflected in the expansion of branch networks as the four major clearing banks in Britain consolidated their positions and established a coherent, hierarchical national banking system (Pratt

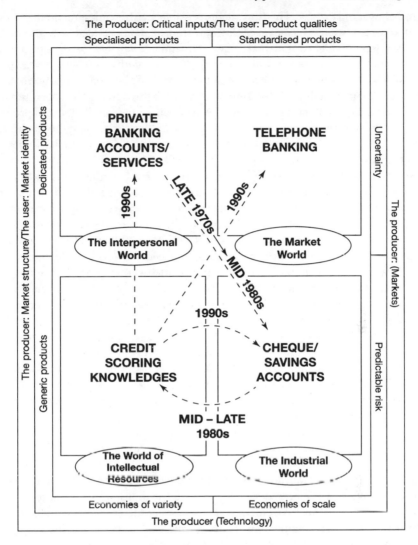

Figure 9.2 Changing worlds of retail banking

1998). Bank branches occupied a pivotal role in the banking division of labour. They were key sites for gathering intelligence about markets and customers, the sites at which banks performed services for customers and the sites at which banks processed and settled the day's business. By the end of the 1970s, the growth of a mass market meant that the part of the retail banking industry dealing with products like cheque and savings accounts began to migrate slowly to the Industrial world of production (See Figure 9.2). Over time, many of the inputs to production became standardised rather than specialised, although initially this standardisation tended to be concentrated in the processing end of the industry (for example, automated cheque-processing began to be introduced from the late 1950s onwards) (Leyshon and Pollard 2000). By the 1980s, however, even 'front office' activities such as assessing creditworthiness were being standardised. This was made possible by the widespread adoption of credit scoring systems as a means of distinguishing *a priori* between 'good/profitable' and 'bad/unprofitable' customers (see Leyshon and Thrift 1999 for a longer analysis of the rise of credit scoring in retail banking). This standardisation of inputs has made it possible for banks to strip out layers of managerial staff who were formerly employed to exercise their specialised skills in credit assessment through face-to-face interviews with customers and potential customers. At the same time, retail financial products have become increasingly generic in nature, so that there is actually little variation between the products on offer, which encourages competition on the basis of price, brand quality or both.

Again, the changing knowledge architecture of retail banking leaves its mark on the geographical anatomy of the industry. With the shift to the Industrial world, banks standardised their inputs by, amongst other things, removing credit assessment and other processing work from branches. Through the 1980s and 1990s, as the 'industrialisation' of retail banking (Marshall and Richardson 1996: 1854) has proceeded apace, the data processing for bank products has been stripped from branches and consolidated in regional data centres. Separate loan centres now deal with credit assessment, while Automated Teller Machines (ATMs) and call centres keep routine customer enquiries out of branches (Leyshon *et al.* 1993; Cressey and Scott 1992). Branches – as sites for intelligence gathering, serving and selling to customers and processing the preceding day's business – no longer occupy the pivotal role they once did as these activities have moved beyond the exclusive confines of the branch (Leyshon and Pollard 2000). Another symptom of this trend is banks' closure of branches; between 1989 and 1995 in Britain, 20 per cent of bank and building society branches closed, with those closures occurring disproportionately in low income communities (Leyshon and Thrift 1997).

The ability of the industry to standardise inputs in the ways described above, however, has been made possible by the growing importance within the industry of what Storper and Salais describe as the World of intellectual resources (see Figure 9.2). Credit scoring systems, and the associated use of geodemographics and 'lifestyle' marketing systems, have been developed outside the industry by specialised consultancies and agencies but have emerged in the last decade or so as key competencies to be deployed in the banking industry as a whole. As Storper and Salais (1997: 37) put it, the World of intellectual resources produces 'qualities which can later be recognised by, and . . . ultimately codified and imitated by others'. This world of production produces generic products from highly specialised inputs, which in this case are analytical devices and models which provide retail financial services firms the means by which they can attempt to screen and filter customers 'at-a-distance'. The introduction of such systems has undermined the dependence within the industry upon codified knowledge, which is now drawn upon routinely only at key points of the overall financial service delivery system (such as, for example, when dealing with particularly affluent, and therefore highly valued, customers, or when resolving credit decisions where credit scoring results are considered to fall between the cut-off scores for either granting or denying a product or service) (Leyshon and Thrift 1999).

This does not mean that banks are no longer attentive to issues of service and product quality, but that many have sought to make the link between the quality of service and the cost of such services more transparent. For example, in Britain, where banking has been 'free' to customers in credit since the mid-1980s, some banks have introduced fee-bearing accounts which come with a range of add-on products and services (e.g. travel insurance, a waiver of commission for foreign currency and travellers cheques, lower overdraft charges and so forth). These 'product-rich' accounts are being increasingly differentiated from the basic, 'no-frills' accounts offered to customers who pay no fees.

The standardisation of inputs, combined within important changes in market regulation, have also made it possible for new firms to enter the retail banking market, with the effect of intensifying levels of competition and market uncertainty. As a result, some firms are adopting many of the hallmarks of the Market world, by competing for market share on price and/or on speed or convenience of service (for example, 24-hour banking, internet banking or by guaranteeing to beat the competition's interest rates on deposits or loans).

Finally, there is evidence that some parts of retail banking are, in the 1990s, returning full circle to the Interpersonal world that much of the industry occupied in the immediate postwar period. As banks have undertaken more sophisticated market research to fine tune the packaging and delivery of debt and investment products to different income groups, they have been refocusing their customer bases around high income individuals and their fee-generating potential. High Net Worth Individuals (HNWIs) can now secure private banking privileges, and the attentions of a dedicated member of staff, to manage their financial affairs. For banks, the extra personal service and face-to-face contact characteristic of the Interpersonal world can maintain the loyalty of their HNWIs and dissuade them from switching their allegiance on the basis of price as market conditions and interest rates vary.

In the 1990s then, the changes we have discussed in retail banking are not only linked to the growing influence of the World of intellectual resources, but also to some of the practices prevalent in the Market world and in the Interpersonal world. As such, descriptions of recent changes within the retail banking industry as a revolution are appropriate. As parts of the industry return to the Interpersonal world in the 1990s, it is clear that the industry's passage through the Industrial world, the Market world and, in particular, the World of intellectual resources, means that it is better able to identify those customers for whom it is worth producing the idiosyncratic labour inputs and dedicated products of the Interpersonal world.

An evaluation and conclusion

Storper and Salais' Worlds of Production framework has considerable utility when applied to the retail banking industry. Indeed, we suspect that it will prove to be a valuable analytical device which can be applied to extant understandings of a broad range of industries (see Henry 1999). Yet, we also encountered a number of problems with the approach when applying it to our case study.

The first issue was methodological. Much of Storper and Salais' discussion of their four worlds of production is highly abstract. Although they provide some discussions of developments in the USA, France and Italy, it is not clear how the categories of the framework have been operationalised empirically in that there is little information on how particular firms and industries are assigned to one world or another. Second, and related, although Storper and Salais identify products as their units of analysis, much of their empirical discussion focuses on classifying *firms* and *industries* into different worlds of production. Given the nature of available statistical data this concession is understandable. It is extremely difficult to undertake any exercise in aggregate empirical description in a framework in which products, and not firms or industries, are the preferred units of analysis. Nevertheless, in this chapter, we have focused on some retail banking products and how they have changed since the 1950s. In so doing, what becomes clear is the difficulty of assigning a firm, let alone an entire industry, to a particular world of production, most especially if it produces a range of different products. Different retail banking products draw on different conventions and worlds of production.

These methodological and empirical queries in turn generate another set of issues. Although the framework arguably provides a useful heuristic for tracing how some retail banking products have evolved through different worlds of production, is it useful to employ such a framework to try to locate firms and industries with respect to different worlds of production? There appear to be some tensions evident in Storper and Salais' discussions of, on the one hand, the richness, diversity and evolutionary nature of production systems, and, on the other hand, the sense of order achieved by classifying firms and industries into one of four worlds. Storper and Salais recognise that real worlds involve 'major and minor keys' (p. 198), that firms are meeting points for dominant action frameworks fused with aspects of other worlds of production. Moreover, they are careful to argue that theoretical parsimony, rather than any claims to comprehensiveness, dictates their choice of four – and not ten or a hundred – possible worlds of production. Yet Storper and Salais go on to describe what they consider to be more and less likely evolutionary trajectories for firms between different worlds.

> the evolutionary pathways between the Interpersonal and Market Worlds and between the Intellectual and Industrial Worlds are wide and deep; weaker channels of evolution exist between the Interpersonal and Intellectual Worlds and the Market and Industrial Worlds; and it is highly improbable that the Market World could be transformed into the Intellectual World (or vice versa) or that the Industrial World and the Interpersonal World could change places.
>
> (Storper and Salais 1997: 198)

In essence, Storper and Salais invoke the imagery of order (their four worlds) while at the same time developing concepts like conventions and untraded interdependencies (and indeed a dynamism in those concepts) that seek to transcend such order.

The examples we have discussed in this chapter suggest that there are a variety of evolutionary trajectories being adopted in the networks which we frame and describe as 'the banking industry'. The banking industry is evolving in a number of worlds of production as its competitive, regulatory, technological and marketing circumstances shift. Understanding these various trajectories and their geographies – the relative stability of the trajectory of products for High Net Worth Individuals (HNWIs) compared with the 'industrialisation' of cheque accounts in the 1970s and 1980s, for example – necessitates an understanding of another ordering framework, namely the broader political economy of finance. Banks in the UK and elsewhere are striving to grow and be profitable in financial services markets which are in the throes of rapid globalisation and a host of technological and regulatory changes (see Martin 1999b). These changes have shifted banks from a Keynesian, regulated, functional role to one in which they are more exposed to market forces. Mergers and acquisitions are producing a financial services arena with fewer, larger players, who are operating in increasingly competitive retail markets, often against non-bank providers of financial services. There is, then, a political–economic arena in which any discussion of forms of economic co-ordination in retail banking – and their likely evolution – needs to be located.

In summary, Storper and Salais' framework is useful in that it provides a template for tracing some of the different intersections of knowledge in, and of , different retail financial products since the 1950s; from the Interpersonal world, to the relative stability of the Industrial world and, more recently, to the world of Intellectual resources. Their framework prioritises flows of knowledge within and between firms, recognises the diversity and unevenness of forms of economic co-ordination and is dynamic in recognising the evolutionary nature of worlds of production. It also draws attention to the complex nature of production within any industry, the need to focus upon evolutionary processes, and that any attempt to describe a particular ensemble of economic co-ordination as an 'industry' is something of a chaotic conception. But in using the framework to assess retail banking a number of methodological concerns become apparent, together with an uneasiness about the sense of order invoked by the four worlds of production. There are difficulties in operationalising the framework and it requires careful political–economic contextualisation to comprehend the challenges of economic co-ordination peculiar to the banking (or any other) industry.

What is clear is that there is now a web of disparate knowledges concerning customers, products, technologies, marketing, strategy and regulation required to compete in the UK retail banking industry. While all knowledges are made 'local' when they are used (Leigh Star 1995), retail banking is now drawing on and localising knowledges generated in other industrial sectors (for example, the management consulting and airline industries) and in other countries (notably the United States, see Leyshon and Pollard 2000). In essence, the web of knowledges that constitute 'retail banking' have become more porous to a variety of knowledges from other industries and places, knowledges that are able to travel in the form of texts, artefacts and people (see Latour 1987).

A number of implications follow from this argument. First, while it is clear that processes of knowledge creation, management and transfer are becoming increasingly significant in the construction of competitive advantage in retail banking, it is difficult to locate retail banking in any one world of production. With changes in the production of its core products since the 1950s, retail banking can straddle several worlds of production.

Second, these changing knowledge webs that constitute the practice of retail banking are shaping the evolution of retail banking geographies. Thus far in economic geography, the spatialities of knowledge production and circulation have been invoked largely to explain the existence of agglomerations or clusters of, say, high technology industry. Retail banking, as we argued earlier, is a very different creature because it is a widely diffused activity with a very different geography to that which characterises high technology industries. Moreover, the shifting knowledges of, and in, retail banking are generating a rescaling of banking operations which is centralising power within organisations and causing banks to abandon those parts of their organ-isation that once operated to gather local, tacit economic knowledge (Leyshon and Pollard 2000). Retail banking's passage from the Interpersonal world to the mass market of the Industrial world has been accompanied by the standardisation of inputs and the consequent development of regional processing centres, followed later in the 1980s by the standardisation of 'front office' activities, such as assessing credit worthiness, and the establishment of separate loans centres. The initial expansion of branch networks that accompanied the development of mass markets has now been reversed, with branch networks being trimmed as data centres, loans centres and call centres have been opened. This standardisation of inputs responsible for downgrading the importance of bank branches has been facilitated, in part, by the credit scoring systems, geodemographics and lifestyle marketing products of the World of intellectual resources.

Finally, as the knowledge architecture of different retail bank products change, banks may find themselves offering products with lower barriers to entry for competitors than previously. Clothing and grocery retailers, for example, are familiar with and have longer histories of working with particular forms of customer and marketing knowledges that are now becoming essential to retail banking products. In sum, as banking knowledge networks become more porous to technical and strategic knowledges generated in other industries and often outside the UK, banks are facing new competition from other firms and industries.

10 Spreading the message

Management consultants and the shaping of economic geographies in time and space

John R. Bryson

The management knowledge . . . of the US . . . like Coca Cola and Levi jeans, . . . is exported worldwide.

(Jacques 1996: 6)

Introduction

Why do companies change? How do new economic geographies evolve? How are new labour practices introduced into firms? How is knowledge transferred between companies as well as countries? All of these are key questions for economic geography and all involve understanding the relationship between knowledge, space and economy. At the centre of this set of relationships are the actions of people, as it is through social relationships that knowledge flows between and within spaces and places. The geography of economic activity is the consequence of the interaction of relational activities operating in time and space. Economies can thus be conceptualised as relative, relational, temporal and spatial. It is difficult to keep track of the mutually constituted nature of economic activity and it is comparatively easy to agree with Doreen Massey that this way of conceptualising time-space 'can make your head hurt' (Massey 1999: 262). The geography of economic organisation is the result of a set of social relationships which have evolved over time and which have produced particular types of production systems. This operates at the level of the individual organisation as well as production system. There are two important points to consider. First, central to this process are the activities of key individuals, managers and consultants as well as a series of texts, for example newspapers, the business press and manuals of best practice produced by professional associations. Second, economic change requires exposure to flows of knowledge either in the form of the direct or indirect movement of ideas between companies and places or through knowledge situated in manufactured items that out compete existing products. A good example of this form of knowledge was the introduction of the Dyson vacuum cleaner that out competed existing cleaners through technological innovation inspired by James Dyson's decision to rethink the conventional cleaner.

Flows of knowledge produce change either leading to the decline of existing economic activities or to alterations in organisation, geography and labour practices. Central to these changes are the activities of management gurus and consultants; those individuals who develop new ways of organising work or are involved in the spread of new management models and theories. The activities of consultants and the importance of management models have only recently been considered by geographers (see Bryson 1997; Thrift 1998). Management theories should be deemed to be central to the geographical project as it is precisely these theories which influence and in certain cases control organisational change as well as altering the conditions of work and employment (see for example Jacques 1996). Central to the geography of economic activity should be an understanding of the transfer of management ideas and techniques into and between companies. In economic geography this is a history which still needs to be authored. This is a history which might begin in Birmingham (UK) as Matthew Bolton and James Watt, Jr., directors of the Soho Engineering Foundry, are considered by some commentators to be the first scientific managers (Pollard 1965: 251–72). To the management consultants Urwick and Brech: 'Neither Taylor, Ford nor other modern experts devised anything in the way of plan that cannot be discovered at Soho before 1805' (1953: 24).

Some of the most recent attempts to explain the spatial reorganisation of production note that there has been a blurring of the organisational boundary of the business enterprise (Bryson 1997). Thus, the trend towards out-sourcing and externalisation has created a situation in which the knowledge or information required to manage an enterprise is controlled and owned by other companies. I would argue that this is not a new trend. There has always been a division of managerial knowledges (Origo 1988: 116–17) and this has, in many instances, been a spatial division of managerial knowledges. To understand the restructuring of British industry involves identifying the sources and locations of the ideas that are driving alterations in the spatial structures of production and the internal organisation of business enterprises. The latter may not appear to be the concern of the economic geographer, but there are important differences in the way in which ideas are localised into the cultural environment of a country (see Guillen 1994, and Jacques 1996 on this point) or business enterprise (Schoenberger 1994, 1997). Thus, an analysis of the production of management knowledge and its diffusion by management consultants, for example, must be situated in an understanding of the ways in which knowledge is consumed by business organisations and institutions.

It must be stressed that this is a first[1] and very partial attempt to address this problem and that further research is required into the impact, spatial diffusion and transformation of a whole series of management ideas which have had an impact on the nature and condition of employment as well as on the structure of economic activity. There is a geographical literature which explores the activities of enterprises and individuals which spread management ideas (see Bryson et al., 1993; Bryson 1997; Bryson and Daniels 1998a), but a curious absence of geographical studies which explore, for example, the activities of Frederick Winslow Taylor (1856–1915), the father of Scientific Management, on the evolution and history of Fordism. The economic geography of Taylorism or Fordism is one of employment statistics and regulation theory and not one grounded in an understanding of the ways in which a particular set of management models or knowledges were created and diffused throughout the USA and Europe. Taylor's name is well known within economic geography, but usually only as a signpost to indicate that the nature of capitalist activity was altering and that something known as 'Taylorism' (Knox and Agnew 1994: 184; Dicken 1992: 116) had occurred. There have been numerous individuals like Taylor who have had a significant impact on the geography of economic activities but like Taylor they have either become signpost descriptors incorporated in a complex theoretical structure such as regulation theory, or are unknown to economic geographers.

The argument is divided into three parts. The first part applies concepts developed in the sociology of translation literature to management knowledge. It highlights the social, cultural and political aspects of knowledge production and consumption. The second part explores the role played by management experts in the diffusion of management models and technique. The third section attempts to construct a history of F. W. Taylor and his disciples (for example, Gantt and Gilbreth as well as the Taylor Society) in order to demonstrate their role as the creators and diffusers of many of the ideas that have been subsequently labelled 'Fordism' by economic geographers.

Gurus, consultants and the sociology of translation

Management consultancy is a knowledge industry that is founded upon the transfer of management models, theories and procedures that are created in universities, large consultancy companies or within client organisations. This type of knowledge flow is complex and dynamic but it usually is the result of active and passive relationships between a series of social actors. Such knowledge flows can be identified and understood using Actor-Network Theory (ANT) (Latour 1987). This approach is not explicitly followed here given the tendency for social scientists to deploy a prostituted version of ANT in order to simplify reality and to remove the complexity that lies at the heart of socio-economic systems (Law 1999: 9–10). Instead attention is focused on some of the earliest papers produced by some of the leading protagonists of what began as a sociology of translation and later became labelled as ANT. These papers provide an important insight into the relationship between management theory and organisational change. The most important paper to explore management consultants using a sociology of translation was that written by Bloomfield and Best (1992) informed by Callon's seminal paper on the fishermen of St Brieuc Bay (1986). They argue that management theories and solutions provided by management consultants can be 'viewed as a process of "translation" or "problematisation" in which solutions are not simply matched to problems; rather, problems are redefined (translated) in terms of existing "solutions"' (Bloomfield and Best 1992: 535–6).

'Translation' is a process, strategy or method by which actors (individuals, organisations) attempt to enrol others into a network. It is via this process that 'the identity of actors, the possibility of interaction and the margins of manoeuvre are negotiated and delimited' (Callon 1986: 203). Problematisation is a specific form of translation in which actors construct and reconstruct the boundaries of a problem. Drawing upon Callon (1986), Bloomfield and Best argue that consultancy involves two aspects of Callon's problematisation. First, 'one actor can make itself indispensable to another by translating a problem of the latter in terms of a solution owned or within the orbit of the former' (Bloomfield and Best 1992: 541). Thus, a consultant can persuade a client that a problem with productivity and profitability can only be addressed by using a particular consultancy specific or even copyrighted solution. The secret of good consultancy is to redefine a problem so that it matches a readily available solution.

Second, problematisation involves an active process of identity construction and repositioning by different actors. In the case of management consultancy and management theorists this involves the establishment of a consultant as 'an obligatory passage point' (Bloomfield and Best 1992: 541). Consultants define themselves as experts possessing specialist knowledge while the client role is constructed as one of dependency. Problematisation occurs both before and during any consultancy assignment and is central to some of the most important aspects of knowledge work: work as role-play, performance and display. In the case of knowledge work the display involves an individual or company convincing others that they alone can supply or have access to a particular form of information or knowledge.

The consumption of management knowledge rests upon perceptions held by potential consumers of the knowledge (management theory, recipe or fad) and of the knowledge producer and supplier. The geographical literature on management consultancy emphasises the role of the reputation of the individual consultant in the acquisition of consultancy contracts (Bryson 1997). Perceived expertise and a visible reputation may be more important than actual expertise. The dialectical relationship between knowledge and reputation is a fundamental part of the flow of knowledge into and between companies. It is also fundamental to understanding the consumption of knowledge in general. Thus, in the world of academic journals, conferences and book publishing the greater an individual's position the more readily will their ideas be accepted. Many academics might protest at this statement, but it can be supported by drawing upon the work of some eminent academics (Latour 1987). To Podolny and Stuart (1995) the status of an individual in the technological community influences the likelihood of an innovation or invention being accepted. They cite the example of IBM whose entry into the personal computer market stimulated software developers to write programs for these machines. The same point is made by Merton who observed that the reputation of a scientist distorts the manner in which their work is received by their peers. He labelled this phenomenon the 'Matthew Effect' on the basis that according to the gospel of St Matthew: 'for unto every one that hath shall be given' (Merton 1973: 445). In his paper Merton suggests that material presented by a scientist with an outstanding reputation may have a greater impact than the same material presented by an unknown scientist. The work of outstanding scientists is read by their peers, whilst that of unknown individuals is not. Jointly authored papers are remembered as the work of the author with the best reputation. Knowledge consumption is thus a social and political process that draws heavily upon the projected, and maybe not actual, life histories of the knowledge producers or distributors.

Systematic versus local knowledge

The primary aim of this edited collection is to explore the relationships that exist between knowledge, space and economic activities. Two of these concepts – knowledge and space – are difficult to define. Definitions of knowledge have already been explored in the chapter by Howells (see Chapter 4). For the purposes of this chapter, however, a distinction needs to be made between 'systematic knowledge' or knowledge which appears to be supported by an informed and maybe scientific or pseudo-scientific analysis, and 'local knowledge' which is knowledge which has developed in a particular cultural, environmental or enterprise environment (see Shiva 1993: 10). This argument is similar to that made by Marglin in his distinction between two types of knowledge: 'episteme' and 'techne'. 'Episteme' knowledge is based on scientific management, analysis and empirical variation of theory. In contrast, 'techne' knowledge comes from use (for example learning on the job) and it tends to be tacit and emotional rather than conceptual. Marglin's terminology, however, only confuses the analysis by the introduction of new terminology. Local knowledge, thus, appears to be less conceptual and theoretical than systematic knowledge and is also less visible. It may be a verbal or tacit form of knowledge rather than being part of a written discourse.

Innis (1951) and Hirst and Woolley (1982) draw attention to the relationship between the communication of information and social processes. To Innis all forms of communication possess an inherent bias. The transition from information written on stone tablets to letters inscribed on portable papyrus is associated by Innis with a transformation of information from that which is 'time-binding' to that which is 'space-binding'. Information (laws, etc.) inscribed in stone will be permanent and long-lasting while more portable media will permit the spread of information. Developments in telecommunications can thus be conceptualised as technologies that increase space-binding effects by permitting the rapid transfer of ideas, information and knowledge through space. Systematic knowledge is more portable (space-binding) than local knowledge (time-binding) and may be deemed to be more legitimate; legitimacy in this case being constructed by its written nature and its diffusion by the scientific and popular media. Although neglected by geographers (but see Barnes 1996: 206–27) Innis' work is useful for understanding the way in which space-binding knowledges are increasingly over-writing local knowledges (see Thomas 1978, for an historical example of this process), or at least combining with local knowledge to produce a different combination of local and systematic knowledge in each country and business organisation (Jacques 1996).

Systematic knowledge, of course, is also the product of a particular cultural environment, but it is knowledge which is increasingly rapidly spread between countries and in most cases it is also knowledge which is rooted in a Western tradition, and in the case of management theory, American tradition (Saurin 1993). Management gurus and management consultants specialise in the replacement of local knowledge(s) with forms of international systematic knowledge.

Consultants and the cadre class

Drawing upon the work of Hirszowicz (1980), the Amsterdam International Political Economy project highlights the importance of a 'cadre function' or social cohert (van der Pijl 1998: 136) in the organisation and supervision of business organisations. The twist in the analysis, however, is that an economy with an advanced division of labour requires new forms of organisation and that these increasingly depend upon 'a set of second-order organisations' (Hirszowicz 1980), for example management consultants. The French term cadre is used by the Amsterdam school to link cadre formation with processes of socialisation (see Hanlon 1994: 198–212). The cadre class can also be described as a professional elite (Perkin 1996) or as a professional-managerial class (Ehrenreich and Ehrenreich 1979). The argument is that a cadre class develops to plan, propagate and monitor economic activity and that such a class is subordinate to the ruling class or owners of capital. The cadres are a product of the higher educational system and multinational corporations. They are a progressive class; open to new ideas and ways of managing and organising economic activities. They are also highly mobile.

Management gurus and consultants are key members of this new class. According to the *Financial Times* '[c]onsultants are in many ways the advance guard of the globalising trend in world business' (*Financial Times*, 24 February 1995, cited in van der Pijl 1998: 160). There is an important caveat to this statement. Management consultancy is not a global industry. Most consultancy is performed locally by small and medium-sized enterprises (Bryson and Daniels 1998b) and there are only a small number of international consultancy companies (Table 10.1). In 1991, these large consultancy companies employed 52,837 consultants. Only a small proportion of this number are, however, full members of a globalised cadre class; the majority of conultants operate within the confines of one nation state. This is not to imply that the majority of consultants are using management models that are developed locally; most of the large consultancy companies run in-house universities which are designed to develop core competencies in the standard and most recent organisational and management models. Thus, Andersen Consulting educates its new professional staff into the Andersen way of doing consultancy by sending them to its own 'university' housed in a former girl's school in Saint Charles, Illinois.

It is the large management consultancy companies which provide a world-wide service, charge high fees and develop entry barriers to the profession based on reputation and the creation of in-house universities and expertise centres. They are also heavily networked organisations with former employees placed in positions of control and power, for example the Boston Consulting group currently has 605 professional employees, but 2,400 alumni (the language this company uses to describe former employees). The alumni are encouraged to consider themselves as part of the company. Staff turnover amongst the large consultancy companies is extremely high; Andersen Consultant's staff attrition rate is 16 per cent per year (O'Shea and Madigan 1997: 79). Consultants from the large consultancy companies are considered to be highly desirable employees by other consultancy companies and by client organisations. A final point is that the majority of the large consultancy companies were established in the USA, and thus are propounding a set of management models that are constructed to meet the requirements of American corporations (Table 10.1). For the geographer it is important to remember that the four bastions of management education: 'the MBA, the business school, managerial publishing and consulting are all US export products' (Jacques 1996: xiii).

Management gurus and consultants transfer the latest management and production techniques or methodologies between companies and countries. They are also involved, along with financial analysts, in setting expectations about normal or expected rates of profitability and returns on capital employed. The technique of 'benchmarking', developed in the 1990s, is a good example of the way in which consultants transfer profitability norms between companies. Benchmarking involves comparing some elements of a client's business to best practice elsewhere. It is based on the acquisition of detailed operational and financial statistics which are difficult to acquire, but which consultants have access to through their normal business activities. Consequently, consultants are heavily involved in encouraging and sometimes forcing companies to restructure, merge and transform their activities.

Table 10.1 Major international consultancy companies, 1991

Consultancy firm	Country of origin	Date founded	Number of offices	Number of professional staff	Countries of operation	Typical clients
Alexander Consulting Group, INC	USA		90	1,400	Western Europe, USA, Latin America	
Alexander Proudfoot	UK holding company, but consultancy subsidiaries have been acquired from the USA		55	1,100	USA, Western Europe, Australia, Japan, New Zealand, Hong Kong, Taiwan, Singapore, Brazil, Mexico	1,500 corporations, of which 200 are part of the Fortune 500 list
American Management Systems	USA	1970	29	2,400	USA, Canada, Belgium, Germany, UK	
Andersen Consulting	USA/UK/Japan. HQ based in Geneva	1989	157	20,000	45 countries	Moscow City Council, New Zealand Inland Revenue, London Stock Exchange, British Rail, Kraft General Foods
Arthur D. Little	USA	1886	25	1,400	60 countries, USA, Latin America, Far East	1,700 clients
A. T. Kearney	USA	1926	60	2,700	30 countries, USA, Europe, Asia/Pacific	General Motors
Bain & Co	USA	1973	25		16 countries	Polish government, Guinness, Dun & Bradstreet, Russian Government

Table 10.1 continued

Booz Allen & Hamilton, INC	USA	1914	30	3,000	75 countries	70 of the top largest companies in the world, half the world's largest banks, governments of 40 countries, 5,000 assignments per year
Bossard Consultants	France		16	360	USA, Western Europe, Singapore, Estonia, Poland	
Boston Consulting Group	USA	1963	40	605	USA, Western Europe, Japan.	Reuters, Harvard Business School, Save the Children Fund
Gemini Consulting	USA	1991	15	800	USA, Western Europe, Japan	75 major transnational companies, several governments
Hay Group INC	USA	1943	79	1,000	32 Countries, USA, Europe, Asia/Pacific, Latin America	
Hewitt Associates	USA	1940	52	2,100	20 countries, Europe, Latin America, Asia/Pacific	
McKinsey	USA	1926	48	2,600	25 countries	75% of the Fortune 500 companies
Mercer Consulting	USA	1900	35	7,857	21 countries, USA, Western Europe, Asia/Pacific, Hungary	Pepsi Co, IBM, Hewlett Packard, GE, UN
PA Consulting	UK	1943	80	1,615	20 countries, USA, Scandinavia, Europe, Asia /Pacific	
Roland Berger & Partners	Germany	1967	14	400	Argentina, Europe, USA, London, Tokyo, Prague, Moscow	
Towers Perrin INC	USA	1917	60	3,500	USA, Canada, Europe, Latin America, Asia/Pacific	

Sources: United Nations 1993; Grosvenor 1989.

Management consultancy should be conceptualised as an extension of Berle and Means (1932) well-known separation of ownership from control in large American corporations. Stocks and shares provide their owners with a financial interest in an enterprise, but they are not necessarily associated with any significant control of a company. Managers with shares have 'nominal ownership' rather than 'effective ownership' of part of a company's assets (Scott 1987: 30). Management consultants will usually have no 'nominal ownership' of a client's company's assets but may have a greater impact on company strategy and activity than managers holding shares in the enterprise.

Relationship capitalism and the flow and distortion of management knowledges

To Richard Dawkins, the biologist, ideas behave like genes in that they are capable of replicating themselves to ensure their survival. Replication is the result of copying and imitations and occurs as a result of gossip, rumour, discussions, conferences and publication. Ideas are also spread on the basis of their status or origin; who has developed a concept may be more important than the concept itself. The diffusion of ideas is also associated with a process of distortion or translation of ideas. This distortion process may be offset by drawing upon ideas obtained through established networks of contacts; trusted direct personal contacts or old-boy – and -girl – networks (Dunbar 1996). Such networks becoming increasingly important as the total population available from which to draw ideas and knowledge expands as a consequence of time-space compression or globalisation. The escalation in the availability of expertise, information and knowledge is increasing the importance of what can only be described as *relationship capitalism* in which personal contacts determine or partially determine the flows of information and knowledge between individuals and companies (Bryson *et al.* 1993; Bryson 1997; Bryson and Daniels 1998b; Bryson *et al.* 1999). Social relationships influence an individual's assessment of the validity of knowledge, as knowledge validity is closely tied to the reputation of the knowledge supplier. To Schoenberger people who run firms have histories and such histories provide them with 'revisable prior knowledge' (1997: 98); revisable via extant networks of social relationships.

The flow of consultancy-deployed knowledge may be conceptualised as an inverse triangle with management gurus at the apex (Figure 10.1). It is management theorists like Mike Hammer (Hammer 1990; Hammer and Champy 1993) and Peter Drucker (Drucker 1994) that are at the apex of this triangle. Above this level are academics who develop techniques in their own right, but also propound ideas created by the gurus. Wider dissemination of these ideas comes via taught MBA programmes, popular magazine articles (Hammer 1990), the 'how to' type of business guides (Norton and Smith 1999; Wilson 1999) and the business press in general. The conversion of a management idea into a fashionable management tool or technique is consequent upon the activities of these semi-related institutions as well as individuals operating in relative isolation from each other. Taken together, these agents produce a management fashion, for example Taylorism, benchmarking or Michael Hammer's business process reengineering (BPR) (Hammer 1990). BPR is a superb example of this type of knowledge flow, but also of a translation of an academic idea.

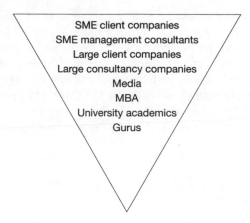

Figure 10.1 The flow of consultancy-deployed knowledge

BPR was first developed by Hammer in a short paper published in the *Harvard Business Review* (1990). This was followed by a book co-authored with James Champy, *Reengineering the Corporation* (1993). BPR is a very simple idea, but very difficult to apply in practice; all that one has to do is answer the question 'if we could start the company again what would be the best way to organise the production process?' It is basically an attempt to return to first principles by identifying and removing forms of organisational, tooling (machines) and geographical inertia. This book became one of the most popular management guides and it encouraged most of the Fortune 500 companies to engage in forms of BPR. The Japanese version of Hammer and Champy's book sold 250,000 copies in its first three months (Micklethwait and Wooldridge 1997: 29). BPR was rapidly adopted by management consultants as a standard technique, but the technique was distorted and was rapidly equated with downsizing. BPR, however, is about efficiency and productivity, not just about redundancy. The majority of BPR projects failed because the focus was on downsizing rather than on reengineering. The reason for this failure can be explained by a sociology of translation; it is much easier for consultants to sell BPR as a technique for downsizing an organisation rather than as a reengineering technique. Downsizing is a relatively simple and quick process, reengineering is a complex and time consuming activity.

Hammer and Champy lost control of their management technique even though they both wrote articles, books and presented lectures in an attempt to distance BPR from downsizing. However, the translations and distortions of their ideas were impossible to stop with the result that BPR was rapidly identified not with innovative management and technologies but with older more 'mature' industrial sectors (Hamel and Prahalad 1994). By 1995, BPR had become big business. The commercial success of BPR for the large consultancy companies was associated with newspaper and magazine articles that associated the technique with company turnarounds, profitability increases and downsizing. The BPR media driven management fad died around 1996; the technique is still extant but marketed under other names. Marketing, branding and the spatial diffusion of a management theory is central to the successful development of a management fad. In many cases consultants are not selling better ways of organising production, but different ways.

Frederick Taylor and Fordism

The history of economic geography over the last twenty years has been closely associated with attempts to explore what has been termed the crisis of Fordism and the development of Post-Fordism (Amin 1994). There are a whole series of problems with such bipolar descriptions of economic change or transition (Thrift 1989). This literature tends to ignore the role played by key individuals and networks of individuals in creating organisational and management models which ultimately impact on the spatial organisation of production. Alterations in the economic system are the result of a whole series of unrelated and related changes made by independent private and public sector organisations. Central to these alterations are the activities of management gurus and management consultants. The other actors in the process are the media and trade and professional associations that simplify, popularise and spread management ideas. It is these organisations which can create a climate of change not as a result of necessity but as the consequence of media created fashions. Fordism in the geographical literature is closely associated with the work of F. W. Taylor, an American consulting engineer who was also both a management guru and management consultant. It is these aspects of Taylor's career that are neglected in the geographical literature. To understand the history of Fordism or Taylorism requires an understanding of the ways in which Taylor's ideas were created and spread amongst business organisations, both in the USA and Europe. Three issues need to be examined: Taylor's theory of scientific management; Taylor as management guru and consultant; and the spatial diffusion and translation of Taylor's management theory.

Taylor and scientific management

Taylor's contribution to management theory was both a set of techniques and an ideology. It is the techniques that have been explored in geography (Cooke and Morgan 1998: 47–53). The ideology of Taylorism concerns a search for efficiency; for the best way of undertaking a task. It was this ideology which spread throughout the USA encouraging manufacturers to experiment with alternative working practices. Taylor provided American capitalism with a theory of management which was absent from Europe until the 1920s and from some countries until the 1940s. As a management guru it is this ideology which is most important as it provided opportunities for a whole series of management consultants to adopt some of Taylor's ideas. These consultants, however, modified Taylor's ideas to make them more acceptable to client companies. It is, thus, correct to identify Taylor's ideological impact on American capitalism, but incorrect to identify a single set of techniques; there was not just one type of Taylorism, but a whole series of different types which were created by Taylor's 'band of disciples' or Taylorites.

Taylor's contribution to management theory rests on the combination of four related principles (Braverman 1974: 85–123; Guillen 1994). First, time and motion studies demystified skills by standardising work tools and practices and by dividing production processes into their simplest constituent tasks. Second, each task should be undertaken by employees best suited to the task: by the cheapest worker. Third, foremen (*sic*) and an incentive system based on differential rates ensure that the 'scientifically determined' task is matched to the right employee. Finally, the execution of work is separated from its conception. Tacit knowledge is identified and codified and transferred from the shop floor to a planning department. Taylor was one of a group developing revised ways of managing shop floor activity, and Taylor acknowledged ideas which he had taken from other peoples' work (Urwick and Brech 1945: 33).

To understand the development of Taylor's ideas it is necessary to situate them in his life history. Taylor was born in 1856 and came from a middle class background. He was well educated and obtained a place to read law at Harvard. Problems with his eyesight meant that he had to find an alternative profession that would not involve long hours reading. In 1875, he started an apprenticeship as a machinist and turner. In 1878, he joined Midvale Steel Works as a labourer in the machine shop. Taylor's principles developed from direct experience of working as a machinist in a steel works. As a machinist he had to adopt existing shop-floor traditions, or work geographies, which restricted output. He was promoted to the position of gang boss and later to the post of foreman. Once on the side of management he became interested in ensuring that every worker under him did a fair day's work. This made him very unpopular and he received threats of personal violence. His four principles developed from his shop floor experience. Taylor's ideas were not readily accepted and many American managers considered that his ideas needed to be developed and refined. Taylor left Midvale in 1890 having climbed from the position of labourer to chief engineer and at the same time obtained a Master's degree in Engineering via evening study.

Taylor as management guru and consultant

From 1890 to 1893 Taylor was employed by the Manufacturing Investment Company to manage two new paper mills in Madison, Maine and Appleton, Wisconsin. This would provide Taylor with an opportunity to apply his management approach to plants that had not established working practices. It would also provide an opportunity to demonstrate the effectiveness of his techniques. This attempt, however, failed for a variety of external factors beyond his control as well as a variety of people related factors (See Kakar 1970: 124, for a discussion of this problem). The workers resented and mistrusted a city man with no experience of running a paper mill. In 1893 Taylor resigned from this company and established himself as a consulting engineer.

In September 1893, Taylor went to Boston to work for his first client, the Simonds Rolling Machine Company in Fitchburg, Massachusetts, a manufacturer of steel balls for bicycle bearings (Kakar 1970: 131). The company was aware of Taylor's work as he had acquired some company stock in exchange for surrendering some machinery patents he held. Taylor's reputation as a consultant cannot be separated from the social capital he obtained from his inventions. This small project involved the introduction of a new accountancy system. In 1894, Taylor's new career as a consultant charging $35 a day for his services was severely tested when he was asked to reorganise the machine shop of Cramp's Shipyard in Philadelphia (Copley 1923). This project failed as the management was not prepared to sack employees who were not prepared to change, and Taylor was 'unskilled in . . . diplomacy or compromise' (Kakar 1970). The early years of Taylor's career as a consultant were not successful. Nevertheless, Taylor's professional reputation amongst other consulting engineers was developing as he began to present papers before the *American Society of Mechanical Engineers*

(ASME). He was beginning to establish a reputation amongst potential clients as well as other consultants.

In 1898 Taylor gave up consultancy and took up full time employment with the Bethlehem Steel Company. Taylor's work at Bethlehem was to introduce his management principles into the yard and machine shop (Urwick and Brech 1945). Like his other attempts to apply his theories this failed and in 1901 Taylor was sacked. At Bethlehem, however, Taylor was able to develop one of his other interests – technological innovation. In 1898 he discovered a method for heat treating tool steel so that machine tools were able to run at unprecedented speeds without any apparent loss of cutting edge. Taylor sold the rights to this idea to Bethlehem. In 1901, at the age of 45, Taylor retired from consultancy as well as everyday contact with the shop-floor. The rest of his life was spent promulgating his principles of scientific management.

The development of Taylor's reputation as a management theorist is closely associated with his development of high speed tool steel. The reception by European engineers of Taylor's first three papers highlights the importance of his high speed steel work. In 1895 Taylor read his first paper to the ASME (Urwick and Brech 1953: 88–107). This paper on the piece-rate system was only documented by one of the three leading British engineering periodicals. His second paper on *Shop Management* (1903) in which Taylor developed the basis of scientific management was ignored by the European technical press. The development of high speed steel enhanced Taylor's reputation amongst engineers and finally led to his election in 1905 as President of the ASME. His presidential address entitled *On the Art of Cutting Metals* was published in 1907 and it is this work which established Taylor's reputation in Europe. This paper drew attention to Taylor's earlier work on *Shop Management* which was only now translated into European Languages. Taylor's work as an engineer and his development of high speed steel cannot be separated from the development of his career as a management consultant and guru. His reputation as an innovative engineer gave his management ideas credibility in the eyes of fellow members of the ASME. Nevertheless, in 1910 the ASME decided that a paper submitted by Taylor should be rejected as engineers would not be interested in its content and that it contained nothing new (Haber 1964: 18). This paper was published in 1911 under the title *The Principles of Scientific Management*. It is this work which played an important role in the development of Fordist production systems.

The spatial diffusion of Taylor's scientific management

As a management consultant Taylor was notoriously ineffective. None of the companies he worked for or with successfully introduced scientific management. According to Nadworny (1955), by 1915, when Taylor died, the scientific managers and their associates had introduced their techniques into 140 establishments. In this analysis only 63,000 employees were identified as working with Taylor's methods. Similarly Nelson (1974: 490) attempted to assess the importance of Taylor's ideas and was only able to identify 48 establishments that had introduced his ideas between 1901 and 1917. Littler (1982) suggests that these studies should not be used to show that Taylorism had a limited influence in the USA. Rather that Taylor and his close associates (Taylorites) were only interested in introducing 'pure Taylorism' to manufacturing plants. Even in 1923, the Taylorites were concerned with their failure to widely introduce full Taylorism, and that one reason for this failure was 'the pressure of unfair competition from illegitimate relatives [consultants]' (Nadworny 1955: 142). In many respects full Taylorism was unpopular as it contained an implicit depiction of the human being as machine (Schon 1983: 237) and it also had a history of conflict between managers and employees.

To understand the diffusion of Taylor's ideas involves recognition of three points. First, Taylor became the visible figurehead of a widely dispersed movement that was concerned with increasing the efficiency of production in manufacturing plants. Second, most applications of his ideas were achieved through consultancy projects undertaken by Taylor's close circle of disciples and friends. Finally, the wider diffusion of some of Taylor's ideas and their translation into 'user-friendly' management recipes was undertaken by a group of neo-Taylorists that were not concerned with promulgating pure Taylorism.

The first and second points can be explored together. During the time he spent working on the shopfloor Taylor developed a group of close followers many of whom became practising consultants using Taylor's theory of scientific management. There are two important individuals, Henry Lawrence Gantt (1861–1919) and Frank Bunker Gilbreth (1868–1924). Some commentators even describe Taylor, Gantt and Gilbreth as the 'trinity' of scientific management (Urwick and Brech 1953: 103). Gantt refined and developed aspects of Taylor's ideas especially in relation to the Gantt chart, a chart designed to permit the pre-planning of production on a daily basis. Frank and Lillian Gilbreth improved the methodology of time-and-motion study by introducing the chroncyclograph, or motion picture camera designed to identify the actions required to undertake any task. The 'disciples' kept in touch with Taylor and discussed cases with him. Gantt and Gilbreth developed scientific management to such an extent that they both argued with Taylor who considered that they moved away from full scientific management to concentrate on only one of the four principles.

In 1911, Taylor's disciples also organised themselves informally into an exclusive club called the Taylor Society. The purpose of this society was to ensure that Taylor's work remained highly visible amongst American managers as well as to co-ordinate Taylor's followers. Organised labour disliked The Taylor Society as it believed like Taylor that scientific management determined wages, hours and working conditions rather than trade union negotiations. The history of Taylorism is, until the late 1930s, a history of management in the USA and Germany and of the activities of a group of Taylor disciples and neo-Taylorites. Taylor's ideas were ignored and neglected by British managers for a variety of reasons: the scarcity of engineers; the conservatism of management and the opposition of trade unions. Scientific management was adopted by German manufacturers like Bosch, AEG, Siemens, Krupp and Lowe shortly before or during the First World War (Guillen 1994: 119). German companies employed American consultants to transfer Taylor's ideas to Germany. Thus, in 1914 Gilbreth was employed by the Auer Electric Company (AEC) in Berlin.

Taylor's ideas were not well received in Britain, yet it is ironic that the only visit Taylor made to Europe after developing scientific management was to Birmingham (UK). In July 1910 Taylor, Gilbreth and Gantt were amongst the delegates attending a Joint Meeting of the Institute of Mechanical Engineers and the ASME held in the Midland Institute, Birmingham. The subject of the meeting was High Speed Tools and everyone expected Taylor to speak on this subject. However, when Taylor rose to address the meeting it was on the subject of scientific management and his belief that 'it is possible to double the output of the men and the machines just as they stand now, and I believe the same is true throughout this country' (Taylor, quoted in Urwick and Brech 1953: 94). In reply to Taylor's six and a half page address the chair of the meeting noted that Taylor's paper 'did not deal with that subject [high speed tools] and he would very much have valued any remarks Dr Taylor might have made on tools and machines' (Urwick and Brech 1953: 96). From this it appears that Taylor had failed to redefine (problematise) the role of engineers in Britain from a concern with machines to a concern with management issues.

Up to 1910 Scientific Management was an approach to shop floor management which was known to a small circle of American engineers. During the years when Taylor was developing and applying his system directly on to the shop floor he was uninterested in publicity. Like all consultants Taylor, the Taylorites and neo-Taylorites obtained clients by word of mouth and especially by the movement of managers between companies (Bryson 1997). Two events occurred which led to Scientific Management's problematisation into the only solution to the problems of low productivity, low wages and unemployment. Adopting Scientific Management would provide the employer with 'a lower labour cost, so that they might be able more successfully to compete at home and abroad' (Taylor's Birmingham Address, cited in Urwick and Brech 1953: 94). Both events involved the acceptance or translation by the business media of Scientific Management into the perceived solution for the problems of American Industry.

The first event was the Eastern Rate Case (1910–11), a famous court case between the railways and business associations from all over the East Coast. In 1910, the railways raised employees wages, but immediately asked the Interstate Commerce Commission to endorse an increase in freight charges. Business groups combined together to try to prevent such an increase and asked Louis D. Brandeis, a popular 'people lawyer' to represent them (Kakar 1970: 174). Brandeis argued that the railways had not proved their case for an increase in rates and that even if they had proved their case the solution did not lie in increasing the rate, but in introducing scientific management. In court he argued that: 'We will show you . . . that these principles [scientific management] are applicable to practically all departments of all businesses, and that the estimate which has been made that in the railroad operation of this country an economy of one million dollars a day is possible is by no means extravagant' (Kakar 1970: 176). The 'million dollar a day savings' captured the imagination of the media and Taylor was hunted by the press. Hundreds of newspaper and magazine articles were published and America entered a craze for business efficiency. To Haber the efficiency craze 'hit like a flash flood, at first covering the entire landscape' (1964: 52). There is a sociology of translation which needs to be authored concerning this period of American management development. Taylor had been provided with a media platform to expound his views, but the media's attempt to problematise Taylor as the management guru did not go unchallenged as many engineers began to resent 'what appeared to be Taylor's virtual monopoly of the subject of management' (Kakar 1970: 177).

The second event was the decision in 1909 by the Ordnance Department of the United States Army to adopt the Taylor system for use in the manufacturing arsenals at Watertown (Aitken 1960). The Chief of Ordnance, General William Crozier, was requested by a congressional committee to compare the costs at Watertown with prices paid to firms in the private sector. Such 'benchmarking' led Crozier to Bethlehem Steel Works and the Midvale Steel Company. Both these companies were able to undercut Watertown and still make a profit and both companies had employed F. W. Taylor. The decision to employ Taylor at Watertown considerably raised the profile of scientific management. However, the Watertown experience ended badly with a major strike and the decision by the House of Representatives to outlaw time and motion study in the arsenals. This was a direct consequence of trade union lobbying and organised labour's success in removing Taylorism from the arsenal undermined the credibility of Taylor's system. Taylor was very much aware that the Watertown crisis would undermine the scientific management movement and he even urged Crozier to 'get the press of the country to take it up [the Watertown case], either by paying them money or without doing so' (Taylor cited in Aitken 1960: 231).

The key problem with Taylorism is that it rapidly became associated with a whole series of negative connotations. Employees were concerned with its explicit anti-trade union stance, and employers were reluctant to introduce all of Taylor's four principles as this involved a complete management overhaul. Consultancy companies drew upon Taylor's ideas, but developed related systems (neo-Taylorite) that enabled them to uncouple 'scientific management' from the negative stigma associated with Taylorism. During the 1930s scientific management began to spread in the USA and Europe as a result of the activities of neo-Taylorite consultants. One of the most important of these consultancy companies was established by Charles Eugene Bedaux (Littler 1982: 99–140; Lash and Urry 1993: 180–1). According to Guillen, Bedaux's 'impact in the United States as well as worldwide justifies retrieving him from the dustbin of history' (1994: 56).

From Taylorism to the Bedaux system

The history of Scientific Management in Britain up to 1940 is largely a history of the Bedaux company. Unlike Taylor, Bedaux was not concerned with justifying his ideas academically rather he was concerned with selling them to engineers and managers (Littler 1982: 107). Bedaux was born in France, but between the ages of 20 and 41 lived in the USA. He worked in a furniture store in Michigan, and it was here that he developed his management system. During the 1930s he returned to France where he purchased a chateau and entertained members of the National Socialist Labour Front (Downs 1990: 57). The Bedaux system developed from Taylor, but unlike Taylor tried to determine the relationship between worker fatigue and the strain imposed by any given type of work.

At its simplest the Bedaux system was a method which standardised the amount of time spent on a task, taking into consideration fatigue and providing for a rest allowance. The Bedaux system like many consultancy techniques tried to obscure or problematise its methods so that workers would be unable to convert actual work undertaken into Bedaux units. Unlike Taylor's system the Bedaux system was designed to be confusing as this worked to the advantage of employers and to the disadvantage of workers and trade unions. A Bedaux unit (BU) comprised a minute of work and relaxation balanced in proportions suitable to a particular task. Sixty BUs was the minimum acceptable hourly effort required by an employee. Unlike Taylor's system, the Bedaux system could only be implemented by Bedaux's own staff of time-study engineers. These men entered the firms as 'neutral' (*sic*) observers and calculated the BU for each task. The Bedaux system was able to identify underperforming workers and trade unions correctly suspected that employers were using the system 'as an instrument for the Darwinian weeding of the workforce' (Downs 1990: 59). Employers using this system publicly displayed individual BU attainments, but with employees highlighted in red with BUs below 60. According to Guillen (1994: 57) over 700 American Companies, employing about 675,000 workers, employed the Bedaux company between 1918 and 1942. In 1937 of the 1,100 or so firms using the Bedaux system, over 225 were British and 150 French (Thurley and Wirdenuis 1973; also see Littler 1982: 112–15). In fact the best cinematic deconstruction of Fordist factory work, Chaplin's Modern Times (1936) includes a scene in which a Mr Beddoes tries to persuade the factory owner to purchase a machine which will feed workers while they continue working (Guillen 1994: 57). Chaplin's depiction of Bedaux as Beddoes highlights the newspaper, trade union and employee distrust of this consultancy company. The BU could be made into a figure of fun, but it was not a laughing matter for employees working on Fordist assembly lines.

Like Taylor, Bedaux tried to problematise his system so that it was perceived to be the only user friendly version of scientific management available. Both the history and geography of the Bedaux company urgently need to be constructed. The Bedaux system was translated by the company and its clients into a perceived panacea for dealing with productivity problems. British trade unions, however, targeted the Bedaux system as a management technique which should be resisted at all cost. In the 1930s the attempt to introduce the full Bedaux system resulted in major strikes in British companies: Rover car works at Coventry (Downs 1990), and the Wolsey Hosiery company in Leicester (Guillen 1994: 224). The problem with the Bedaux system was that the full system automatically removed pay inequality between men and women and this was something which trade unions and employers were reluctant to permit. To Downs the survival of the majority of car companies in the West Midlands depended on 'a wage hierarchy which held women's pay at half that of men' (1990: 72).

Discussion and conclusion

The history of capitalist production is one of new management panaceas appearing to solve old problems. It is also a history of the professionalisation of management, of the development and evolution of a science of management, and the growth of management consultancy (Chandler 1977: 466). Consultants have played a central role in the transfer of management theory and techniques between companies and countries. They operate as bridges between competing companies as well as countries. The geography of Fordism was constructed around the transfer and translation of Taylor's ideas; and it is important to note that the processes involved have played, and are still playing, an important role in economic, workplace and employment change.

Taylorism introduced order and 'scientific rationality' into business enterprises, but at the same time it challenged the individuality of workers. The human relations movement developed in the late 1920s and 1930s at the Harvard Business School as a reaction against Taylorism (Schon 1983: 239). Human relation's proponents explored the monotony of work, absenteeism, employee turnover and low morale arguing that all of these reduced productivity (Guillen 1994: 17–18). The history of the human resource movement mirrors that of Taylorism. To understand the similarities between these two movements is to explore the ways in which knowledge is created and disseminated. What ties these movements together is an understanding that in the social sciences of which management science is one as well as within business organisations 'meaning is not discovered; it is imposed' (Gillespie 1993: 4).

There are a number of superb histories of the development of management knowledge which are largely unknown within the geographical literature. The best of these is Gillespie's study of the development of human resource management through the well-documented Hawthorne experiments directed by Elton Mayo between 1924 and 1933. Two important points come from this study which are crucial for understanding the spatial division and diffusion of management knowledge. First, to understand the Hawthorne experiments it is necessary to understand 'how a body of complex and uncertain data, tenuous hypotheses, and conflicting interpretations was transformed into a relatively dependable body of social scientific knowledge' (Gillespie 1993: 5). What occurred at Hawthorne was the construction and dissemination of an official interpretation of a set of difficult-to-interpret experiments into the relationship between productivity and conditions of work. The way in which this knowledge was produced was influenced and shaped by a complex network of institutions and actors ranging from the Harvard Business School, the National Research Council, The Rockefeller Foundation and Western Electric, the owners of the Hawthorne site. The key actor in this process was Elton Mayo who enabled the research to discount the interpretations provided by employees at the Hawthorne works and replace such interpretations with those which existed in the established academic literature.

Second, the actors involved in this process were part of a network of academics and companies which provided the interpretation of the Hawthorne experiment with credibility as well as ensuring that the official version was widely disseminated. Thus, Gillespie notes that a 'network of academics and corporate interests crystallised around Mayo, which gave him the power to develop an authoritative interpretation of the experiments at Hawthorne and ensured that the findings would exist not just at Hawthorne, but also at the AT&T head office in New York, at the Harvard Business School, and then rapidly in management and social science circles' (Gillespie 1993: 265–6). The same type of network, but composed of different actors, crystallised around Taylor.

Both of these points emphasise the social construction of knowledge in general and of management knowledge in particular. The spread of Taylor's ideas in the USA and to Europe and the translation of his ideas by neo-Taylorite consultants all depended upon two factors. First the life histories of the individuals involved. Thus, Taylor acquired clients as well as disciples as a result of established friendship networks; social relationships bind individuals together and it is through these bindings that knowledge and information flow. Second, the business media, or more correctly business journalists, play an extremely important role in constructing the reputations of management theorists or theories. Thus, Taylorism as well as BPR developed on the back of significant media exposure.

To understand the geography of Taylorism and of Fordism requires an understanding of the social, cultural and political constructions of the knowledge(s) which lie behind these production systems. The same is true for all production systems. The difference between F. W. Taylor and the consultancy cadre is one of scale. There are more consultants and more competing management theorists and theories. The escalation in the number of competing theories has resulted in a transformation in the behaviour of large and multinational corporations which has largely gone unnoticed. In 1994, as part of an ESRC project, I was involved in 40 in-depth qualitative interviews with large companies. One of the most memorable aspects of these interviews was the emphasis placed on change. This emphasis on change is partly caused by intensified competition, but is also the consequence of conflicting and competing management theories. There is always another way of running and managing a company, and of course in reality perhaps no single right way.

Note

1 This chapter is part of a larger project consisting of four related papers. The first paper (Bryson 1997) explores the flows of management knowledge in to large corporations. The second (Bryson and Daniels 1998c) constructs a recipe-based analysis of management knowledge and suggests that such knowledge is localised by the activities and translations of management consultants. This chapter is the third paper in the sequence of analysis. The final paper explores the consumption of management knowledge by client companies.

11 The free and the unfree

'Emerging markets', the Heritage Foundation and the 'Index of Economic Freedom'

James Sidaway and Michael Pryke

The purpose of Newspeak was not only to provide a medium of expression for the world-view and mental habits proper to the devotees of Ingsoc [as the official ideology is known], but to make all other modes of thought impossible. It was intended that when Newspeak had been adopted once and for all and Oldspeak forgotten, a heretical thought – that is a thought diverging from the principles of Ingsoc – should be literally unthinkable, at least so far as thought is dependent on words. Its vocabulary was so constructed as to give exact and often very subtle expression to every meaning that a Party member could properly wish to express, while excluding all other meanings and also the possibility of arriving at them by indirect methods. This was done partly by the invention of new words, but chiefly by eliminating undesirable words and by stripping such words as remained of unorthodox meanings, and so far as possible of all secondary meanings whatever. To give a single example. The word free still existed in Newspeak, but it could only be used in such statements as 'This dog is free from lice' or 'This field is free from weeds'. It could not be used in its old sense of 'politically free' or 'intellectually free', since political and intellectual freedom no longer existed even as concepts, and were therefore of necessity nameless.

(George Orwell 1954: 241–2 (originally published 1949)).

Introduction: following freedom

'Developing countries have entered a new age of globalized private capital.' So proclaims the World Bank in a recent report *Private Capital Flows to Developing Countries* (1997: 9). The proclamation is relevant to this chapter as it provokes questions about the processes involved, and language used, to chart the knowledge maps that help to steer the rising (and turbulent) flows of private capital into so-called developing countries. In our approach to this issue we look first at how one influential bundle of knowledge, represented in the workings of the Washington DC based Heritage Foundation and its *Index of Economic Freedom*, seeks to help private sector investors make sense of the investment opportunities opening up in developing countries. The accompanying aim of such knowledge production is, needless to say, to nudge these same countries into accepting what, for the likes of the Heritage Foundation, is now an 'undeniable truth': that freedom, including freedom from poverty, lies in engendering efficient market mechanisms. As the Foundation's website boldly states, 'Poverty is caused largely by ill-conceived and repressive economic policies.' Freedom and free markets are one and the same thing.

Our route then takes us through the rapid 'enfolding' of geo-politics into geo-economics which has followed the end of the cold war, and which has produced today a *geo-political-economic* discourse regulated by free market international institutions such as the World Bank and the World Trade Organization. The final section explores the scripting of important elements of this discourse.

Yet this is not an abstract process removed from everyday life or of import only to 'developing countries'. More and more people in the West (the so called 'industrialized countries' in official World Bank literature) are being drawn into and implicated in the outcome of this move in a number of rapidly evolving ways. For example, the growth of the 'new age of global capital' since the 1980s has witnessed and prescribed the faster integration of financial markets. This has enabled institutional investors, such as pension funds and mutual funds, to invest a growing proportion of their investments internationally. Between 1989 and 1994, for instance, total international investments by pension funds rose from US$302 billion to US$790 billion. In the USA, open-ended mutual funds increased their exposure to 'emerging markets' (as developing countries have come to be renamed in financial investment speak), from US$1.5 billion in 1990 to US$35 billion in 1995, as the number of emerging market country funds rose from 19 to 505, and the number of global funds jumped from 9 to 773 over the same period (World Bank 1997: 14, 106). The same trends are apparent in most industrial countries and growth potential is even greater. Feeding this expansion – and this is where more of us are being drawn in – is the demographic shift under way in industrial countries. Amongst other things, the ageing of the populations of these countries has spurred governments to begin to deregulate both pension *provision* – notable through privatization – and pension (and other savings related) *investment* to allow allocation of investment flows across expanding swathes of the globe. How these new spaces of investment opportunities are *made knowable*, to both institutional investors and those buying into pensions and savings schemes, is the subject of the remainder of this chapter. As will become clear, the appeal to a particular interpretation of *freedom* – with its slippage between self-reliance and equality, and the free market (Love Brown 1997: 99) – has been central in presenting the potential investment geography of the new age of global capital.

'Where in the world?'

For just $2.95, anyone in the US with the inclination and access to a moderate-sized newsagent or bookstore can take out a copy of the monthly magazine *Individual Investor*. Of course it is also available on-line or by postal subscription.[1]

> *Individual Investor* is one of an array of popular financial-media that have grown along with the greater prominence of financial capital in the 1980s and 1990s and the American bull-market share boom of the 1990s. Reflecting on the latter in April 1998, *Newsweek* (which describes itself at *The International Newsmagazine*, yet is very obviously marked by its North American origins) carried a cover story about 'Why we're all married to the market', noting how: 'We [Americans] have become a nation of stock junkies – splurging on mutual funds, covering market tables and studying personal finance columns.'
>
> (Samuelson 1998: 33)

A few pages further on *Newsweek* reminded readers that Wall Street has perme-ated daily life as never before. Bars play CNBC; truckers can talk about the Hang Seng Index. Online investing with companies like E* Trade is eating into traditional brokerage, and it has created great fortunes, at least on paper. The market has stirred a new frenzy of anxiety and envy. After all, as any investor knows, it isn't about how much you've made, but how much you *might* have made. More players – including more women, minorities and young investors coming into the market – simply means more ways to miss out on 'Warren Buffet-size profits' (Leland 1998, 37–8). This expansion of interest is the context in which *Individual Investor* becomes a profitable prospect for its publishers.[2]

It is the April 1998 issue of *Individual Investor* (the same month that the Newsweek story on the American bull market appeared) that we choose to concentrate upon here. Under an Editorial entitled 'Following Freedom' the smiling, smartly dressed Jonathan Steinberg (1998: 8) who is the Chairman and Editor-in-Chief of *Individual Investor* notes how his magazine will this month be dedicated to what he called 'the perennial question, Where in the world should I invest now?'. What this turns out to be, under the banner of the cover-story 'Freedom Markets, Open Economies, Rich Returns', is a fifteen-page celebration and explanation of uses of something called 'the *Index of Economic Freedom*, a comparative guide to 160 economies, published for the past four years by the Heritage Foundation in Washington DC, and *The Wall Street Journal*' (Vestner 1998: 47).

According to *Individual Investor*, the Index provides 'a chart to help them ["inter-national investors"] navigate the choppy waters of international investing'. It is held to be particularly useful to those 'seeking the promise of modern-day El Dorados, awaiting discovery and about to leap from the third world to the second, and from the second to the first' (Schlegel 1998: 61).

The unfamiliar form of reference here to 'first', 'second' and 'third' worlds is inter-esting. The conventional model of 'three worlds' was developed after the Second World War and followed the end of most European overseas empires, and the rise of both Soviet and American power and conflicting American and Russian (Soviet) pretences to global leadership. For more than 40 years this model was a common way of interpreting the world's great ideological and socio-economic divides (whilst at the same time rendering these divides meaningful and 'real'). 'First world' referred to the West (led by Washington), 'second world' to the Eastern bloc communist countries led by Moscow and 'third world' to the rest, that is the former colonies or quasi-colonies many of which sought a non-aligned (neither East nor West) position. Whilst this enduring and arguably rather simplistic tripartite division of global space has been thoroughly disrupted by complex economic and political shifts in the last decade or so, Schlegel's apparent misreading of the standard 'three worlds' model is notable for at least one reason. It is a reading which revives an idea frequently present in certain versions of development and one codified in the West (especially America) during the Cold War. Reworking prior colonial discourses – in which the colonial power was held to be an agent of 'civilisation' and 'progress' for benighted and 'backward' 'races' of Africa and Asia – American policy makers (most famously Walter Rostow who was assistant national security adviser to the Kennedy Administration) developed 'modernization theories' which purported to describe and prescribe *the* path from 'underdevelopment' to 'modern', 'developed' status. Simple and immensely appealing, both to Western policy makers concerned to show that alliance with the USSR and communism was a flawed path to 'modernity' and to many third world elites, modernization became the core ideology of development. Moreover, as Rist (1997: 103) notes, the theory of modernization as codified by Rostow 'has not ceased to nourish the hopes as well as the illusions of the rulers of both North and South'. Rist adds that it now forms part of the framework through which post-communist countries and the Peoples Republic of China codify and present 'development' policy. Approaches and policies do depart and divide, however, in terms of key practical and ideological questions, such as the role of the state and the place of regulation and capital controls. Indeed it is evident that the identification of the most direct and appropriate path to 'modernity' and the sense of what combination of social, economic and cultural forms such modernity may amount to, form contested themes.

Whilst the eclipse of the communist model of state-led accumulation and comprehensive planning has imposed a certain limit on the apparent range of such trajectories (i.e. the orthodox communist model is no longer seen as a viable reference frame or possibility), questions around what is the most effective way to regulate and sustain capitalist accumulation continue to be prominent. These frequently take the form of contrasting policy agendas. So the form of capitalism and policies formulated in, for example, the Peoples Republic of China look rather different to those of Russia, New Zealand, Brazil or South Africa. Such differences are not absolute nor are they reducible to 'cultural' variations or 'differences in level' of 'development'. They are formed in part out of an ideological arena. That is they embody different policies concerning the stuff of economic theory and policy; issues such as the roles of the state, currency controls, fiscal and monetary policy and so on. The accelerated turbulence in the world economy (most notably in South-East Asia and other places designated as 'emerging markets') during the late 1990s has reinforced the sense of such differences. Whilst the IMF and World Bank and the United States seek further deregulation as a response to a crisis whose main roots they locate in the countries affected, Wade and Veneroso note that in response

> [many] Asian countries are no longer prepared to accept that there is only one game in town: the one that gives banks, hedge funds and pension funds of the West freedom to enter and exit their markets at will. Their moves are finding echoes in Latin America, Russia, Europe and elsewhere. China's stock has risen. We now have a live case-study of a sharp alternative to the IMF model: Malaysia. If successful, it will have a powerful demonstration effect.
>
> (Wade and Veneroso 1998: 40)

In such a context the form of the model celebrated in Individual Investor can be seen to represent an unusually pure form of the so-called 'neo-liberalism' – one that has been hegemonic (in various forms)[3] at the World Bank and IMF and in London, New York and other Western centres of intellectual and of economic power. Neo-liberalism advocates a (re)configuration of the nexus of relations[4] between the social powers of 'state' and 'market' in such a way that the sense of the role of the 'market' is maximized. Amongst other things neo-liberal policy amounts to freely convertible currencies; complete openness to foreign capital and its repatriation; the maximum commodification of all aspects of society; land and property markets; minimum regulation in labour markets, marketization of health, education, social services – the list is almost endless and by now very familiar. Significantly, neo-liberalism is a global(izing) project, institutionally embedded in transnational corporations, banks and the international institutions of the World Bank and IMF. Indeed part of the power of neo-liberal visions is in the way that they mobilize a certain notion of the inevitability of 'globalization' and a related sense of progress. Such pretensions to universality and inevitability disguise the way that the origins of neo-liberalism are geographically and historically specific products of the West. The reiteration of Freedom Markets in *Individual Investor* has some direct affinities with the prior colonial discourse of progress and Western leadership. According to *Individual Investor* there is a *linear progression* from 'undiscovered [markets]', through 'emerging [markets]' to what it calls 'grown-up economies'. This is firmly within a [colonial-derived and Cold War] knowledge-vision that the West is the best and that it shows the path to the future – of 'progress', 'development' or 'civilization' – to the rest of the world. Yet it is also symptomatic of something more recent: that is the diminishing credibility of the post-Second World War geo-political division of the world into third (developing), second (industrial-communist) and first (industrial-developed). Therefore, beyond the Cold War, in a 'New World Order', the old poles of North and South and East and West seem passé. Places once usually designated as belonging to a third or second world now achieve the status of 'emerging market' in the latest order of classification. And as investors seek the next best investment ('the perennial question' exercising the minds of those with capital to spare) a geo-political imagination of East-West/South (three worlds) is being overwritten by a more complex geo-economic script.

What this amounts to is a new or revised way of classifying countries. It is a process in which the importance of a political designation (is the country pro-Western, pro-communist or non-aligned?) has been replaced by a classification which records first and foremost a country's investment potential. Certainly a variety of geo-political designations persist. Yet the classification of countries according to a characterization of their investment potential has grown greatly during the 1990s and seems set to continue. So alongside the idea of 'failed states' requiring some form of Western intervention (Bosnia, Somalia, etc.) and places projected as 'rogue states', requiring Western surveillance and discipline (Iraq, Libya, North Korea, Yugoslavia, etc.), are places designated as 'emerging markets': the next best investment opportunity.

Therefore the authority granted by *Individual Investor* to the *Index of Economic Freedom* represents one of the ways that in the contemporary world geo-politics is being folded into geo-economics. A process of enfolding that forms part of a neo-liberal vision of what the world is and could be; a vision that is also connected to, in Jacques Derrida's (1994: 82) words: 'mechanisms [that] are starving or driving to despair a large portion of humanity. They tend thus to exclude it simultaneously from the very market that this logic nevertheless seeks to extend.'

Geo-politics contra geo-economics?

The idea of something called 'geo-politics' has had a complex and varied history since it was coined a century ago by Rudolf Kjellen, a conservative Swedish academic and politician. Tracing this history, O' Tuathail explains that:

Back in the early years of the twentieth century, Kjellen and other imperialist thinkers understood geopolitics as that part of Western imperial knowledge that dealt with the relationship between the physical earth and politics. Associated later with the notorious Nazi foreign policy goal of Lebensraum (the pursuit of more 'living space' for the German nation), the term fell out of favour with many writers and commentators after World War II . . . [But] During the later years of the Cold War, geopolitics was used to describe the global contest between the Soviet Union and the United States for influence and control over the states and strategic resources of the world. Former US Secretary of State Henry Kissinger almost single-handedly helped to revive the term in the 1970s by using it as a synonym for the superpower game of balance-of-power politics played out across the global political map.

(O'Tuathail 1998: 1)

Knowledges, practices and expertises that had in the 1950's and 1960's been grouped under the designations of 'International Relations', 'Strategic Studies', 'Foreign Affairs' or 'Area (that is Asian, Slavic, Latin American and so on) Studies' were thereby re-inscribed as 'geopolitics' by Kissinger (see Hepple 1986) and other American 'statesmen' such as Zbigniew Brzezinski (see Sidaway 1998). Such Cold War geo-political visions were always entwined with and constructed from cultural, social and economic references. Indeed the entire Cold War confrontation was in part a clash of social systems (capitalism versus statised socialism) with their associated systems of ideological, political and cultural systems of difference. Although conceptions of 'economic' difference were present within much geo-politics – in references to free enterprise, strategic resources and an 'American way of life'; to private homes, mobility and mass-consumption contrasted with centrally administered and statized 'communism', for example – American Cold War geo-politics essentially foregrounded a version of space as military–strategic distance that American Cold War geo-politics foregrounded. In this vision places were classified according to their military security value, naturalized and legitimated through a code of containment. Writing under the name of 'Mr. X', the American diplomat George Kennan had argued that to 'contain' what he saw as the inherently expansionist USSR, American policy must take the form of: 'adroit and vigilant application of military counter-force at a series of constantly shifting geographical and political points, corresponding to the shifts and manoeuvres of Soviet policy' (Kennan [Mr. X] 1947: 575).

After a wave of successful Third World revolutions in subsequent decades, the spectre of containment failing and communism spreading led to the elaboration of a metaphor of 'dominoes'. In the 1950s, when America was beginning its involvement in Indochina and reportedly contemplating nuclear strikes to relieve the retreating French colonial forces from advancing (communist) Vietnamese nationalists, US President Dwight D. Eisenhower (1953–61) had claimed that: 'the loss of Indochina will cause the fall of Southeast Asia like a set of dominoes' (quoted in Aspery 1975: 708). Subsequently the idea of dominoes falling (culminating perhaps in communists advancing to the boundaries of the USA) haunted the US geo-political imagination (see Johnson 1994; O Sullivan 1982). To many conservative American commentators, the advance of revolutionary nationalism and the emergence of a whole new set of pro-Soviet Third World governments in the 1970s, seemed to confirm the worse fears of Mr X, Eisenhower and other earlier 'Cold Warriors'. From Afghanistan via Angola to Ethiopia, Nicaragua and Vietnam, a phase of successful Third World revolution was interpreted in Washington as the culmination of the 'prophecy' of Mr. X and others – those who had painted the USSR as inherently expansionist and the US as in mortal danger should its responses be inadequate.

By the late 1970s the voices calling for the United States (once again) to go on the offensive, regain its resolve not only to contain the Soviet Union but to roll-back Soviet strategic advances, were becoming increasingly strident and organized (see Dalby 1990). It is in this context that the Heritage Foundation as a codifier and popularizer of a new Cold War geo-politics came on to the Washington scene. James Ridgeway notes how the Heritage Foundation had been

> Organized in the early seventies by Paul Weyrich and Edwin Feulner, two young Congressional aides, with the prodding and support of beer magnate Joe Coors, Heritage caught fire with the Reagan presidency. It became Reagan's brain trust, and its Mandate for Leadership, prepared in 1980, became the 1,100-page blueprint for taking over and refashioning government . . . Abroad, Heritage argued for much more consistent application of commando-style guerrilla activities in places like Nicaragua and Afghanistan to counter what it perceived as the threat of communism, while at home it called for the rapid privatization of government programmes and reductions in social programs.
>
> (Ridgeway 1997: 12)

Aside from its right-wing domestic agenda, Heritage also joined other right-wing organizations that had been calling for a geo-political strategy that would not only halt but reverse perceived Soviet advances. Under President Ronald Reagan (1981–89), these hysterical voices found official approval. Reagan's domestic policy focused on income-tax and welfare cuts (so called Reaganomics) and was combined with attacks on organized labour conducted in the name of Christian, American and Family values. Reagan's foreign policy was, like a number of earlier ones before him (variously, Truman, Eisenhower, Nixon and Carter, all concerned with refining containment strategies) designated as 'doctrine' and bearing his name. The Reagan Doctrine contained a number of strands. In the *first* place it was about the USA decisively re-gaining the advantage in terms of nuclear war-fighting capability. The biggest ever 'peace-time' increase in US military budgets saw the development of new nuclear weapons programmes (such as MX and the Neutron bomb) and the much vaunted Star-Wars anti-missile 'defence' systems, and the development of new means of 'delivery' – cruise-missiles, stealth-bombers and new submarines and aircraft carriers. The *second* strand of the Reagan Doctrine was about the provision of aid and logistical support to insurgent movements waging war against a number of those revolutionary states that had come to power in the 1970s with various levels of Soviet military and diplomatic support. These anti-communist insurgents included the Contra in Nicaragua, the Mujahadeen in Afghanistan and the União para a Independência Total de Angola (UNITA) in Angola. The CIA and branches of the US military intelligence system funnelled aid, advice and weapons to these organizations. The consequences for those countries on the receiving end of this doctrine were widespread dislocation, and in at least two cases (Afghanistan and Nicaragua), the effective collapse of the revolutionary governments. What role the Reagan Doctrine played in bringing about the wider global collapse of communism at the end of the 1980s is a subject of ongoing debate. What is clearer, however, is that the collapse of a consistent enemy generated a certain dis-orientation in the United States. The sense of a loss of mission intensified, given that the United States' identity had been so marked by the Cold War. During the epoch of Cold War the US had a special sense of mission, a common enemy which at the same time allowed a collective sense of identity.

Following the collapse of the Soviet Union, the old geo-political certainties of East and West no longer applied. In this context a variety of visions of 'New World Order' have emerged. While it would be beyond the scope of this chapter even to begin to map the variety of versions of what the New World Order is or should be[5] we would note that prominent amongst them is the idea of a (geo)political threat to American 'security' being replaced by new (geo)economic dangers. This 'geoeconomic' discourse takes a variety of formats. In the early 1990s it was often a concern with the threat to America from aggressive Japanese capitalism. In Cold War geo-politics, Japan was subservient to the US. This threat-discourse was widely disseminated in business, strategic and popular culture, cropping up in Hollywood movies such as *Rising Sun* (1993), Philip Kaufman's adaptation of Michael Crichton's best-selling pulp-novel, *Rising Sun*, about the 'cut-throat' actions of Japanese corporations buying-up America. In more codified and conventional academic form, Edward Luttwak argued that:

> Everyone, it appears, now agrees that the methods of commerce are displacing military methods – with disposable capital in lieu of firepower, civilian innovation in lieu of military-technical advancement, and market-penetration in lieu of garrisons and bases.
>
> (Luttwak 1990: 17)

Part of this geo-economic idea is the notion that the world is in danger of 'splitting' into giant (and antagonistic) trading blocs. For example, an article in *Time* magazine warned its readers to 'Beware the three-way split' (Talbott 1992) noting the 'danger' of revived competition between Europe, North America and East Asia. For all the over-statement in such alarmist accounts and in Luttwak's own work, there is a recognition that the vision of open markets and transnational liberalism guided by regulatory institutions of the World Bank, the International Monetary Fund (IMF) and the World Trade Organization (WTO), is open to contestation and must actively be reproduced and (re)constructed. It is within this contested arena that the *Index of Economic Freedom* operates.

A new front-line?

As Vestner's (1998: 48) article in *Individual Investor* explains:

> The Index was designed for policy makers in Washington, far from the trading floors of Wall Street. The idea, says Johnson [one of the three compilers], was to document how overbearing governments squelch growth. But taken one step further, the Index becomes a way of sizing up a market as a potential place in

which to invest. It's a simple leap: economic freedom leads to economic growth, and that drives stocks.

The logic here is every bit as selective as that of Cold War stories of dominoes. It is deceptively simple. Freedom for the Heritage Foundation signifies a quantifiable combination of private property rights, low taxation, openness to capital flows, a limited black market, absence of what they term 'wage' and 'price' controls (e.g. regulation for minimum wages and subsidized necessities) and 'government intervention'. The resulting ranking produces a strange hierarchy: from free to repressed. As Table 11.1 shows, in this vision, Hong Kong is the 'most free', followed by other well known bastions of 'freedom' such as Singapore and Bahrain. Most 'repressed' is North Korea, but places such as Mozambique, Congo, Bosnia and Haiti are also prominent amongst the 'repressed'. It does not take long to uncover the selective partiality and tragic-comic nature of this list.[6]

Singapore for example is one of the mostly highly regulated societies in the world – a 'strong state' seeks to discipline not only the conventional 'political' scene and key aspects of economic and social life, but also makes a range of close interventions in issues such as appropriate family size, cleanliness and deportment of citizens and their sexual relations. Selective but enormous state intervention has underpinned Singapore's trajectory for the past 30 years and continues to be central today. Such interventions cut-across what may be categorized (even in quite narrow terms) as the 'economic, the social and the political', for example in respect of compulsory savings schemes for all employees. According to the Index, Bahrain is number three in the annals of 'freedom', a designation that Joe Stork, an advocacy director at Human Rights Watch, might beg to disagree with. Writing in the quarterly report of the Washington-based *Middle East Research and Information Project*, he notes how

> Since early June 1997, an upsurge of crude firebombings, street demonstrations and heavy repression has added some nine deaths and an unknown number of arrests and injuries to the toll of ongoing unrest in Bahrain. The troubles erupted there three years ago with demonstrations over unemployment, discrimination and the refusal of the ruling family to modify its *monopoly* over the state and the public purse.
>
> (Stork 1997: 33, our emphasis)

Neighbouring Saudi Arabia appears in Heritage's category of the 'Mostly free'. But we might ask 'mostly free' for whom? The men of the ruling family completely dominate the power structure to the extent of the country representing a kind of private alliance between an extended family and foreign oil companies, backed up by a strategic relationship with the United States. Even the 'nationality' of their subjects is the name of the family (Saudi). In his exposé the Saudi family's dependent relationship on the West, Saïd Aburish notes how

> Fahd and his family monopolize governance; realize billions out of corrupt practices; dismiss ministers at will; imprison, torture and murder their people and discriminate against Christians.
>
> (Aburish 1998: 224)

Table 11.1 Index of Economic Freedom

Free		39	Argentina	77	Slovak Rep.	117	Niger
1	Hong Kong	39	Barbados	77	Zambia	120	Albania
2	Singapore	39	Cyprus	80	Mali	120	Bangladesh
3	Bahrain	39	Jamaica	80	Mongolia	120	China
4	New Zealand	39	Portugal	80	Slovenia	120	Congo
5	Switzerland	44	Bolivia	83	Honduras	120	Croatia
5	United States	44	Oman	83	Papua New	125	Chad
7	Luxemburg	44	Philippines		Guinea	125	Mauritania
7	Taiwan	47	Swaziland	85	Djibouti	125	Ukraine
7	United	47	Uruguay	85	Fiji	128	Sierra Leone
	Kingdom	49	Botswana	85	Pakistan	129	Burundi
		49	Jordan	88	Algeria	129	Surinam
Mostly free		49	Namibia	88	Guinea	129	Zimbabwe
10	Bahamas	49	Tunisia	88	Lebanon		
10	Ireland	53	Belize	88	Mexico	**Repressed**	
12	Australia	53	Costa Rica	88	Senegal	132	Haiti
12	Japan	53	Guatemala	88	Tanzania	132	Kyrgyzstan
14	Belgium	53	Israel	94	Nigeria	132	Syria
14	Canada	53	Peru	94	Romania	135	Belarus
14	U.A.E.	53	Saudi Arabia	96	Brazil	136	Kazakstan
17	Austria	53	Turkey	96	Cambodia	136	Mozambique
17	Chile	53	Uganda	96	Egypt	136	Yemen
17	Estonia	53	Western	96	Ivory Coast	139	Sudan
20	Czech Rep.		Samoa	96	Madagascar	140	Myanmar
20	Netherlands	62	Indonesia	96	Moldavia	140	Rwanda
22	Denmark	62	Latvia	102	Nepal	142	Angola
22	Finland	62	Malta	103	Cape Verde	143	Azerbaijan
24	Germany	62	Paraguay	104	Armenia	143	Tajikistan
24	Iceland	66	Greece	104	Dominican	145	Turkmenistan
24	South Korea	66	Hungary		Rep.	146	Uzbekistan
27	Norway	66	South	104	Russia	147	Congo
28	Kuwait		Africa	107	Burkina Faso		(formerly
28	Malaysia	69	Benin	107	Cameroon		Zaire)
28	Panama	69	Ecuador	107	Lesotho	147	Iran
28	Thailand	69	Gabon	107	Nicaragua	147	Libya
32	El Salvador	69	Morocco	107	Venezuela	147	Somalia
32	Sri Lanka	69	Poland	112	Gambia	147	Vietnam
32	Sweden			112	Guyana	152	Bosnia
35	France	**Mostly unfree**		114	Bulgaria	153	Iraq
35	Italy	74	Colombia	114	Georgia	154	Cuba
35	Spain	74	Ghana	114	Malawi	154	Laos
38	Trinidad and	74	Lithuania	117	Ethiopia	154	North
	Tobago	77	Kenya	117	India		Korea

Source: Heritage Foundation, web site http://www.heritage.org/

Note: For four years the Heritage Foundation of Washington DC, and The Wall Street Journal have published the Index of Economic Freedom, which ranks countries according to the level of government intervention in the economy. Limited restrictions on such factors as prices, wages, and capital flows, the authors of the Index say, provide a strong platform for economic growth. Here's how the countries of the world rank today (2000).

In detailing the enormous corruption, hypocrisy and extravagance of the higher male members of the Saudi clique, Aburish goes on to note what he calls

> the hideous way in which the average Saudi is treated. Amnesty International, Middle East Watch, the Minnesota International Lawyers Association and other human rights organizations have documented endless cases of imprisonment, torture and elimination within the kingdom. Solitary confinement for years on end is a regular happening (former Saudi Ambassador to Switzerland Abdel Aziz Al Muammar), kidnappings and disappearances are common (Saudi writer, Nasser Al Saïd, and dissident, Muhammad Al Fassi), political executions without proper trial are frequent (Abdallah Al Mutheif) and even women are not spared torture and humiliation (Alia Makki).
>
> (Aburish 1998: 242–3)

So Saudi Arabia's 'mostly free' status might have been strange news to them, as it would be to women in Saudi Arabia, seeking to engage in such mundane activity as filling up a car with petrol (difficult as all women are banned from driving in the Kingdom).

Blind to such discordant signs, let alone ambiguities, the Index claims a kind of objectivity from which in turn are deduced a whole set of conclusions following a simplistic logic. As Vestner claims

> In the end, what emerges from the pages of the Index is a handy framework for international investing, a Baedeker [travel guide] for the financial traveller.
>
> (Vestner 1998: 48)

The reference here to a travel guide is interesting. For as Roland Barthes has shown (1993 [1957]) travel guides are highly selective in what they choose to represent. While the organization of the world that they present seems natural, the set of differences and distinctions that they select as 'obvious' tourist sights/sites, however, are based on ideological myths as to exactly what constitutes 'history'. Barthes comments on the way that the mid-1950s Hachette World Guide to Spain (*Guide Bleu* in French) contains a list of Christian monuments to see, whereas the inhabitants of the land (then suffering the brutal repression of the Franco regime) 'constitute [only] a charming and fanciful decor, meant to surround the essential part of the country: its collection of monuments'. Yet

> To select only monuments suppresses at one stroke the reality of the land and that of its people, it accounts for nothing of the present, that is, nothing historical, and as a consequence, the monuments themselves become indecipherable, therefore senseless. What is to be seen is thus constantly in the process of vanishing, and the *Guide* becomes, through an operation common to all mystifications, the very opposite of what it advertises, an agent of blindness.
>
> (Barthes 1993: 75–6)

In similar ways, it is suggested, the 'objectivity' of the Index like that of 'real travel guides', is wholly phantom. For what is left out, as well as what is regarded as causing what, are all highly selective. Such is the case of course with any narrative. However what is notable with respect to the Index and similar schemes that purport to guide investors in 'emerging markets' is the claim of universal objectivity and knowledge with universal applicability. In this they share a feature with the mode of thought conventional to traditional geo-politics. That is, they purport to be a kind of view from nowhere. Yet below the surface claims and appearance they turn out to be highly dependent upon a certain set of assumptions that are closely associated with the specific culture of Western society.

Writing about this aspect of what he terms 'the geo-political gaze', O'Tuathail notes how

> By gathering, codifying, and disciplining the heterogeneity of the world's geography into the categories of Western thought, a decidable, measured, and homogeneous world of geographical objects, attributes, and patterns is made visible, produced. The geopolitical gaze triangulates the world political map from a Western imperial vantage point, measures it using Western conceptual systems of identity/difference, and records it in order to bring it within the scope of Western imaginings.
>
> (O'Tuathail 1996: 53)

Therefore, in a similar way to Cold War geo-political rhetorics that classified the 'objective' strategic value of places the idea of freedom-repressed is a binary classification scale into which places can only be codified if factors that might disrupt the categorization are excluded or written out. Our reference to the complex socio-political and economic conditions in Singapore and Bahrain shows how, for them to appear as objective, 'freedom markets', some things must be excluded. A closer look renders the classification in terms of economic freedom-repression unstable – likely to be disrupted by (the return of) those aspects of each country that the Index tries to forget. In a sense therefore, the Index is symptomatic of the way that the discourse of which it forms a part – the idea of places as 'emerging markets' – is a variation of that classifying (Western) imagination that underwrote Cold War and imperial geo-politics. In a sense it is partly comprehensible as a form of geo-power, closely associated with the frames of geo-politics rather than as a wholly new and 'different' geo-economic discourse.

In so far as it takes the form of institutional regulation and thereby dogma, we may detect it in the discourses of emerging markets – including the Index, as well as in the operation of regulatory agencies such as the World Bank. The increased power of the latter institutions is part of a complex story of shifts in the global political economy which must lie largely beyond our scope here.[7] However, for the anthropologist Daniel Miller

> these institutions may be said to embody a 'pure capitalism' to contrast with what might be called the 'organic capitalisms' of actual commerce. Pure capitalism is that which is envisaged in economic theory, and is dependent on the construction of certain ideal conditions that would facilitate capitalist enterprises working 'efficiently'. All geographic, cultural and institutional specificity is swept aside in a text-book style model of economic logic and rationality.
>
> (Miller 1997: 42)

Such 'pure capitalism' has some broad roots in a certain moral code in which the best kind of person is seen as entrepreneurial, independent and self-reliant, an associated conception of what 'human nature' is and how it is best realized via the 'market'. This was evident in the formation of modern Western economics during the Scottish enlightenment. Adam Smith's *The Wealth of Nations*, published in 1776, the same year that the American Republic was founded, is the clearest and most famous exposition of this moral judgement of the 'market'. In turn, Adam Smith's account of the power and moral worth of 'free markets' reinforced those notions of individual autonomy, sovereignty and freedom that the American Declaration of Independence encoded (Derrida 1986). Subsequently, as Love Brown (1997: 99) explains:

> The Free Market is a powerful icon in Western civilisation, especially for Americans because it incorporates and integrates the three important core principles of American culture: self-reliance (individualism), equality and resistance to authority.

Freed from the feudal restrictions of Europe, America became the land of the free market, and the ideology of 'free enterprise' a part of the construction of American identity. And in so far as conceptions of the West also incorporate ideas of capitalist enterprise and market potential (which are at least as important as, and cannot be dissociated from, the Western sense of itself as 'developed' and 'democratic', America became its natural leader. The development of the market is part of what differentiates a sense of the West from the rest. In Carrier's words

> the market model is more than just an idealization of economic activities in particular times and places. It is also an idealization that itself idealizes the modern West . . . Such idealizations start with differences between what goes on in the modern West and what goes on in non-Western societies or in pre-modern Western societies. These differences, invariably partial and relative attributes of particular times and places, are transformed into absolute and defining essences of types of social life that stride with their capital letters, across the page and through the mind – the West, the Orient, Communism, Capitalism and so forth. In this process of constructing distinct and opposed types of life, the attributes that differentiated the modern West themselves become purified and exaggerated, become essences.
>
> (Carrier 1997: 31–2)

It is in the form of such an essence that market freedom, of the abstracted form celebrated in the Index, may acquire a certain authority. There is much more to the complex story of global finance than appears in the market-triumphalist accounts of Heritage. Indeed as we have charted here, the Index obscures the complex and messy social reality of places and the market relations in which they are embedded. In this context we should recall that there is a long tradition of inciting 'freedom' within a system of control, regulation and discipline.

In their 'counterculture' interpretation of American history (from which we have drawn part of the title for this chapter) Carrol and Noble describe a paradox at the cultural heart of American society in terms of its internal geo-politics. They note how a policy of racial separation, brutally enforced on native peoples and African slaves,

> revealed a striking paradox in American attitudes toward space. It was only by imposing a rigid territorial order on non-WASP [White Anglo Saxon Protestant] peoples that white Americans could preserve a sense of geographical mobility for themselves. . . . This paradox – the institutionalization of space and the advocacy of geographic mobility – stimulated the territorial expansion of the United States during the nineteenth century.
>
> (Carrol and Noble 1977: 16)

To a degree this has been common to all states, particularly colonizing ones like America. Beyond the nineteenth century, as the United States became the embodiment of and leader of the West (or, as its leaders preferred to say, of 'The Free World'), this paradox, or dialectic of control (of others) in the name of (selective) freedom underpinned a wider American strategy. At times, in geo-political discourse and practical strategy it led not just to the imposition of control, but the physical elimination of subject populations – all done in the name of defending this selective notion of freedom. In the Vietnam war for example, military spokesmen declared that some places 'had to be destroyed in order to be saved' [from falling to the communists]. In terms of some of the consequences, specifically the dismantling of social welfare provision, free or subsidized health care and education services as part of the strategies of maximizing 'market freedom', we are not so far away from the same logic.

Conclusion

In this chapter we have centred on some of the key moments that have signalled a transfer in emphasis away from geo-politics to a geo-political economic discourse following the ending of the Cold War. In doing this we have noted how over a relatively short period of time a free market discourse has become woven into our everyday knowledge and its relation to space and the economy. Moreover, in focusing on the workings of the Heritage Foundation and its Index, we have suggested how a geo-investment knowledge is being produced that makes the spaces of the new age of global capital both familiar and calculable to a growing proportion of the West's population.

We would locate the rhetoric of Freedom Markets and the Heritage Foundation's Index on Freedom amongst a wider array of disciplinary apparatuses that have appropriated and invoked a language of freedom. Freedom is a powerful word (in its capacity to mobilize and demand action), but one with an enormous range of applications. It is a highly flexible sign, subject to appropriation by a wide range of discourses of power. We should recall that in perhaps the most extreme case, Nazi concentration camps were marked by the slogan over the gates (through which millions entered, but few returned) *Arbeit macht frei* (Work makes freedom). The apparent absurdity of this claim betrays the way that 'freedom' has been invoked by even genocidal systems. Recall too that the most brutal of communist regimes have eliminated millions and deprived many more of basic liberties in the name of (amongst other things) a greater freedom. These extreme cases are, of course, a long way from the rhetorics of freedom markets. Yet we should recognize the duplicities and inconsistencies of the freedom market rhetoric. Beyond the claim to 'objectivity' present in the latter is an attempt to reconfigure a certain kind of 'freedom' spatially. For in the circulation of a free market discourse is an inseparable process of re-presenting the unfamiliar space of developing countries as the now natural hunting grounds of the free market global investor.

Notes

1 US subscriptions $22.95 a year; Canadian and foreign, add $13 a year. *Individual Investor Group, Inc.*, 1633 Broadway, 38th floor, New York, NY 10019, www.iionline.com

2 Today about 100 million Americans (approaching 40 per cent of the population) own shares, either directly or indirectly through mutual funds. In Britain, data from the mid-1990s indicate that nearly a third of the population own shares. However, in both countries, and elsewhere in 'developed' economies, exposure to stock markets is considerably wider than these figures suggest. This is because pension and endowment funds (savings schemes with insurance, commonly used in the UK as a mortgage repayment method) invest in shares. For more data on the expansion of stock market exposure and a critical assessment of the potential consequences, see Bygrave (1998).

3 This is not to say that the rise of neo-liberalism at these most prestigious and significant codifiers of development discourse has been a simple story. Indeed neo-liberalism forms a constellation of discourses with considerable internal heterogeneity which has found expression at the World Bank and IMF and between them (Berger and Beeson 1998).

4 'State', 'market' and 'society' should not be seen as stable fixed components that come together in some mixture or other to form a regulatory nexus, but rather as expressions of power whose meaning is itself constructed through and rendered effective through the nexus.

5 For a beginners guide, see Part 3 of O'Tuathail *et al.* (1998) or Chapter 3 of Derrida (1994) for a more difficult, but rewarding diagnosis.

6 According to the Heritage Foundation (quoted in Johnson *et al.*):

> The Heritage Foundation/Wall Street Journal 1998 *Index of Economic Freedom* measures how well 56 countries score on a list of 10 broad economic factors: The higher the score on a factor, the greater the level of government interference in the economy and the less economic freedom. A total of fifty independent economic variables were analyzed. These variables were grouped into the following 10 economic factors: Trade policy, Taxation, Government intervention in the economy, Monetary policy, Capital flows and foreign investment, Banking, Wage and price controls, Property rights, Regulation policy and Black Market. . . . An objective analysis of these factors continues to demonstrate unequivocally that countries with the highest levels of economic freedom have the highest living standards. Similarly, countries with the lowest levels of economic freedom also have the lowest standards of living.
>
> (Johnson *et al.*: 1998)

7 For accounts, see Berger and Beeson (1998) and Corbridge (1992).

12 Rethinking 'useful' knowledge

Co-operative science and the new genetics

Morag Bell

Introduction

Notwithstanding the success of technoscience in improving life expectancy, human communications and environmental knowledge on a global scale, the search for collective and personal control in an unstable and uncertain world have been consistent features of Western modernity. This chapter focuses on some dimensions of the debate underpinning this search for control, notably a recurring concern to improve the quality of life. The chapter concentrates on the meaning and application of 'useful' knowledge which are implicit in this debate, including the sources, authority, ownership, access and 'proper' use. Whilst these issues are of recurring importance in intellectual and public arenas, the chapter examines their contemporary significance within a context of supposed doubt and uncertainty. With reference to examples, it explores the significance attached to boundaries, both moral and material, in shaping the discourse of 'useful' knowledge and how concepts of space, place, time and agency are implicated in their definition and manipulation.

The discussion is divided into three main parts. The first outlines some contemporary notions of uncertainty and anxiety. The second considers their association with the 'pathology' of the global. As a way of overcoming this pathologisation of space, I consider the significance of relational approaches to place, space and time and their role in reshaping human identities. Some implications for the meaning and authority of 'institutionalised' knowledge are also considered. The third part of the chapter examines these concepts and approaches with reference to the new genetics. In focusing on the human form with a view to enhancing individual well-being, the new genetics is widely regarded as among the most significant breakthroughs in biological science with profound implications for human identity and life quality. Notwithstanding its supposedly humane objectives and universal applicability, the new genetics and the Human Genome Project in particular, represent a highly controversial form of technoscience. In examining some dimensions of the controversy, I consider the importance attached to boundaries in structuring the debate, the implications for broader relational approaches to space, place and time and some tensions which arise in the application of these approaches.

Living with uncertainty

Numerous contemporary critics claim that these are times of unprecedented change, of doubt and anxiety. The optimism and exhilaration widely associated with global interconnectedness is set against an equally powerful mood of doubt and anxiety. Many different explanations have been offered. For some commentators this mood is a function of 'manufactured uncertainty' (Giddens 1994: 184). It marks the end of old beliefs, once approved with trust and confidence that through advances in human knowledge, the good society could be conceived, planned, and implemented. Whilst the directions of human life may be no less predictable today than in the past, it is argued that the origins of this unpredictability have changed as the growth of human knowledge generates new uncertainties (Smart 1992). Other writers focus on the contribution of 'globalisation' to this mood of anxiety. Granted, in an era of space–time compression, a global consciousness has been interpreted as individually liberating, socially desirable and technically possible. Equally, in undermining dualist categories like North–South, nature–culture, and in questioning supposedly fixed identities associated with nation, class and 'race', its positive qualities have been emphasised. However, in disrupting established interpretations of time and space, 'global' times may bring with them new instabilities (Light 1997; Williams 1997). According to some writers, the elimination of regulated diurnal patterns of work, and of seasonality in food production and consumption, creates within many societies a disjuncture between the temporalities of nature and culture with supposedly negative human and environmental effects (Adam 1996; Adam *et al.* 1997; Grove-White 1997).

A mood of uncertainty has also been related to the bewildering effects of new mutations in space, referred to by Fredric Jameson (1991: 44) as postmodern hyperspace. Stimulated by new visions of and technologies applied to design, according to Jameson, they appear to transcend 'the capacities of the individual human body to locate itself, to organise its immediate surroundings perceptually, and cognitively to map its position in a mappable external world'. In connecting this disorientation at the level of individuals to the 'global', Jameson suggests that the disjunction between the body and the built environment also stands as a symbol of 'the incapacity of our minds, at least at present, to map the great global multinational and decentred communication network in which we are caught as individual subjects'. Beyond these accounts, contemporary uncertainty has been associated with the insecurities which follow from a shift in relational ethics, notably the removal of constraints on human liberty. As part of the folklore of Western modernity, values of harmony, cleanliness and the routine of order have, it is argued, constrained natural, disorderly, human impulses. Unsurprisingly, by inhibiting personal freedom in the interests of individual and collective security and social cohesion, the discontents arising from these constraints have stimulated frequent challenges. For Zygmunt Bauman (1997: 2), however, 'deregulation' and with it a loss of security, are central features of our times; they form part of the collective consciousness, in countries of the North at least. Commenting also on relational ethics, Ulrich Beck (1994a: 10) argues that principles such as responsibility and guilt are not suited to comprehending or legitimating what he sees as 'the return' of uncertainty and uncontrollability associated with the turbulence of 'mutually contradictory, global and personal risks'.[1]

For many commentators, these anxieties are all features of the postmodern condition the roots of which lie in the 'crisis' of modernity. As descriptions of the contemporary world, it is argued that they mark a decisive historical shift in terms of beliefs, material conditions, relations between nature and culture, orientation in space and time and in relational values. Other commentators are much less convinced about this representation of our times. In favouring the concept of risk society, or institutional reflexivity, they claim that it is based on a misreading or misrepresentation of modernity itself as an epoch (with a beginning and end) of certainty. In forming part of an inherently modernist impulse, a tendency to discredit the old in order to privilege the new, it is argued that the search for both collective and personal control in an unstable and uncertain world has been a consistent feature of Western modernity (Lash *et al.* 1996; Osborne 1995). Equally, there is a tendency in these accounts to marginalise the countries of the South. They seemingly mask the conditions of uncertainty, risk and insecurity which have been, and remain, for the majority of the world's population an essential part of the daily struggle for survival (Simon 1998).

Posing new questions

These debates over the boundaries of past and present, local and global, are significant not only in stimulating critical reflection; they also provoke normative questions. For some theorists all criticism stimulates debate about values. However, as Sayer and Storper (1997) point out, it is a terrain which many social scientists have been unwilling to enter. This unwillingness can be related to a current mood of not only public but also intellectual anxiety and scepticism which influences some approaches to the boundaries of space and time. The ethics and practices underpinning global processes have contributed to what is, for many cultural critics, a particular example of the pathologisation of space, the pathology of the global. It should be emphasised that concepts like global and international have meant different things at different times and do not necessarily refer to the planetary scale. Indeed, Stuart Hall (Chen 1996: 393) argues with reference to cultural studies that it is not about globalisation, it is about 'being "globalized" – a very uneven and contradictory process'. Roy Porter (1997: 6) expresses similar caution about the 'global' in his single-volume history of medicine. He states boldly that human health and disease should be discussed 'from a global viewpoint' since 'no other perspective makes sense'. But, in seeking to avoid narrow ethnocentrism, he writes not only about the 'historic power of "western" medicine as uniquely global', but also of the many other medical systems which have profound influence 'the world over'. Intellectual memories of the past compound a contemporary reluctance to engage in global theorising. The tarnished associations of such Enlightenment principles as humanism with the practices of colonialism have served to highlight the western male bias implicit in the 'universal' (Rogers 1992). They have also exposed the ethnocentricity implicit in Europe's imperial assumptions that the cultural infirmity of others required the healing power of eurocentric transformations (Vaughan 1991).

Granted, challenging grand narratives produces new intellectual doubts and global sensitivities. But, as Derek Gregory (1994) suggests, they also provoke theoretical reflection on ways of describing and representing the world. Indeed, they may offer a context for, and a validation of, renewed interest in normative theory. Recent critiques of the human and environmental effects of global processes have stimulated this more positive intellectual reflexivity.[2] The need has been recognised for critical enquiry not only to question the values which underpin these processes but also to shape social practices in more 'positive' ways. These critiques emphasise that intellectual anxieties do not merely produce ambivalence. Nor, to use David Livingstone's (1998: 15) words, do they lead universally to a 'postmodernist certainty' that 'everything is *just* representation and that no judgements can be made about adequacy or inadequacy, accuracy or inaccuracy, truth or falsity'. Scepticism as a characteristic of contemporary intellectual consciousness may indeed stimulate an equally powerful impulse, the desire to make a difference. It may provoke debate about the relations between values, the meaning of 'useful' knowledge and power, where power is conceived not only as an ability to criticise, to challenge and to resist those with power to oppress. It may also be interpreted in therapeutic/redemptive terms; to heal, to enable and to transform in a manner which extends beyond the limited association of therapeutics with medical enterprise and the human body, or of redemption with religion and the human soul (Bynum and Nutton 1991). A cultural environment of empowerment and enablement supposedly characteristic of the present would appear to afford new possibilities – the possibilities of a 'better' life (Strathern 1996).

Relational approaches to space, place and time

I should like to highlight two themes which underpin this search for alternative modes of socially useful knowledge. These themes centre around alternative ways of looking outwards, and of looking back in time, as a basis for reshaping human and environmental relations. Collectively they point to forms of spatial and temporal consciousness which seek to transcend the exclusivity frequently associated with established social, intellectual and institutionalised boundaries. They emphasise the importance of relational approaches to place, space and time, and suggest the kinds of ethics which should underpin these relations. In her book *Moral Boundaries*, Joan Tronto (1993) uses the metaphor of boundaries to address two sets of questions; the first strategic and the second visionary. She argues that by noticing boundaries (a strategic issue) we identify the inclusions and exclusions and that this knowledge provides the basis for debating alternative visions of life (an imaginative issue) and how they might be achieved. Within this context, my first theme concerns modes of looking outwards.

For many writers, a sense of place as an intrinsic part of human subjectivity has been interpreted as an appropriate lens through which to approach the world and to comprehend global processes. Whilst simultaneously looking outward and inward is important in defining any form of social, cultural and place identity, it is also argued that this 'way of seeing' has either positive or negative qualities depending on how otherness is interpreted. It is negative when the external world is read as a threat or an arena to be controlled, and positive or progressive where interrelations across boundaries and mutual benefits are acknowledged. In the aftermath of European imperialism, approaches to 'the geographical beyond' (Massey 1995: 183) are widely advocated which are not hegemonic in the sense of threatening to others, nor supportive of a postmodern tendency towards cultural relativism, nor myopic and defensively supportive of parochial favouritism. As a contribution to resolving what in many countries of the North is interpreted as the contemporary crisis of identity, perspectives are favoured which, liberated from notions of the outside as threatening, recognise the fluidity and permeability of cultures and question the significance of established, institutionalised boundaries.[3] As Stuart Hall (1996a) argues, the concept of identity needs to be grounded within the historically specific developments and practices which have disturbed the relatively 'settled' character of many populations and cultures, above all in relation to the processes of globalisation. As an alternative way of seeing the world, a conception of space as reciprocal relations is advocated and of place as neither fixed nor immutable but defined through these relations. Doreen Massey (1995: 183) emphasises that 'places are always already hybrid'; they have shifting identities which are shaped by the dynamic nature of social and cultural relations across space. Implicit in this relational approach to space, place and identity are notions of 'identification with' and 'cooperation across' to which I shall return later, and which, lodged in contingency, are not intended to obliterate or undermine difference (Hall 1996b; Harvey 1996).

Closely related to this spatial reflexivity is the second theme, that of critical reflection on our temporal consciousness. Similar issues emerge. Recent reassessments of space and place have stimulated a re-reading of the past and the boundaries of the present. As an intellectual metaphor, reflection has been associated with the Enlightenment idea of 'seeing with an added eye', and with a modern logic of temporality which interprets change as ignorance transformed to knowledge. More critically, it provides scope to identify the unconscious categories of our knowledge and the ruptures as well as the continuities within modernity (Beck 1994b: 174). It has, for example, highlighted the naïve certitudes implicit in the discourse of progressive evolutionary change with its emphasis on technical solutions to social and environmental questions over political and economic action. In offering a challenge to those for whom history is threatening, either because of the embarrassing evidence it exposes, or in view of the interests that it serves, as David Livingstone (1998: 15) argues, by re-presenting past representations of peoples and places, history can be criticised for its inauthentic characterisations. For many writers from a postcolonial perspective, re-reading the past forms a necessary basis for assuming the responsibilities of cultural translation (Bhabha 1994). In exposing 'the comforts and continuities' implicit in invented traditions, it is suggested that these revelations can play a positive role. They can facilitate the shift from a carefully controlled and familiar order based on comfortable certainties to a position of greater openness (Morley and Robins 1995b).[4] They can guard against approaches to history which focus merely on the need for censorship or on appropriation in the interests of the powerful. As part of this critical reflection, numerous recent studies of Britain's imperial relations are exposing the subtle variations which existed within dominant ideologies. In offering a critique of the imperial claim to speak for a common humanity or of a binary opposition between coloniser and colonised, they emphasise how the metropole and colonies were deeply implicated with each other and that these interrelations were shaped by diverse, complex and conflicting imperial and colonised voices (Rattansi 1997).

Rethinking 'useful' knowledge

The discussion so far highlights the importance of approaches to identity within the countries of the North that are liberated from the dualisms implicit in inside and outside, past and present. In opening up new spaces of knowing based upon relational approaches to place, space and time, sentiments like insecurity and threat are contrasted with calls for a greater openness to the complex webs of relations which link people, cultures, places and nature around the world and which, with their own space–time geometry, render inadequate any notion of fixed or stable identities. But support for these approaches does not imply that all judgements should be suspended about the validity of different positions. Rather, it favours the making of decisions based upon more open forms of discourse and critical reflection which have the capacity to combine redemptive and transformative qualities. The implications of these sentiments for the meaning of 'useful' knowledge are profound. They provoke judgements about good and bad and therefore the authority and legitimacy of differ-ent forms of knowledge. They raise questions about ownership of and access to these different forms, their 'proper' use and the validity of the sources from which they emerge.

Forms of knowing and human identities are produced within many different social and material spaces. Conventionally, within the spheres of science, these include laboratories and scientific institutions. In assessing the processes which construct us as human subjects, it is helpful to explore the extent to which calls for a greater openness in social and environmental relations are reflected also in scientific inquiry, and within the spaces of 'institutionalised' knowledge. The very meaning of the concept is worthy of close examination, specifically the significance of institutionalised boundaries in shaping its production, communication and application. Donna Haraway (1997: 41) suggests that, 'For . . . the true believers in the church of science, a cure for the trouble at hand is always promised. That promise justifies the sacred status of scientists, even, or especially, outside their domains of practical expertise.' In their ability to uncover the secrets of life, for many commentators the natural and biological sciences continue to benefit from our anxieties. Through their capacity to shape human identities and the relations between humans and the environment, they provide 'useful' knowledge beyond the boundaries of the laboratory and the academy.[5] But in offering insights not only of the public but also for the public, they have relevance beyond the limited realms of public policy. In acquiring an optimistic twist within the public domain, scientific knowledge remains for some, liberating, empow-ering and enriching. This 'sacred status' does, however, disguise the debates, the controversies and the uncertainties which lie behind the production and translation of 'institutionalised' knowledge in different social and cultural settings.[6]

Critiques of the natural and biological sciences from numerous different perspectives examine the production of scientific knowledge, its sources, for whom and for what purposes it is produced and disseminated. In questioning the values underpinning scientific inquiry and their relations to broader social values, these critiques frequently implicate 'Western' science and medicine in the hegemonic interpretations of space and the progressive readings of time which normative theory now seeks to challenge (Macleod and Lewis 1988). Feminist and environmental critiques, for example, have linked scientific research and science-based technologies with gender inequalities and environmental degradation in different cultural contexts (Fox Keller and Longino 1996). Equally, biological discourses of 'race' and environment have been shown to be complicit (wittingly or unwittingly) in modern misreadings of cultural identity and in practices of imperial control (Livingstone 1992a; Poole 1997). Institutionalised knowledge with its roots in the North has also been criticised for the persistent undervaluing of alternative knowledge systems within the context of unequal North–South relations (Slater 1997). In distinguishing between normative claims about how science should proceed and evaluative claims about how successful science has been, many of these critiques go beyond an impulse to challenge by offering constructive assessments of the implications for the natural sciences of the adoption of alternative social and environmental ethics. As part of what Tiles (1996: 220–32) refers to as the 'logic of domination', she focuses on the historic interest of modern science within western cultures in prediction and control as shaping criteria for success. But, in emphasising that scientific goals are also subject to contestation, she argues for a shift in epistemological values away from 'power over' through prediction in favour of the 'development of science with a human face, science which aims more at cooperating with than at conquering nature'.

This advocacy of alternative principles and approaches for the natural and biological sciences is not, of course, new. It echoes broader critiques of a culture of individualism and corporate capitalism which persisted through the twentieth century. A review of Rachel Carson's recently published biography reminds us that 'it is impossible to overestimate the historical significance of Silent Spring or its author' (Porritt 1998: 29). Published some forty years ago as a challenge to the ineptitude and moral inadequacies of the scientific establishment, it highlights the profound contribution made by this social critique of a gospel of technological progress, to modern health and environmental movements. Whilst an ability to predict may offer a powerful antedote to uncertainty, it also provokes persistent debate both within and beyond scientific institutions over the human and environmental implications of this scientific and technological power. Within a turn-of-the-century context in which knowledge is widely seen as partial and contingent, these social movements have been interpreted as posing significant epistemological and political challenges to institutionalised knowledge (Jamison 1996; Williams 1996). In scientific and market cultures, public consultation with a view to acting upon, as opposed to 'improving', lay people's knowledge and opinions is interpreted as marking a significant shift within academic and policy arenas from the dominant 'deficit' model of public understanding of science (Wynne 1991; 1995). Some convergence can be identified here with countries of the South where concepts of 'indigenous' knowledge, popular participation and partnership across public, private and voluntary agencies have occupied a central place in much development discourse for over two decades.

I want now to draw together these debates surrounding the boundaries of space, place, time and knowledge by examining how they are being applied in practice. I shall refer to one contemporary example which relates directly to quality of life. The new genetics involves the application of biological science and science-based technologies to the human body with a view to enhancing individual well-being and also the social good. Bryan Turner (1993: 10) argues that in Western society 'the question of the body' remains a dominant and linking theme in contemporary thought about the nature of the individual, social organisation and the natural environment. The new genetics has profoundly important implications for the transformation of the human form and, by implication, for the meaning of life quality. It is nevertheless a highly controversial form of biological science both within and beyond the academy. My purpose is to examine the geographical discourses implicit in the debates and controversies surrounding the production and application of this 'new' scientific knowledge. I shall examine the importance attached to notions of uncertainty, co-operative science, human rights and the interactions between nature and society in these debates. Particular attention will be given to how the boundaries of space, place, time and knowledge are employed and manipulated.

Mysteries of life, life quality and the 'new genetics'

In her discussion of technology and enabling identity, Marilyn Strathern (1996: 49) suggests that 'Perhaps it was always a condition of modernist Euro-American knowledge practices that uncertainties increase as fast as certainties are established; each horizon exposes more terrain.' The new genetics provides one powerful example. The gene is 'the essence of identity and the basis of human difference'; genetic research offers the promise of enhanced 'prediction' and greater 'control of behaviour and disease' (Nelkin 1994: 26). As an example of international scientific cooperation, the Human Genome Project is 'the Holy Grail of future health' (Shakespeare 1995:8). These confident and authoritative views are reflective of much of the promotional language underpinning the new genetics. Comprising a body of knowledge and techniques arising out of the discovery of recombinant DNA in 1973, the new genetics has a long and complex genealogy. It can be traced to laboratories within the North in earlier moments of the twentieth century. Profoundly important in this spatial genealogy is the discovery in 1953 of the structure of DNA by James D. Watson and Francis W. C. Crick in the Cavendish Laboratory, Cambridge, in which, with the help of the X-ray crystallograph, they saw the 'secret of life' (Watson 1968: 23 and 155). Two interrelated objectives underpin the new genetics. It seeks 'ultimate answers to the chemical underpinnings of human existence' (Watson 1990: 44) with a view to facilitating research into the genetic components of a range of disease, illness and behaviour and its application in the clinic in the form of testing, screening and treatment. New genetics can therefore be interpreted as central to the traditional responsibilities of science and medicine. It seeks both to understand and to enhance human life by reducing distress and suffering. But the technologies that are enabling of genetic certainty and prediction bring with them new uncertainties about the knowledge with which they are associated, its sources, for whom and for what purposes it is produced, communicated and applied. As Strathern (1996: 48) suggests, 'the case is a classic one – not too much technology but too much knowledge'.

Notwithstanding its humane objectives, genetic research and its application are the focus of widespread academic and public debate, central to which is the meaning of human identity, useful knowledge and life quality. In creating new visions of life through scientific analysis of and medical interventions in the body, the new genetics bestows on biological science and medicine the possibility of unprecedented powers to heal and to transform human life. In doing so, it generates new choices and risks in decision-making in which questions of individual liberty, human rights, social accountability and responsibility irrevocably arise. Methods of improving humankind and life quality by the use of science-based technologies and medical intervention form part of an intellectual heritage which interprets the human body as a site of health and normality, a deviant site invaded by disease and a location for 'humane' interventions. They are therefore concerned with the body not only in its material but also representational forms in terms of the identities attached to it and the ensuing politics of difference.[7] But genetics research also illustrates how the body is interpreted not only as a site for manipulation but also for human thought and action. It is implicated in debates of national and global significance which transcend a concern with the construction of knowledge in the laboratory and its application to the human form. Reference will be made to three interlocking debates with which it is associated: the role of biological science in international co-operative endeavour, the relations between heredity and environment in shaping human health, and the meaning of individual choice in demographic, social and environmental engineering in the interests of a 'good society' or the ideals of a 'perfect health culture'.

Reshaping boundaries: space/place, past/present, nature/nurture

The Human Genome Project draws extensively on the language of space and time for its institutional identity and public legitimacy. Identification with a broad scientific community across national boundaries is central to the initiative. Launched as an international programme in the late 1980s, it is reported to be the largest ever collaborative research effort in the life sciences and has frequently been referred to as biology's equivalent to the moon-shot (Cantor 1990). By mobilising the resources and expertise of scientists in laboratories across the world, not least because of the vast costs involved, its primary objective is to map and order the entire complement of human genetic material, the genome. Likening the scientists involved to 'explorers venturing into unknown territory', Wilkes (1993: 85) indicates that they 'decided first to equip themselves with maps'. Continuing the analogy, according to Balmer (1996: 534) the maps, 'like geographical maps, can be of varying type and resolution'. They can range from 'large-scale linkage maps of genes in relation to other genes based on frequency of co-inheritance, through various types of physical maps that locate "landmarks" in the DNA, and eventually to the highest resolution, the sequence of chemical base-pairs which make up the DNA molecule.' During the 1980s a series of mapping projects had begun to evolve in different countries. Notable among these was the USA where, as the Cold War was coming to an end, a new agenda for science provided an effective antidote to the volatility of national defence research and policy (Kevles 1997).[8] Writing in 1990, a key architect of the project, James D. Watson (who, with Francis W. C. Crick had discovered the double helical structure of DNA), emphasised the importance of openness in terms of ownership and the benefits which would accrue. In highlighting what he regarded as a fundamental principle of all scientific knowledge, namely, that it should transcend institutionalised boundaries, he argued that 'the nations of the world must see that the human genome belongs to the world's people, as opposed to its nations' (Watson 1990: 48).

Notwithstanding this internalist rhetoric and the supposedly global significance of a project designed to reduce the uncertainties of life, genetic research highlights how the human body has also become a site of struggle and resistance, not least to interventions in what are perceived to be natural processes. Scientific and public anxiety over the ethical implications of human genome mapping, and its association with 'disease-gene hunting', highlight these sensitivities. They centre around the relations between heredity and environment in shaping human health. For many critics of the new genetics, a re-reading of history emphasises this point. Memories of past manipulations of genetic research and fears of a new eugenics underpin their concerns. By contrast, among its professional advocates the definition of boundaries between past and present serves to preserve a safe distance between the two. Recent investigations by Kerr *et al.* (1998a: 175) among a 'small but important' network of British scientists and clinicians involved in the new genetics and concerned with the social and cultural impact of their work, indicated the significance attached to temporal boundaries in the meaning they assigned to eugenics. Like other supporters of the new genetics, eugenics was widely characterised as an 'outdated and discredited' ideology associated in particular with an era of prominence from the 1920s to the 1950s. It was interpreted as 'unscientific' and associated with attempts at the coercive manipulation of the gene pool of the future (White) race in the interests of imperial control rather than with a concern for the autonomy and rights of the individual (pp. 176–85). Coupled with this censorship of history, the importance widely attached to environment-society relations in shaping human health and well-being would also appear to challenge the notion that the new genetics, like eugenics, is crudely deterministic. That genome mapping cannot be a medical panacea is reflected in a recognition within the specialist press of the complex interaction between genes and the environment, rather than to the privileging of nature over nurture which genetic research necessarily involves (Cantor 1998).

This manipulation of boundaries between past and present, and between nature and nurture, is reinforced in public discourse by a distinction between nationalism and universalism. The international co-operative approach implicit in the Human Genome Project would appear to distance further the new genetics from eugenics in the importance it attaches to mutual scientific support and shared benefits. A world-wide survey of genome mapping activities conducted in 1991, identified eight countries with established national projects, namely Denmark, France, Germany, Italy, UK, Japan, former USSR and USA. Measures to instigate national programmes had been taken in a further seven, namely, Australia, The Netherlands, Canada, Chile, Sweden, Korea and New Zealand. International programmes had also begun or had been proposed by the EC, UNESCO, Latin America and the Nordic countries. World-wide co-ordination of the initiative, described as 'a loose confederacy of programmes', was the responsibility of the Human Genome Organisation (HUGO) (Balmer 1996: 535). Established in 1988 at the instigation of a leading American geneticist, despite a lack of resources, by 1992 the Organisation had offices in London, Washington, Osaka and Moscow (Anderson and Aldous 1992). In its inter-continental reach with implications of universal applicability, the Genome Project might appear to allay the fears of sceptics that the research could be used to benefit only narrowly defined political and cultural interests.

Tensions between universality and particularity are nevertheless implicit in these anxieties. Distinctions made by some scientists between the application of supposedly 'neutral scientific knowledge' within democratic states and the 'politically distorted bad science' associated with totalitarian regimes in both past and present, notably Nazi Germany and China respectively (Kerr *et al.* 1998a: 185), serve to highlight the significance widely attached to the boundaries of place in shaping the 'proper' use of scientific knowledge. Underpinning this propriety is a further distinction between 'individual informed choice' supposedly characteristic of democracies in contrast to the coercive social politics and infringements of human rights associated with totalitarian states. Reports in western publications during 1995 in connection with the introduction in that year of China's Law on Maternal and Infant Health Care, highlighted its eugenic implications. Building on the country's existing one-child-family policy intended to limit population growth, an Editorial in The Lancet (1995: 131) characterised the new legislation as designed 'to reduce the perceived burden of disability to Chinese society' by making compulsory a premarital medical exami-nation for serious genetic diseases, some infectious diseases and 'relevant' mental disorders. Notwithstanding its coercive nature at individual level and the widespread international condemnation of it, this same editorial adopted a less strident tone. In emphasising that 'China has population problems on a scale inconceivable to the west', it cautioned against the arrogance and ethnocentrism implicit in hasty condemnation of the new law. Rather, it continued, 'eugenics may be a product of the west, but what right does the west have to judge elsewhere? China has a unique history and culture. It is perilous to impose western morality on this country.' In seeking a more 'constructive' approach, the value of 'dialogue' between western and eastern doctors and medical scientists was highlighted with a view to encouraging 'collaboration in research' and to reducing 'misunderstanding of each other's cultural differences'.

Situated knowledge: Individual choice, human rights, social context

Whilst the moral relativism implicit in cross-cultural comparisons has been fiercely criticised (Clarke et al. 1995; Tsering 1995), in highlighting the interplay between science, politics and culture, they raise questions about the extent to which certain values can be universal even if the practices with which they are associated are culturally specific (Smith 1997b). Drawing on the work of Donna Haraway (1989), they also highlight the significance of 'contexts' and 'situated knowledges' beyond the specific cases of coercive regimes. An emphasis on the objectivity of 'good' science as opposed to politically distorted 'bad' science ignores the interacting social and spa-tial worlds within and beyond the walls of the laboratory which shape the nature of scientific knowledge and its usefulness. Writing in 1990 (p. 48), James Watson indicated that, notwithstanding the formation of HUGO, 'how to ensure that we, as nations, work together instead of indulging in costly competitive races for the same chromosomal objectives, is not yet settled'. Links between the funding of genetic research and the patenting of particular gene sequences of interest to the international biotechnology industry, points to the blurring of boundaries between basic and applied knowledge and to the difficulties of ensuring that, in terms of ownership, the genome remains public as opposed to private property (Kay 1993; Fujimura 1996).[9] Equally, whilst cross-national openness is implicit in the Human Genome Project, the provin-cial nature of the international is clear in the distribution of scientific participation with its emphasis on laboratories in the rich countries of the North.[10] Links between selectivity and professional usefulness have also been identified. Concerns within the scientific community that only the larger groups or those engaged in extensive collaboration would have the resources to produce what is likely to 'count as worthwhile science' have been coupled with fears over the diverting of funds from other smaller-scale, 'creative' research in the life sciences (Balmer 1996: 560; Kevles 1997).[11] Further concerns over the provincial nature of the international have been expressed in terms of the 'useful' application of new genetic knowledge. Notwithstanding its supposedly humane objectives, alternative interpretations of international humanitarianism have been emphasised. Suzuki (1989: 347) questions the ethics of large-scale resource allocation to genetic manipulation, often focusing on 'extremely rare hereditary illnesses', when existing scientific knowledge applied to environmental conditions could prevent the diseases suffered by 'millions of children in Third World countries'.

This geographical partiality has temporal parallels. Whilst the construction of ethical divisions between contemporary genetic research and past 'malpractices' offers a particular strategic reading of history and the present, alternative interpretations challenge these temporal boundaries. Evidence of the complex and diverse nature of early-twentieth-century eugenics sharply contrasts with a tendency towards stylised representations of it as merely coercive and deterministic pseudoscience (Kerr *et al.* 1998a). Equally, recent attempts to link genetics with particular behavioural traits, including intelligence and aggression, are not necessarily dissimilar to earlier biological explanations of differences among ethnic groups and between the sexes, used subsequently to justify inferiority (Alper and Natowicz 1992). The supposed universal applicability of the new genetics also has many antecedents in the annals of science-based technologies applied to human populations within the context of humanitarian discourses. Feminist writers, for example, have demonstrated how, during the late nineteenth and early twentieth centuries, within a range of scientific fields, including bio-medicine, the view that sex and reproduction were more fundamental to woman's than man's nature resulted in the emergence of gynaecology as a separate branch of medicine devoted to the diagnosis and treatment of the female body (Oudshoorn 1996). In the development of medical technologies, one consequence of this process of naturalisation was the quest for universally applicable contraceptives without regard to the different cultural and social contexts which shape women's experiences (Whitworth 1994). The negative human consequences are well known and persist in many countries of the South where population control is deemed to be a precondition for the solution to environmental problems (Hartmann 1987; Oudshoorn 1996).

As new ways of seeing life through scientific investigation change the nature of life itself (Fox Keller 1996), uncertainties over the communication and use of genetic knowledge in screening, treatment and care, reflect an awareness of its power to change the nature of identity and the meaning of life quality. These uncertainties provide further evidence of the complex range of interactions between individuals and the environment (beyond the genes–environment relation) which define and shape human health and well-being. These centre around notions of choice and the auton-omy of the individual. Whilst individual informed choice as opposed to coercive social policy is central to much medical discourse, in democratic states at least, this emphasis has a tendency to separate the individual from the contexts within which choices are made, and from the broader implications of the decisions reached. Consultations with lay people over their views and knowledge of the new genetics and the implications of these for policy making, highlight this point. As Kerr *et al.* (1998b) emphasise, the decisions of those who participate in genetic testing are embedded in an array of cultural, economic and social experiences and pressures associated with the meaning of normality, marginality and life quality which may vary profoundly within and between countries. As part of a strategy to minimise human differences and impairment around a socially constructed norm, for some commentators the Human Genome Project, with its implications of stigma and discrimination, is not radically different from traditional approaches to the 'disability problem' (Shakespeare 1995). Furthermore, as a consequence of screening, the creation of new medical distinctions between those 'at risk' and 'not at risk' to particular diseases has implications for 'informed choices' over possible lifestyle changes, access to medical treatment and the provision of health care. It raises new questions over the significance of 'place' in defining the meaning of individual autonomy, access to 'useful' knowledge, and the responsibilities of the state (Clarke 1995; Harper 1995; House of Commons 1995; Kerr et al. 1998a).

Conclusion

Contemporary intellectual reflexivity and a mood of uncertainty with which it is associated, provide scope to reconsider some of the interrelations between, and meanings attached to, human identity, life quality and useful knowledge. Within the North, relational approaches to space, place and time are widely interpreted as progressive alternatives to established hegemonic identities, the impulse to cultural relativism and the myopic leanings implicit in a tendency towards the pathologisation of history and the global. In emphasising both the strategic and visionary significance of boundaries, these relational approaches are designed to offer more open ways of thinking about peoples and places and the forms of knowledge on which they are based. The practical application of these approaches in strategies to enhance human well-being and life quality point to some difficulties. The new genetics provides an important twist to the notion that identities are always hybrid and dynamic. In creating new visions of life and new possibilities for its further transformation, it also illustrates how the human form is not only a site of manipulation and control but also of resistance to interventions in what are deemed to be natural processes. Implicit in the controversies surrounding this new science and its associated technologies, are debates of national and global significance which transcend a concern with the construction of knowledge in the laboratory and its application to the human form. These debates are not merely about disagreement and dispute but also uncertainty. The technologies that are enabling of genetic certainty bring with them new uncertainties about the knowledge with which they are associated, its sources, for whom and for what purposes it is produced, communicated and applied.

Central to the search for, and use of, this genetics knowledge is a shared concern with the intricate connections between different domains of life, the biological, environmental and social. The geographical discourses employed in the debates which surround it, highlight the complex ethical questions involved and how, in addressing these questions, boundaries can be consciously manipulated and controlled. Much is made of the internationalism involved in the production, ownership and application of this knowledge. An emphasis on newness provides a means by which to distance it from past 'malpractices'. Equally, fears of genetic determinism and coercive social policy are challenged by an emphasis on the interactions between nature and society in shaping human well-being and life quality, and the importance of individual rights and choices as independent decision-makers. These interpretations do, however, overlook the partiality of the international and the cultural power on which this co-operative science is based and which it serves to reinforce. They also disguise somewhat superficial readings of the past and the complex social, cultural and economic contexts which constrain the power of human agency, shape prevailing interpretations of disability, attitudes to difference and collective responsibilities for care. In effect, as material and moral boundaries are constructed, manoeuvred and blurred between space and place, past and present, nature and nurture, the individual and the social context, some of the tensions and contradictions which may arise in the application of relational approaches to space, place and time become clear.

Notes

1 These themes are widely debated. They range from global environmental risk (Prins 1993) through forms of working life (Pahl 1996) to the relations between science, public policy and public opinion (Stern and Fineberg 1996).

2 It forms part of a broad contemporary interest in the relations between geography and ethics (Proctor 1998; Sayer and Storper 1997; Smith 1997a).

3 Iriye (1997) argues that the cultural side of international relations has received less attention than geo-political power and the relations between states. He emphasises the need to consider, for example, the environmentalists, journalists, scholars, artists, musicians and students who flow between cultures, mitigating the rhetoric of national leaders and the menacing accumulation of troops and arms.

4 The need to avoid superficial re-readings is also emphasised (Kennedy 1996).

5 Latour (1987) accounts for science's power in the world by focusing on the networks which transcend and dissolve the dualism between 'inside' and 'outside' the laboratory.

6 The importance of reflexivity is recognised in the social studies of science which stresses the constitutive power of knowledge generally and of scientific knowledge in particular (Knorr-Cetina 1981; Latour 1987; Murdoch 1997).

7 As a subject of common interest to the sciences of nature and society, it provides further evidence of the degree to which, following the work of Michel Foucault, people's lives in western cultures have been medicalised by science and government.

8 Kevles (1997) notes that as a project funded in the United States by the American Government, it has many antecedents. These include the programme of the United States Coast Survey established in 1807 to map the country's coasts, and which was eventually extended to include a geodetic map of the country. Similarly, in 1879 Congress established the United States Geological Survey.

9 Notwithstanding the initial co-operative impulse, scientific competition associated with gene patenting has become increasingly prominent in all forms of genetic research. Within Britain, under the Comprehensive Spending Review to the year 2002 announced by the Labour Government in July 1998, a combination of public money and resources from the Wellcome Trust for science and engineering includes a sum for equipment, reported by *The Independent* as required 'to keep the UK in the race to decode and understand the function of all genes in our DNA – the Human Genome Project' (Arthur 1998: 7).

10 A similar point has been made within cultural studies in which the universalising tendencies implicit in its self-presentation as an international or global movement overlooks its own particularity (Stratton and Ang 1996).

11 In charting the dynamics of the UK's Human Genome Mapping Project from the late 1980s as part of the international programme, Balmer (1996: 537 and 540) noted that, when separated from analysis of gene function, many scientists regarded gene mapping as

'boring and repetitive'. Parallels were drawn with taxonomy or map production which, on its own, was perceived to have no significance or little explanatory value. The gradual erosion of professional autonomy was also identified as the Project became increasingly directed in order to fulfil its stated mission as 'a collective exercise' designed to 'meet the needs of the "genome mapping community as a whole"' (p. 548).

Part III

Becoming in the (k)now

Spaces of identity

13 Space, knowledge and consumption

David B. Clarke

The central issue of philosophy and critical thought since the 18th century has been, still is, and will, I hope, remain the question, *What* is this Reason that we use? What are its limits, and what are its dangers?

(Foucault 1989: 342)

Introduction

Towards the end of the second millennium, in the wake of the information revolution, the long-established alliance of science and commerce has seemingly taken on a new intensity, now defining a globalized 'space of flows' capable of sweeping away much of what went before (Castells 1996). Whatever the merits of such a conception, however, the quiet backwaters of everyday life appear not so much unperturbed by such a momentous sea-change, as eminently capable of adapting to it, even as it interrupts and transforms the established rhythm of their own daily ebb and flow. Thus, whilst 'technoscience' has undoubtedly set a new pace for global change, much of the hyperbole regarding the 'information revolution' neglects to consider the way in which social life has always been profoundly affected by knowledge.

Successive shifts in the status of knowledge have, periodically, brought about the most dramatic discontinuities in the figuration of society. Knowledge today, for instance, exhibits all the signs of the dissolution of the alignment of knowledge and power first forged at the Enlightenment. Whilst this alignment – which saw the establishment of what Foucault termed the 'disciplinary societies' – permitted the initial development of the productive system of industrial capitalism, capitalism has since outgrown the conditions of its birth, rendering the power/knowledge syndrome inherited from the Enlightenment of ever diminishing relevance to its own reproduction. Displacing the modern form of power/knowledge, it is above all the market that has assumed primacy with respect to the systemic reproduction of capitalism. That this role accords to a sophisticated *consumerist* logic is the chief proposition I seek to explicate here. For in the light of such an understanding, I argue, it becomes evident that the current information revolution is but the epiphenomenon of a far more fundamental change in the way in which capitalist society reproduces itself.

Two aspects of contemporary social change stand out as particularly appropriate conceptual lenses for bringing this complex picture of change as a whole into focus. First, the interconnected transformations in capitalism, knowledge, power and everyday life have entailed corresponding changes in the nature of *space–time*. Second, these transformations have, *in toto*, been responsible for gestating the conditions under which the vexed issue of identity has raised itself with increased frequency and concern. In short, the relations between knowledge, space–time and identity are co-dependent and co-determinant, such that changes in any particular field are transmitted and reinforced throughout the others. Furthermore, and insofar as '*the ludic dimension of consumption has gradually supplanted the tragic dimension of identity*' (Baudrillard 1998: 192, italics in original), it can be shown that current concerns over identity – concerns intimately related to transformations in power/knowledge and space–time – equally pertain to a fundamental change in the nature of capitalism. These anastomosing trajectories of capitalism, power, knowledge, space–time and subjectivity, I propose, provide the most suited means with which to comprehend the often puzzling array of social transformations characterizing the latter part of the twentieth century and the end of the second millennium.

Knowledge, legislation, legitimation – and beyond

> The narrative function is losing its functors, its great hero, its great dangers, its great voyages, its great goal.
>
> (Lyotard 1984: xxiv)

It may seem trite to point out that Lyotard's (1984) *The Postmodern Condition* is, as its sub-title plainly states, *A Report on Knowledge*. But the point is easily overlooked. Lyotard's is nothing if not a rigorous examination of what amounts to a changed context for knowledge; of knowledge in a changed context. The surfacing of ideas of knowledge and truth as incommensurate, relative, situated, positional, etc., seems to signify the end of a certain way of viewing the world; one that pinned its hopes on the power of reason. This is, of course, an inevitably retrospective historical formulation. As Bauman (1992a: 24) remarks, the sense of a 'postmodern condition' *necessarily* rests on a definition of modernity 'reconstructed ex-post-facto as an era possessing the selfsame features the present time feels most poignantly as missing, namely the universal criteria of truth, judgement and taste seemingly controlled and operated by the intellectuals'. And it is the perceived absence of such criteria that animates current debate over knowledge – whether this absence is denied or wished away (Harvey 1989); lamented as having been lost too soon (and regarded as recoupable) (Habermas 1983, 1985); or seen as the final relinquishing of what always amounted to a vain hope (Lyotard 1984; Bauman 1987).

Others have commented on the difficulty of treating of 'knowledge' in the singular, of the distinction between 'knowledge' and 'expertise,' of the 'situated' quality of knowledges, and so on (Giddens 1994b; Habermas 1978; Seidman 1998; Thrift 1985). Whilst not divorced from these issues, they can be circumvented for our present purposes by focusing on 'intellectual practices' or the 'intellectual role' (Bauman 1987: 3). This is no arrogant affirmation of the supposed superiority of intellectual knowledge. It is, rather, a case of focusing on a vision of knowledge that took on a singular importance during the Enlightenment, insofar as it held out the promise of *ordering* a turbulent world. Such a focus, moreover, provides ample room for such notions as 'situated' knowledges. Indeed, it helps to reveal something of their context of emergence, as the power of knowledge declared at the Enlightenment is judged afresh, with the all-important benefit of hindsight.

As Bauman (1987) notes, the term 'intellectual' was coined only at the beginning of the present century. It was above all a rallying cry. It signalled an attempt to rise above the diversification of specialist discourses of knowledge – different fields of 'expertise' – produced by the ongoing, fissiparous development of modernity. It did so, moreover, with a nostalgic backward glance towards the status mythically afforded to knowledge at the time of the Enlightenment. For

> [i]t was in that era that the power/knowledge syndrome, a most conspicuous attribute of modernity, had been set. The syndrome was a joint product of two novel developments which took place at the beginning of . . . modern times: the emergence of a new type of state power with resources and will necessary to shape and administer the social system according to a pre-conceived model of order; and the establishment of a relatively autonomous, self-managing discourse able to generate such a model complete with the practices its implementation required.
>
> (Bauman 1987: 2)

If knowledge and power were initially brought into such an alignment during the Enlightenment, however, innumerable circumstances have since conspired to ensure their disengagement. And it is the 'subsequent divorce between the state and intellectual discourse, together with the inner transformations of both spheres, [that] has led to an experience articulated today in a world view and associated strategies often referred to under the name of "post-modernity"' (ibid.). The grounds for the divorce are undeniably complex, as were the initial conditions that led to the marriage. Let us begin to examine these with a look at Foucault's notion of disciplinary power.

'Foucault', Deleuze (1992: 3) notes, 'located the *disciplinary societies* in the eighteenth and nineteenth centuries; they reached their height at the outset of the twentieth.' Such societies represented the eventual response to the breakdown of the localized mechanisms of self-reproduction that characterized the communities of pre- and early-modern Europe. According to Foucault,

> in the seventeenth and eighteenth centuries, we have the production of an important phenomenon, the emergence, or rather the invention, of a new mechanism of power possessed of highly specific procedural techniques, completely novel instruments, quite different apparatuses, . . . which is . . . quite incompatible with relations of sovereignty.
>
> (Foucault 1980: 104)

This 'invention' of disciplinary power marks out a fundamental discontinuity in the figuration of society. Disciplinary power, as Foucault conceived of it, bares a 'capillary' form, being of the order of a *dispersed technology* – rather than a possession, capable of being seized. As such, it differs fundamentally from the sovereign power that preceded it. Its chief mechanism is surveillance, again marking a contrast with sovereign power. Pre- and early-modern societies were left much to their own devices, subject to secular power predominantly in the form of the demands made by the sovereign for a share of the surplus such societies produced. The role of the sovereign was thus strictly circumscribed, being, as Deleuze (1992: 3) notes, 'to tax rather than organize production'. Modern society, in contrast, sought to deploy the dispersed technology of power-through-surveillance to organize and discipline heterogeneous bodies, in a manner designed to achieve a certain productive *efficiency*. If sovereign powers 'did not consider the production of the surplus as their responsibility, fully satisfied as they were with the enforcement of its repartition' (Bauman 1983: 36), this changed fundamentally with the advent of the disciplinary societies.

Bauman (1983), however, questions Foucault's characterization of disciplinary power as an 'invention'. To all intents and purposes, the community-based repro-duction characterizing pre-modern Europe already exercised its own proto-form of disciplinary 'power', and its own manner of 'surveillance'. The seventeenth and eighteenth centuries did not so much witness the 'invention' of disciplinary power, therefore, as its *discovery*. Something formerly practised routinely, unquestioningly and matter-of-factly, was brought into view as a problem; and once problematized, the way was opened to a new 'solution'. Pre-modern communities had engaged in a kind of general, reciprocal 'surveillance', ensuring forms of behaviour suited to the community's own reproduction (together with the production of an economic surplus sufficient to placate the demands of the sovereign). Viewed against the highly visible form of sovereign power, such mutual 'surveillance' was hardly recognizable as a form of *power*. Only when redeployed, in the characteristically asymmetrical terms of a power relation, did surveillance become apparent as a form of power; marking the transformation of the substantive content of sovereign power into the capillary form of a dispersed technology.

Disciplinary society amounted, therefore, to a distinctive form of social organization – one that was to afford to knowledge a new centrality. It rested, as Bauman (1992a) emphasizes, on a new way of viewing the world; one centred on the discovery of *culture*. Whereas differences in the forms of human life had previously been recognized but regarded as essentially pre-ordained and ineluctable – each conforming to its own position within the 'great chain of being' (Lovejoy 1961) – the rearticulation of difference as *cultural* difference problematized what had previously seemed unproblem-atic. Once framed as specifically cultural differences (rather than, say, *natural* or *given* ones), different forms of life could not but be regarded as essentially malleable – as something that need not be as it is, something amenable to change by the self-conscious deployment of human design. So it was that, for almost a century up to the last quarter of the eighteenth century, the term 'culture' carried a strong 'actionist' implication – expressing a *task* with respect to a *problem*.

> To culture (cultivate) land, meant to select good seed, to sow, to till, to plough, to fight weeds and undertake all the other actions deemed necessary to secure an ample and healthy crop. This was exactly the shape of the task in relation to human society, as it appeared in the wake of the bankruptcy of self-reproducing mechanisms.
>
> (Bauman 1987: 94)

The emergent *ideology of culture*, then, owed more than a little to the breakdown of the previous, localized mechanisms of social reproduction. It spoke of a situation where 'some forms of life were becoming problems for some others' (Bauman 1992a: 5), where a formerly unproblematic co-existence – in no need of 'solution' – was no longer deemed tenable. It was the strength this ideology gave to the vision of human diversity as malleable and manipulable that made possible the emergence of the disciplinary societies.

Arguably the most crucial factor in their emergence was the knowledge essential to the formidable task of stemming autonomous behaviour – in favour of a heteronomous behaviour appropriate to the modern vision of an ordered, rational society. A unique series of historical and geographical circumstances, which came together in France in the third quarter of the eighteenth century, was finally responsible for galvanizing the new kind of power. Although various contributory factors preceded this moment, and although its development was to evolve significantly from this point on, these unique circumstances, 'in a flash, short-circuited knowledge and power' (Bauman 1987: 25; cf. Hazard 1968a, 1968b). The circumstances in question – most notably, the rise of absolutism and the new role carved out for knowledge by *les philosophes* – were to provide for a fundamental 'redeployment of political power in the wake of . . . the waning of the feudal principle of association between land-owning rights and administrative duties' (Bauman 1987: 25). It was not so much any particular unity of thinking that marked out the significance of *les philosophes*, as the role created by *le monde* within which they and their ideas circulated; by the self-professed autonomous status captured in the very idea of *la république des lettres* (Habermas 1989). The problems arising from the bankruptcy of community self-reproduction – often long-standing problems such as vagrancy, which had ingrained a profound sense of insecurity in early-modern Europe (Fébvre 1968; Muchembled 1978; Fletcher and Stevenson 1985; Beier 1985) – were now to be self-consciously overcome by the mobilization of the power of knowledge over the dead weight of tradition.

The Enlightenment, whatever its undoubtedly messy and complex realities (Darnton 1984), was to impose upon knowledge a new sense of certainty; a condition that had remained absent since the fracturing wrought, most notably by the Reformation, on a pre-modern certainty underwritten by divine authority. The preceding protracted period of sceptical thought, associated with a tolerant humanist culture, was now to be relabelled as the 'Pyrrohnian crisis' (Popkin 1979), signalling the new certainty underpinned and guaranteed by knowledge. Efficient power demanded efficient knowledge, and the power of reason proclaimed at the Enlightenment was marked by a universalist, proselytizing ambition. Such a self-conscious proclamation – characterized by d'Alembert, in his preliminary discourse to Diderot's *Encyclopédie* of 1751, as the 'intensification' of the '"light of reason" first lit in Greece, which had been "rekindled" in the fourteenth century after almost a millennium of darkness' (Hume and Jordanova 1990: 1) – placed the men of knowledge in the metaphorical role of *legislators*, giving them 'the right to command the rules the social world was to obey' (Bauman 1992a: 11). Such a role was, moreover, self-legitimating. It could now seem only reasonable that those with privileged access to right knowledge should assume the right to instruct others, 'deprived of such an access, what to do, how to behave, what ends to pursue and by what means' (Bauman 1992a: 9). Enlightenment, however, inevitably cast its shadows. The limits imposed by the world of patronage and privilege within which the Enlightenment's thinkers circulated would ultimately be responsible for gestating the tendencies that would erupt into revolution. From the start, the autonomous role of *les philosophes* was destined to witness the kind of frustrations and limitations that could only multiply when brought up against the realities of social power. But none of this detracts from the significance of the role granted to an *autonomous* public sphere – in contradistinction to the earlier role afforded to landed interests – and the new centrality this autonomy gave to knowledge (Habermas 1989). As Bauman (1987: 70) insists, the French Revolution was 'nothing else but the end product of the discourse originated by the Age of Reason and continued by the Age of Enlightenment'. And whilst one can see certain of these conditions in other parts of Europe at earlier times (Hill 1940; Corrigan and Sayer 1985), and whilst a number of circumstances had already set the conditions of possibility for the capitalist development which portended the transition from feudalism (Baechler *et al.* 1988; Duplessis 1997; Hall 1985; Mann 1986), it was nonetheless the unique circumstances coming together at the Enlightenment that finally set the cast from which the disciplinary societies were to be moulded.

As Foucault (1980, 1989) emphasizes, moreover, the change from sovereign to disciplinary power witnessed a new 'politicization of space' in the eighteenth century: 'The cities were no longer islands beyond the common law. Instead, the cities, with the problems they raised, and the particular forms that they took, served as models for the governmental rationality that was to apply to the whole territory of the state' (Foucault 1989: 336). Whilst Foucault confesses to a certain 'awkwardness' when stated in such terms – both Livingstone (1992b) and Dodgshon (1987), in different ways, reveal a far more protracted political significance to space than the developments with which Foucault concerned himself – there are good reasons why Foucault should focus on the eighteenth century. The period gave rise to a growing concern with institutions of confinement – 'prison, hospital, factory, school, family' – 'each having its own laws' and being characterized by 'the organization of vast spaces of enclosure' (Deleuze 1992: 3–4). As the reproduction of the lower orders of society changed from a process beyond the concerns of sovereign powers, to be redefined as the direct responsibility of the dominant classes, the resort to 'spaces of confinement' proved the most efficient way of surveilling and disciplining bodies. Reorganization of space was necessary to mould a new social reality; to proscribe autonomous behaviour and ensure the heteronomous behaviour appropriate to the goal of an ordered, managed society.

However, whereas state power had initially appeared fully dependent on knowledge, its gradual disengagement from such a need, as the state developed new governmental technologies of its own, was to render the legitimizing services of the intellectuals increasingly redundant. It is in such a light that the coining of the term 'intellectual' should be interpreted. For what was subsequently to be painted by the intellectuals as a 'legitimation crisis' (Habermas 1976) was, from the point of view of an increasingly robust modern state, nothing of the sort. The development of the new technologies of confinement and surveillance allowed for an increasingly efficient form of social reproduction, less and less reliant on legitimation from those claiming superior knowledge. Indeed, the fissiparous 'creative destruction' of new, specialist forms of expertise came to undermine the original role conceived for knowledge. If this tendency gradually demoted the role of knowledge *per se*, however, it was nonetheless not without its compensations. Expertise was increasingly indispensable to social reproduction, although now the task was primarily to ensure the smooth functioning of the political technologies of confinement, disciplinarity and surveillance (and, increasingly, the needs-creation that would ensure social compliance by means of 'seduction' rather than out-and-out coercion). Such requirements hardly granted the intellectuals the kind of role first envisaged by those *philosophes* originally responsible for carving out such a position. But the swelling ranks of experts, professionals and bureaucrats found adequate consolation, the loss of political autonomy being associated with either a significant level of material comfort, or a significant degree of intellectual freedom, or both.

Accordingly, 'while rendering the "legislative" function of the intellectuals irrelevant, the modern state disposed of any reason why the intellectual discourse should be subject to political control or otherwise externally limited or regulated' (Bauman 1992a: 16). Such freedoms were not, of course, granted always and everywhere with equal magnanimity – as the inglorious histories of state socialism and fascism that swept through the public sphere in the twentieth century amply demonstrate. Such differences – which revealed intellectual freedom to be fragile indeed – made those who had been granted this 'autonomy of no practical consequence' (ibid.) cherish it as all the more precious. To those holding on to such freedom, however, a further blow was yet to be dealt. The intellectual gratification derived from the autonomy to pursue the ideals of truth, judgement and taste for their own sake was largely reliant on the intellectuals' belief that it was theirs by right to hold a monopoly on such issues. The discourse to which they committed themselves held a certain privileged position within the culture whose values it preserved and refined, simultaneously defining, for the intellectuals, the singular importance of their task. But, as the 'interests of the state in culture faded (i.e. the relevance of culture to the reproduction of political power diminished), culture was coming within the orbit of another power the intellectuals could not measure up to – the market' (1992a: 17).

It was, moreover, in relation to this brash new competitor – whose powerful role had already been established beyond doubt by the early years of the twentieth century – that a particularly spectacular turn-around in the intellectual conception of 'culture' occurred. The original ideology of culture was firmly tied to a vision of *universal* values of truth, judgement and taste, which the intellectuals, in the role of legislators, had seen as fulfilling its historical destiny in the vanquishing of heterogeneous forms of life founded on supposedly inferior or backward ways of thinking. The homogenizing tendency now attributed to the market was cast as performing precisely the same kind of role – but in a manner that no longer seemed quite so reasonable. From now on, the vision of culture increasingly promulgated by the intellectuals was one that cherished diversity, one that sought to defend difference against market-driven homogenization. Whilst it is probably far from the truth of the matter to see the market as possessing such homogenizing tendencies, it is nonetheless telling that this *volte face* in the intellectuals' vision of culture was only finally able to take place once the original legislative role of the intellectual had been irredeemably lost. In other words, the new possibility of envisioning culture as *relative* arrived on the scene together with the delegitimation of the 'legislative' role of the intellectual (which rested, of course, on a vision of culture singularly opposed to such relativism).

It is in relation to this new situation that one can finally begin to appreciate the problematic status of knowledge in the eyes of its self-declared guardians today. Certainly, the computerization of society and the attendant information explosion, which sees knowledge ever more entrained within the vicissitudes of the market, appears as anathema to those understandings of knowledge owing most to their Enlightenment heritage. Thus, for instance, the contemporary role of critical theory largely draws its strength, sense and importance from the original status afforded to the ideals of pure reason – a status that appears as being degraded and diluted as the 'mercantilization of information' begins to hold sway (Lyotard 1984). Here, as Bauman (1992a: 20) puts it, 'in a curious twist of mind, this pure, ethereal, cultural ideal is believed to be reality's best chance'. But as Lyotard (1984: 14) notes, the sort of role for knowledge propounded by critical theory rests on an internally contradictory 'partition solution'. It draws a distinction between, on the one hand, knowledge of 'the positivist kind, . . . directly applicable to technologies bearing on men and materials, . . . lend[ing] itself to operating as an indispensable productive force within the system'; and, on the other hand, knowledge of 'the critical, reflexive, or hermeneutic kind . . . reflecting directly or indirectly on values or aims' and resisting 'any such "recuperation"' (ibid.). In so doing, however, it accepts a situation that is 'unacceptable' in the very terms of the vision of universal knowledge from which it draws its strength. Transformations internal to the sphere of knowledge have, in short, resolved themselves in paradox (in the literal sense of that term: *re-solvere*, to re-release; to unfasten again).

The *delegitimation* of universal knowledge such paradoxes have served to nourish has, therefore, spawned a new situation; one presenting the status of knowledge in entirely novel terms. The 'incredulity towards metanarratives' (ibid.: xxiv) – together with a loss of nostalgia for such metanarratives – amounts to a recognition of the relativity of knowledge that is 'not just a transitory state', 'a difficulty to be resolved theoretically *and* practically', but is, rather, 'an apodictically given existential condition of knowledge' (Bauman 1992a: 21). Lyotard's (1984) much-discussed methodology of 'language games' amounts to one strategy for coping with the new situation coming in the wake of the delegitimation of universal knowledge. For Bauman, such a strategy recasts the role of the intellectual in accordance with the metaphor of the *interpreter*. Interpretation 'must make the interpreted knowledge sensible to those who are not "inside"' a particular language game; but, given the absence of anything but the relativity of knowledge, and thus 'having no extra-territorial references to appeal to, it has to resort to the "inside" itself as its only resource' (Bauman 1992a: 22). Lyotard (1986), however, concerns himself above all with the *injustices* associated with the incommensurability of knowledges, and hence with the necessity of *enforcing the exclusivity* of language games or 'phrase regimes'. Indeed, the existence of the 'differend' – that which escapes expression within one genre of discourse but not within another – entails that the intellectual's role must, paradoxically, be to enforce the exclusivity of genres; to insure against one genre legislating against another. The new role for the intellectual is to 'legislate against legislation'; the intellectual adopting the position of the 'great proscriber' (Lyotard and Thébaud 1984). For those who refuse the nostalgic strategies of critical theory, as well as resisting the redefinition of the intellectual role as prescribed by the market (Cohen 1993), the task remaining, however it is envisaged, is to address the apodictically given conditions of knowledge that have been disavowed for so long. It is into these uncharted waters that knowledge now passes.

Space–Time in transition

The past went that-a-way.

(McLuhan 1967: 74)

As the above account has sought to demonstrate, the postmodern condition is one expressive of the loss – albeit without nostalgia – of a state of affairs underpinned by universal precepts of truth, judgement and taste. It brings in its wake a retrospective reconstruction of a modernity guided above all by the presence of precisely such universal precepts of knowledge. What has remained almost entirely implicit in the foregoing account, however, is that modernity *as such* rested on a particular vision of space and time. The modern conception of time, for instance, through a process of metaphorical transference, drew upon certain properties more usually associated with space. It also carried with it certain direct implications in terms of space. The contention to be pursued here is that the composite vision of modern space–time has now begun to break down, in a manner which explains the preoccupation with space that surfaces in all manner of writings on postmodernity (Harvey 1989).

With the benefit of hindsight, it is perhaps unsurprising that modernity, given its pre-eminent concern with the power of reason, found itself committed to a foundational sense of time as history. The precepts of reason, as modernity conceived of them, gave to time a sense of direction, directly comparable with the itinerary of a journey in space. The notion of progress – which, through the sustained application of reason, was to improve upon and displace all preceding forms of culture by the sheer fact of its *self-conscious* recognition of culture as man-made – set the vision of time as historical advancement in sway. Whereas 'for most of the history of Christian Europe, time-reckoning was organized around a fixed point in the slowly receding past', modernity's faith in reason 'set the reference point of objective time in motion, attaching it firmly to [modernity's] own thrust towards colonizing the future in the same way as it had colonized the surrounding space' (Bauman 1987: 110). The faith placed in modernity's self-defined historical centrality was such that it automatically registered as wanting any alternative form of existence, which was thereby characterized as belonging to the 'wrong side' of history.

Moreover, the world as perceived by modernity amounted to a world marked by a sense of *finality*. 'Time progressed from the obsolete to the up-to-date, and the up-to-date was from the start the future obsolescence' (Bauman 1997a: 86). In other words, the modern sense of time projected the inadequacies of the present back from the future – necessarily casting the present as in need of improvement. To paraphrase Benjamin (1968), it is as if our faces are turned towards the past as we are irresistibly propelled into a future to which our backs are turned, able to witness the debris of history piling before our eyes. Or again, to cite McLuhan (1967: 74–5): 'We look at the present through a rear-view mirror. We march backwards into the future.' Such a vision of time endowed certain events with particular qualities and, in so doing, carried particular implications with respect to space. Specifically, it replaced any notion of cyclical recurrence – a notion that, by all accounts, was commonplace before the dawn of modernity, and which has never been fully exorcised from the West – with a *linear* conception of time; a conception concerned with origins and finalities, with the idea of 'no return'. Births or origins are, by definition, endowed with the quality of being *reversible* or revocable; something that is 'here today' can just as well, as the popular adage has it, be 'gone tomorrow'. Finalities or endings, however, are irredeemably *irreversible*: 'a being that has ceased to be will not be again' (Bauman 1993a: 32). In such a world, which accords to Henry Ford's reputed characterization of history as 'one damn thing after another', '[s]pace is at a premium. Beings must vacate the place they occupy if other beings are to appear' (ibid.). Space in its modern conception represents a vessel of finite capacity, within which room has to be made if the irresistible force of progress is to proceed. A key characteristic of modern space was, therefore, its representation as abstract, homogeneous, and empty (Lefebvre, 1991). Thus, space under modern conditions amounted to a clearly ordered, bounded and mappable 'cognitive space' (Bauman 1993b: 146), providing a suitable template for the ordering of a rational society, and for plotting out the trajectory of its progress over time.

One way of understanding postmodernity – which amounts to a re-statement, in different terms, of Lyotard's 'incredulity towards metanarratives' – is in terms of a changed vision of space–time. This new vision issues from the decomposition of faith in History as a progression set in motion by the universalizing power of Reason. And it involves nothing less than the end of the modern vision of linear time. 'What *dies* is annihilated in linear time,' writes Baudrillard (1990a: 32, emphasis added), 'but what *disappears* passes into a state of constellation. It becomes an event in a cycle that may bring it back many times.' Postmodernity, therefore, can be conceived as a kind of re-release of space from the modern vision which held it to be *scarce*, in direct consequence of the meaning put on time as linear, progressive, and irreversible. Hence the association of postmodernity with the untimely, the uncanny, and the unhomely (*unheimlich*) – with the return of everything modernity sought to repress (for fear it would undercut its attempts to impose order and stability). Accordingly, as Lyotard (1984: 81) suggests, the '*Post modern* would have to be understood according to the paradox of the future (*post*) anterior (*modo*).' The postmodern sensibility amounts to a recognition of the modern tendency to stake its hopes on an all too elusive presence; and a reflexive recognition that things only ever *will have been*.

The break-down or decomposition of linear time marks the end of all finalities – the paradoxical end of 'the end' as such; the end of the irreversibility of endings (Baudrillard 1990b, 1994; Clarke and Doel 1995). Unlike death, disappearance carries no certainty of finality; it becomes an event marked by the possibility of its recurrence. Likewise, origins, once seen to be characterized by their reversibility, now take on an uncanny sense of irreversibility. Once born – if they can be said to be 'born' as such; for what seems to be a birth is now just as likely to be a reincarnation – they are here to stay. In a world devoid of the apparent certainties of a space–time structured by linearity and irreversibility, the virtual co-existence of reversible and undecidable beings implies a fundamental non-scarcity and heterogeneity of space; an ecstatic space marked by excess (Bataille 1985, 1988). Bauman (1993a: 32) conceives of it thus: 'space has many levels, its living rooms and its cellars, open stages and hidden limbos. To make room at one level, beings may, and do, just move to another.' It is preferable, however, to conceive of space–time as a multiplicity of foldings, unfoldings and refoldings on a single plane of immanence: '*Everything now returns to the surface*' (Deleuze 1993: 45).

The upshot of all of this, as space–time takes on new qualities in the wake of the decomposition of modernity, is that space–time loses the solidity and durability it once seemed to possess. The possibility of space–time providing a framework within which to plot out unambivalent paths of progression seems now to have all but disappeared. The certainty once ensuing from the robust modern structure of space–time is no longer in place. Again, we face here an unavoidably retrospective definition of modern space–time. Just as postmodernity was seen to take its putative meaning from the absence of universal criteria of truth, judgement and taste once seemingly defined and controlled by the intellectuals, those qualities retrospectively attributed to modern space–time are the ones we now find conspicuous by their absence. 'What we think the past had – is what we know we do not have' (Bauman 1997a: 87).

From identity to transformation: production, consumption and (post)modernity

> We now live . . . in an *open space–time*, in which there are no more *identities*, only transformations.
>
> (Bauman 1992b: 184)

One of the most dramatic growth areas in the social sciences in recent years – one seemingly intimately related to the kind of transformations described above – concerns the question of 'identity'. Superficially, it seems as if identity has only begun to pose itself as a problem in the wake of the fragmentation, relativity, incommensurablity, undecidability, indiscernibility, etc. that characterize our putative 'postmodern condition'. Yet 'at no time did identity "become" a problem; it was a "problem" from its birth – was *born as a problem*' (Bauman 1996: 18–19). Once again, we are facing a particular case of 'the general rule of things being noticed only *ex post facto*' (ibid.: 18). It is only with the eclipse of a certain vantage point that identity reveals itself as having been a problem from the start; and identity, it now seems, was constructed in anamorphic perspective all along.

A definitive quality of a problem is that it calls for action, and this is indeed a fully ingrained quality of identity. 'Identity,' proposes Bauman (1996: 19), 'though ostensibly a noun, behaves like a verb, albeit a strange one to be sure: it appears only in the future tense. . . . Identity is a critical projection of what is demanded and/or sought upon what is; . . . an oblique assertion of the inadequacy or incompleteness of the latter.' Identity is necessarily formulated in relation to the need to reduce the gap between what is and what might yet be. In Giddens' (1994b: 80) terms, 'Identity is the creation of constancy over time, that very bringing of the past into conjunction with the future.' Hence the nature of 'identity' as a process, a project, and a *task*. For Giddens, such 'constancy' is prerequisite to the ontological security that any form of social existence demands, for reducing the unbearable contingency and unpredictability the world would otherwise possess. This ambivalence and contingency, however, has been dealt with in different ways, at different times, and in different societies. Given the above remarks regarding the significance of linear time to modernity, the relation between identity and temporality demands that attention is paid to the connection between modernity and identity.

For the most part of human history, societies structured social life in accordance with 'an orientation to the past, such that the past . . . is made to have a heavy influence over the present . . . and used as a way of organizing future time' (ibid.: 62). A distinctive trait of modern society, however, as we have already seen, was the carving out of the future as a separate time-frame, one whose content and actuality is always still to be made. Thus, under modern conditions, rather than the past providing a guide to the present and a means of anticipating (and therefore discounting) the future, the future represented a place to be colonized anew, overcoming the inadequacies and limitations of the past. To the modern mind, socially ascribed identities of the kind found in Lovejoy's 'great chain of being' were an anathema; they came from the past, without a view towards what the future might be made to be. They foreclosed on the future, discounting the possibility that the future may always be otherwise than more of the same. If modern conditions were to prevail, identity had to take on a new character.

Pre-modern societies comprised of 'self-enclosed ranks', each representing an ideal for identity, such that 'emulation across the boundaries was frowned upon and considered morally morbid – a tinkering with the divine order' (Bauman 1992a: 4). The ideology of culture ushered in by modernity, however, signalled the end of this structure. Whereas once the ideals of identity were 'as numerous as the ranks themselves' (ibid.), the universalizing rationality of modernity instated an ideal characterized by its abstract generality – and hence the task of identity was increasingly delegated to the self. Enjoined to forge its own identity, the modern subject – born of the seventeenth century (cf. Althusser 1972; Dumont 1977, 1986) – experienced a certain freeing of the bonds of tradition. Such a freedom, however, brought with it certain costs. Specifically, it brought us ever closer towards a situation where, to use Giddens' (1994b: 75) phrase, 'we have no choice but to choose' (which is *not* to imply that choice has ever been unconstrained; nor that certain choices were made more accessible to some than to others, being distributed broadly along class lines, in terms of gender, and in terms of numerous other complexly imbricated social dimensions). Choice, it follows from this, necessarily possesses a double-edged character. It encapsulates freedom of action; but thereby, paradoxically, renders such freedom *necessary* – coupling it to a potentially anxiety-ridden responsibility. So it was that, under modern conditions, identity took on the character of a personalized problem, the solution to which lay in the future; or more accurately, in a certain orientation towards the future – without which 'the future' would not be what it is, here and now in the present. In the modern world, therefore, it was always a matter of constantly maintaining an identity; of *identity-building*. Modernity promised the subject personal fulfilment and self-completion, but since the future never actually arrives as such – since such completion can never be finally achieved – modernity founded itself on a *false* promise. Completion only ever appears in the paradoxical tense of the future anterior; it only ever *will have been*. Such a situation, of course, provided ample opportunity for social commentators to expose the rife inequality accompanying the freedoms that modernity offered; but it nonetheless differed profoundly from pre-modern social arrangements, inasmuch as it at least offered the promise of the possibility of self-advancement. (That it did so falsely, that it could not keep its promise, and that it inevitably led to different degrees of advancement – distributed unevenly in relation to class, race, gender and so on – in no way detracts from its difference from all preceding modes of social arrangement.)

In such a situation, to borrow the allegorical language deployed by Bauman (1996), life amounted to a kind of 'pilgrimage'; a search for a destiny necessarily removed from the here and now; a future destination towards which one must aim. Life as pilgrimage was cast as a matter ultimately resting with the individual; as a matter for the volition and good judgement of the individual, and hence as an individual *responsibility*. All of which carried certain strategic implications – if one was to hold out the hope of success. Such a situation entailed that one must select one's destination sooner rather than later, if maximum progress were to be made; and that one must stick unswervingly to the path, deferring gratification today in order that more substantial rewards might accrue when one finally arrived at one's preselected destination in the future. Little wonder, then, that the world of work should have assumed so central a role in the modern social order. And this was so not only in individualized terms, where work 'made the difference between affluence and indigence, autonomy and dependence, high or low social status, presence or absence of self-esteem' (Bauman 1988: 71), but also in terms of social integration: 'it was through the work-place that the most meticulous social surveillance and monitoring of individual behaviour took place. . . . [T]he work-place served, in other words, as the main training-ground for attitudes and actions suitable for the hierarchically differentiated norms of . . . [modern] capitalist society' (ibid.: 72). Work also formed, moreover, the central systemic necessity: 'The maintenance and the reproduction of economic and political structures depended on capital engaging the rest of the population in the role of pro-ducers' (ibid.: 73). The sphere of production encapsulated the logic underpinning the task of identity-formation itself. As Baudrillard (1987: 21) reminds us, 'The original sense of "production" is not in fact that of material manufacture; rather, it means to render visible, to cause to appear and to be made to appear: *pro-ducere*.' However, the validity of such a life-strategy was fundamentally dependent on the world possessing suitable qualities for one's progress to be tracked as a progression; on the world retaining sufficient stability that so arduous a journey might finally yield its eventual rewards.

If conditions where one could make one's tracks towards a pre-defined future destination were once prevalent, as Bauman (1996: 23) curtly remarks, '[t]he world is not hospitable to pilgrims any more'. More accurately, if modernity promulgated the belief that the future would see a progressive increase in the order of the world, making it increasingly amenable to life as pilgrimage – and making life as pilgrimage more and more attractive – the increased entropy inevitably resulting from the sum total of localized orderings always rendered such a belief an illusory one. Modernity was always a self-reinforcing maelstrom, however much it tried to annihilate disorder and ambivalence (Bauman 1991). Hence, rather like the harsh lesson learned from the housing market – that 'negative equity' is always a possibility – the more determined and well-constructed a life-strategy is, then the more *inflexible* it is also likely to prove. This is a matter of considerable importance, given the apparently increasing likelihood of fundamental change in the social environment. Hence, whereas 'the *modern* "problem of identity" was how to construct an identity and keep it solid and stable, the *postmodern* "problem of identity" is primarily how to avoid fixation and keep the options open' (ibid.: 18). Deferring gratification today, for example, might prove not to be so valid a strategy – if the possibilities for gratification might evaporate altogether overnight. In the context of an uncertain future, 'get it while you can' becomes the appropriate slogan for the postmodern problem of identity. Or, more accurately, given that a problem has, by definition, a solution, identity now reveals its formerly hidden *aporetic* nature.

The situation we face today, therefore, is fundamentally different to the world where pilgrimage was not only a viable but a dependably profitable life-strategy. More precisely, such a situation seems increasingly prevalent within certain 'expanding enclaves of "post-modern" existence' (Bauman 1997b: 24). Under such conditions, one can no longer be certain of attaining a sense of security by sticking to the straight and narrow. One can no longer be sure that sheer unflinching dedication will guarantee the kind of future payoff it once promised. The inflexibility or rigidity involved in following such a life-strategy seems to be increasingly risky; a case of putting all one's eggs in one basket. In a society in which risk has become endemic (Beck 1992), danger now seems to accompany every strategy that once promised to reward dedication to the project of building one's identity. The road to success seems to wend its way through a landscape engendering a profound sense of anxiety (Pahl 1995).

Because modern life as a form of pilgrimage sought to render the world such that identity could be 'built at will', the 'space in which identity was to be built [was turned] into a desert' (Bauman 1996: 23) – devoid of all the worldly distractions likely to divert or prematurely halt one's progress. But the desert, whilst 'comfortingly featureless for those who seek to make their mark, does not hold its features well' (ibid.). In such a world, progress becomes increasingly difficult to track across the shifting sands, just as it becomes increasingly difficult to determine in advance whether one's chosen destination – even if one could be sure of making one's way there – will prove to be an oasis or merely a mirage. Increasingly, we might hypothesize, the suspicion is that the latter is a more common outcome than the former. Success from following a strategy underpinned by the logic of production thus seems increasingly a chimera, not least because of the way in which the goal-posts keep moving. The belief that success from sustained dedication could ever be the *norm* appears increasingly incredible. The putative solution to the problem of identity has, accordingly, become detached from the logic of production. The era shaped by the '*tragic dimension of identity*' (Baudrillard 1998: 192) has run its course.

That this is so results, above all, from the fact that competition in the field of production possesses the character of a zero-sum game (Bauman 1988) – generating both winners *and* losers, the winners only possible *at the expense of the losers*. In addition, it possesses the form of an 'elimination contest' (Elias 1982), whereby the established winners serve to constrain and even prevent new competitors from taking part in the game. Such characteristics have fuelled the perception that this is no longer – if indeed it ever was – a game for mass participation. Even in the most literal of interpretations, however, this is not to deny that work is still *necessary*; but it is perhaps precisely as a necessity – for many as a 'necessary evil' – that work increasingly assumes the status of a means for life, rather than an end in its own right. For those inhabiting the expanding enclaves of postmodern existence, 'in which people are consumers first – and workers only a very distant second' (Bauman 1997b: 24), work is increasingly regarded as a means for a life constructed *elsewhere*. One perceives a general unwillingness to form, and act in accordance with, the kind of future goal that an identity based on work was once expected to attain. Whether literally or metaphorically, the work ethic no longer holds sway.

If identity has become detached from the logic of production, however, it has done so only to find a new affinity with a *consumerist* logic, the ludic nature of which affords a greater degree of play within the system. Increasingly, we are presented with a 'market solution' to the 'problem of identity' – a 'problem' which actually amounts, of course, to a supplementary detour for the aporia of identity. 'Identities are not scarce goods,' writes Bauman (1988: 63), bringing home the point that in the sphere of consumption we face the situation of a non-zero-sum game. In such a situation, consumers as 'winners' do not necessarily exclude other consumers from claiming an equal prize (Hirsch's (1976) 'positional goods' represent, here, the exception that proves the rule). This is not to suggest that consumption does not have its social costs – in the form of those excluded from what amounts, perhaps, to a 'rigged' game; those systematically dealt a 'rum deal' – but that, for those engaged in the game, each can potentially come away a 'winner'. The 'market solution' to the 'problem' of identity provides, in principle, the opportunity for one consumer to procure something *without precluding another from doing so too*. There may, of course, be other consequences stemming from this: it may fuel dissatisfaction as a given object becomes too common and indistinctive against the crowd; or stocks may run out before all the *potential* winners have played their turn. One suspects, however, that these counterposing tendencies might well be manipulated by those bringing their goods to market. The characteristics of the market are, moreover, particularly well suited to the concern of keeping one's identity flexible rather than with building a rigidly fixed identity. Consumption is characteristically organized in terms of (or, from the standpoint of the individual consumer, presented in the form of) an immense combinatorial and kaleidoscopic array of commodities – a '*system* of objects', which renders commodities as differential sign-values rather than particular use-values (Baudrillard 1981, 1996, 1998). Such a system is suitably geared towards overcoming the possibility of the market solution to the problem of identity 'undoing itself' by tending towards the provision of precisely the same commodities, such that individual distinctions all but lose their distinctiveness.

This situation has brought about a profoundly important reconfiguration of the *systemic* role of consumption (Bauman 1987, 1998; Clarke 1997; Clarke and Bradford 1998). Market-based consumption, as a mechanism promising the means to tackle the problem of identity, does not exist for beneficent reasons. Given that identity is a problem with an impossible solution – that is to say, since the subject can never achieve the completion it desires, and is destined to experience an insatiable lack – consumption is a never-ending process. To this extent, therefore, and despite its character as a non-zero sum game, consumerism too, like the earlier form of capitalism, makes a promise on which it cannot deliver. The lack the subject experiences cannot be made good, despite the incessant assertions emanating from the advertising industry that it can. The market, moreover, does not act to ameliorate this lack, but finds additional 'problems' for which market 'solutions' can be made available. Whilst advertising has long been recognized as functioning as a kind of retroflex social sanctioning for individual consumers (such that, as Erlich *et al.* (1957) discovered long ago, car-owners tend to take note of the adverts for the brands they *already* own, as a means of confirming the wisdom and foresight of their consumption choices), paradoxically, advertising *as a whole* simultaneously serves to arouse *as well as* curb anxiety – keeping the consumer market in a state of continual flux or dynamic tension. Under such conditions, 'choice' becomes increasingly pervasive and obligatory. As Giddens (1994b: 76) puts it, 'all areas of social life come to be governed by decisions – often, though not universally, enacted on the basis of expert knowledge of one kind or another'. Hence, whilst the production of uncertainty is rife, it is faithfully accompanied by the production of the specialist knowledge or expertise capable of curtailing such uncertainty. In short, symptom and cure are twin-born, as everything from home-improvement magazines to Prozac seems to demonstrate. Thus, although Warde (1994: 890) fundamentally questions whether 'consumption really is a source of chronic anxiety', pointing to the absence of widespread, explicit anxiety-related disorders, as Bauman suggests, consumerism pulls off a remarkable balancing act. In so doing, moreover, it arguably represents the most efficient and potentially the most hegemonic figuration of capitalist society yet seen:

> The consumer market is . . . a place where freedom and certainty are offered and obtained together: freedom comes free of pain, while certainty can be enjoyed without detracting from the conviction of subjective autonomy. This is no mean achievement of the consumer market; no other institution has gone this far towards the resolution of the most malignant of the antinomies of freedom.
>
> (Bauman 1988: 66)

Conclusions

The survival of capitalism owes a great deal to its evident capacity to transform its own systemic structure. The manner in which it has taken on a consumerist guise provides a clear-enough illustration of this. Today, not only has the system of objects assumed an increasingly central role in defining individual identity – or rather, in providing the means for individual *transformations* in an era where the 'problem of identity' is rendered aporetic – it has also taken on an increasingly predominant role in terms of social integration and systemic reproduction. The form of social compliance achieved through the system of objects; through *seduction*, through the 'carrot' of needs-creation rather than the 'stick' of repression (Baudrillard 1990c; Bauman 1987; Deleuze 1992) – seems to guarantee the perpetuation of capitalism. It does so, however, only at the cost of a relentless process of decoding and deterritorialization (Baudrillard 1993; Deleuze and Guattari 1988); always a dangerous strategy for opening up weaknesses in the least suspected of places.

Consumption thus presents itself as nothing more than an individualized form of pleasure, whilst simultaneously operating as a form of social control – a form capable of generating social compliance in a manner very different to the methods of constraint and confinement practised by the disciplinary societies. As Deleuze (1992: 4) puts it, whereas '[e]nclosures are *moulds*, distinct castings, . . . controls are a *modulation*, like a self-deforming cast that will continuously change from one moment to the other.' Such control generates compliance by ensuring the implicit understanding that it is in people's own interest to play by the rules of the consumer society; that it is in their self-interest to accept and internalize the belief that this society is suited to them, and they to it. Thus, as Baudrillard (1998: 78) insists, 'The truth of consumption is that it is not a function of enjoyment, but *a function of production* and, hence, . . . not an individual function but an *immediately and totally* collective one.' But whilst consumerism permits capitalism to project itself in the image of individual freedom, the freedom it promotes is a freedom of an extremely limited kind. As Bauman (1988:

88) puts it, 'such freedom of expression in no way subjects the system, or its political organization, to control by those whose lives it determines'. Consumerism thus bears a thoroughly duplicitous character; presenting itself as the paragon of freedom whilst representing its highly specific reduction to one particular area of life. It simply defines away other questions of freedom – in particular the freedom to influence or determine the nature of the social system in the first place.

Insofar as it relates to the individual, to the selection and subsequent use of commodities (Kopytoff 1986), therefore, consumption represents an increasingly central arena of freedom of choice, ever more closely connected to the 'problem' of identity; to the task of attaining ontological security in the realm of everyday life. But this should not be permitted to hide the fact that consumption is more central than ever to the systemic dimension of society (Baudrillard 1998), which it has begun to redefine in its own image – encouraging the perception that '[t]he "system" has done all it possibly can . . . the rest is up to those who "play" it' (Bauman 1996: 34). Consumerism, as the postmodern guise of capitalism, thus silences its discontents more thoroughly than ever. For whereas 'repression' inevitably produced its own discontents, 'seduction' is a strategy far more effective in ensuring compliance. As Bauman (1993a: 44) puts it, '[l]osers have no less reason than winners to wish that the game goes on. . . . Postmodernity enlists its own discontents as its most dedicated stormtroopers.'

From such a perspective, it becomes increasingly apparent that the so-called 'information revolution' amounts to little more than the market-driven redefinition of knowledge in the wake of the disengagement of the modern power/knowledge syndrome. This is not, of course, to deny the continuing power of knowledge in its fully commodified form – as information, pure and simple. Nor is it to deny the immense social implications issuing from this, as knowledge becomes instrumentalized and commodified to such an extent that it is not only a resource, a productive force, and a speculative source of profit, but also an increasingly invidious dimension of social inequality. Nor is to neglect the saturation of the entire *socius* with an ever-increasing proliferation of specialist forms of expertise – growing at a pace equal only to the increasing levels of anxiety to be found at all levels of the social hierarchy (and only most notably amongst the 'comfortable classes'). To the contrary, it is simply to emphasize the way in which the disengagement of the modern alignment of power and knowledge has evolved to a stage where consumerism, choice and the ideology of the market hold sway. The 'information revolution' is but one consequence of this transition towards a fully fledged consumer society. As Sampson (1994: 37) expresses it, 'such a culture of consumption is quite indiscriminating and everything becomes a consumer item, including meaning, truth and knowledge'. Who could have anticipated such consequences as flowing from the modern marriage of power and knowledge and their subsequent divorce? Who could have envisaged knowledge proving to be such a dangerous thing?

Acknowledgements

This work is supported by ESRC Award No. H52427002294, which supports the Fellowship 'Consumption, Lifestyle and the City'. My thanks to Marcus Doel and Martin Purvis for their comments on an earlier version of this piece. I alone, of course, accept responsibility for its remaining inadequacies.

14 Virtual culture
Knowledge, identity and choice

Frank Webster

Introduction

One hundred and fifty years since they were written, Karl Marx and Friedrich Engels' words are more resonant today than even on first publication of *The Communist Manifesto*: 'All fixed, fast-frozen relations, with their train of ancient and venerable prejudices and opinions', pronounced the pair, 'are swept away, all new-formed ones become antiquated before they can ossify. All that is solid melts into air' (Marx and Engels 1948, in McLellan 1997: 224). A good many commentators would probably have denied this claim back in the 1850s, since even while Empire was expanding and trade growing apace, there were few willing to claim that, for example, the relations between the sexes were not pretty well fixed, or that Great Britain might soon lose its newly established colonies in Africa and India. But nowadays it is simply conventional to acknowledge that everything is changing, that what it is changing into is itself set to change, and that all this is happening at an accelerating rate. All that was once assured, all previous certainties, all axial principles now seem challengeable and indeed are challenged. It is now quite orthodox to observe this sort of thing, and such observations are as likely to come from right wing circles as from those on the left. A heady cocktail of technological development, globalizing tendencies, the breathless pace of industrial innovation and economic adaptation, round the clock exposure to multimedia images, political turmoil, the commonplace experience of international travel, scientific explorations into the recesses of the human mind and to the far corners of the earth . . . , such a brew readily convinces that, indeed, everything is set on change.

So might we say then that the words of Marx and Engels have been confirmed as we reach the millennium? Up to a point we surely can. The ceaselessness of change, the constancy of upheaval, the collapse in confidence that things will continue much as they were, all such lend support to the original assertion. There is, however, a singular difference between the diagnosis of Marx and Engels and the situation which prevails today. The authors of *The Communist Manifesto* were convinced that the social, economic and political turmoil they saw all around would lead ineluctably to clarity of vision amongst those enveloped within it: 'All that is solid melts into air,' declared Marx and Engels and, the same sentence continued, 'man is at last compelled to face with sober sense, his real conditions of life, and his relations with his kind' (Marx and Engels 1948, in McLellan 1997: 224). Who today can resist a knowing smirk, even an embarrassed giggle, at such hopelessly modernist faith? That one might come to see, sharply and unambiguously, the reality of one's circumstances, and this precisely because of their uncertainty?

Theoretically it is now commonplace – after the 'linguistic turn' in social science, post-structuralism, deconstruction, and the rest – to scoff at such naivety (and to be on guard against modernism's authoritarian tendencies), but it is defied too by more substantive developments (Lyotard 1979; Seidman 1994). Thus today a central and widely acknowledged constituent of life is its 'virtuality', and this not only in the obvious sense of routinely participating in relations mediated by electronic media. To be sure, this is what is usually conjured when one imagines the 'virtual society': it is something which involves linking with others via computer communications networks of one sort or another, thus undergoing a less than full-bodied connection which is thereby 'virtual' in being not quite 'real'; or it points to the enormously increased output of media which informs us of what is happening here and there, though it is the only 'reality' we may experience of what is shown; or, more audaciously, it evokes those cases of simulation (of an aircraft in flight, a rocket on take-off, a Formula One racing car) which are experientially indistinguishable from 'reality'. In all cases like this a division between 'reality' and its 'representation' is maintained, and it is the widespread adoption of information and communication technologies (ICTs) in such instances which easily draws attention to the 'virtuality' of a good deal of life today.

However, to speak of the 'virtual life' is to suggest a great deal more than an emphasis on mediated existence, with its implication that there is some dissonance between 'reality' and what is 'virtually' experienced. It extends to an appreciation that everything that we do, whether it be in working, discussing, playing, interpreting, loving, or learning, is 'virtual' in so far as it involves artifice, leading unavoidably to the conclusion that there is no point zero of 'reality' to be found. Everything being humanly constructed, from making sense of one's personal circumstances, shaping one's body, to the design of offices, it follows that 'virtuality' is in everything, and that nothing is any longer really real. Everything is a simulation of one sort or another because everything involves fabrication, from the design of cities, styles of dress, to the writing of history. And more than this: everybody knows that this is so, hence there can be no justification for Marx and Engels' hectoring claim that, at last, through force of circumstances, long-bewildered people will come to appreciate, like the mighty duo themselves, what is real. The upshot is that Marx and Engels, and their eagerness that we come to understand the 'real conditions of life', are to be abandoned as ingénues, though commentators remain happy enough to acknowledge that all does indeed melt into air. Everything is made, reconstituted, and remade over again in a never-ending process. Similarly, perceptions are developed, reconsidered, and reconceived in a never-ending cycle. Things are put together, they fall apart, and are reconstituted in new ways. And so it goes. Such is the 'virtual life' and, as such, there can be no hope of establishing a foundational point of 'reality'.

In what follows I intend to offer a critique of this point of view, one which is widespread in current social theory and beyond. In so far as I do so, I shall be offering a defence of modernists in the manner of Marx and Engels. However, prior to the critique, I shall say more about virtuality today. In the first part of this essay I examine aspects of the 'virtual life', pointing to its characteristic features of malleability, choice and pervasive social construction, all of which oppose foundationalist claims. I then describe how these features lead to a paradox: celebration of the changeability of human life alongside an incapacity to identify (and still less endorse) the direction of change. Later, I charge that enthusiasts of the 'virtual life', while they are correct to refuse foundationalism in emphasizing that the world is socially constructed, they overstate the degree to which arrangements are matters of choices. They do this, I suggest, because they downplay (and even ignore) the constraints that inhere in the historical development of capitalism. I conclude by arguing that the triumph of capitalism, and its extension deeper into everyday relationships simultaneously with its increased global reach, while undeniably a construct of social forces, imposes enormous limitations on the options people might have for alternative futures.

The possibilities of the virtual life

Today everything changes, nothing is permanent. Indeed, the only thing about which one may be confident is that which is here now will sooner or later be gone. It is only to be expected that some will find this disconcerting, desperate to stop the world to get off at some stable point. I will have more to say about this reaction, but for now I would emphasize those who most readily acknowledge the disorientation and unsettling consequences, but who proclaim that the ride is a heady and even an exhilarating one. Futurists are to the forefront in this, quick to reveal techno-utopianism as a motivating force. For instance, Nicholas Negroponte (1995: 183) envisages the 'information superhighway . . . creating a totally new global social fabric' which will turn everything upside down, but empower and improve things nonetheless, while Kevin Kelley (1994: 2) expresses awe at the prospect of a 'neo-biological civilization' which is at once so mind-bogglingly complex and fast-adapting that it is 'out of control' yet is also conveniently self-stabilizing. With all this sort of thinking there is an enormous emphasis placed upon new technologies[1] – ICTs, genetic engineering, biotechnology or a combination of all three figure recurrently (Anderson's (1997) concept of an emerging 'bio-information society' is indicative), and today this finds a particular accord with postmodern reflection. Mark Slouka (1995: 30) considers this alliance of high technology and popular philosophy a 'mating of monsters', but Baudrillardesque enthusiasm for the boundless opportunities digital technologies present to recreate realities, to reconstitute places, bodies, and even selves, runs rife in positive commentaries on the impermanence and artificiality of life today (see Woolley 1992; Rushkoff 1996). Sadie Plant (1997: 46), for instance, is delighted that '(t)here is turbulence at so many scales that reality itself seems suddenly on edge', believing women are especially suited to benefit from the normal turmoil and turbulence of cybersociety.

To thinkers such as this change is full of possibility, and therein is its attraction: everything being subject to change, anything is feasible. Key terms of reference are 'cyberspace' – the electronic information network (no)places where people 'meet'; 'virtual reality' – the 'simulation' that is found in and around cyberspace; 'cyborg' – 'cybernetic organism', the melding of high technology and living things; and 'cyberculture' – the forms of life manifest in a world of cyberspace. That the origins of this terminology may be traced to science fiction, specifically to North American 'cyberpunk' writer William Gibson's novel *Neuromancer* (1984), where the word 'cyberspace' is first introduced, underscores the appeal of the imaginable being fabricable, the changeability of the changed, the possibilities of the future.[2]

This emphasis on possibilities, something which stems from recognition that everything is 'virtual', co-exists with acknowledgement of the impossibility of any and all forms of foundationalism. That is, people may now be free to be whatever they want to be, but no-one can say, at least with any authority, what they ought to be, since to do so is to make claims for a rock-solid morality (or human nature, or organization, or social relationship) which conflicts with the premise that everything changes precisely because everything is humanly constructed and is thereby subject to reconstruction. Hence if someone – perhaps a gauche modernist in the manner of Karl Marx or Friedrich Engels – puts about the idea that they can see the 'real conditions of life' (which others ought also to see, but unfortunately do not as yet), well, he or she will be laughed off the stage as readily as would Mrs Doasyouwouldbedoneby with her Christian creed, since all they may legitimately offer is a construction of reality which is as much a fabrication as the 'reality' itself.

A corollary of this way of thinking is that, though the world around and even within us is malleable to an unprecedented degree, any goals which might guide reconstruction are quite incapable of justification on any ultimate grounds (see Gray 1995). For instance, Francis Fukuyama's (1992) efforts notwithstanding, the suggestion that 'democratic capitalism' is the route on which all societies are set (and where history itself comes to an end) because it combines economic efficiency with acknowledgement of individual worth, is defied by the fact that there are other, equally viable, forms of capitalism than the Western model (for example, the Asian Tigers, Islamic societies, and, perhaps most important, China) which may be economically successful yet authoritarian, market oriented yet culturally sharply different (see Huntington 1996) from the secular states in the West (Gray 1998). Less dramatic, but also telling, are the daily interventions and involvement of governments, pressure groups and business people which subvert the operation of an abstractly imagined laissez-faire capitalism, and by challenges to commercial practices by those committed to different goals, be they religious, political or environmental. In short, there can be no foundational legitimation of the free market since there are practical alternatives that match its efficacy and the abstract model of laissez-faire is nowhere found in practice.

Similar conundrums are encountered at every turn: the world can be made and remade, but it cannot be justified on any solid grounds. Consider, for instance, the matter of sexuality and the body. It is increasingly the case that what were once thought of as naturally bequeathed are now regarded as largely matters of human choice (Shilling 1993). Sexual behaviour has long been disconnected from the fact of reproduction, and even from any supposed natural behaviour, but this especially so in recent decades. Anthony Giddens' (1992) concept of 'plastic sexuality' captures well the position we are in today where one may choose one's sexual preferences and practices. Age, for instance, no longer results in a natural boundary of sexuality, and in consequence we have the prospect of 'sexy' twelve-year-olds while the over-seventies are encouraged to maintain regular intercourse. There are, of course, fears about child abuse, yet these coincide with the release of the film *Lolita* which tantalizes audiences with child sexual display, and, as for the elderly, there is a host of technical procedures available to assist in enhancing performances, notably the drug Viagra. Relatedly, there is by now a good deal of literature which details the malleability of the body, a subject which is worked upon by exercise and health regimes, adjusted by technologies (from hip replacements to new hearts, implanted corneal lenses to electronic pacemakers, dental treatment to plastic surgery), so much so that we have witnessed already the advent of the 'cyborg' (see Haraway 1991b). So we find ourselves today in a situation where one's sexuality and body form is by and large an issue of choice.[3] Yet it is because this choice comes from our capability of intervention in one's sexuality and the body itself that just what one's body and sexuality might be is so agonizingly problematical. Should one discipline the body to be this shape or that? Ought one to have a baby in one's twenties or wait until one is forty? How is it that sixty-year-olds may claim to be 'middle aged'? If there are no pre-given limits, then how are parameters to be established? Having broken free from natural limits, there are left no firm foundations on which to base choice.

Look where one will, one comes across this paradox. Relationships, with neighbours, partners, or colleagues, are, increasingly, matters of choice rather than of compulsion. Yet this freedom has been accompanied by recognition that there are no longer clear bases for forming relationships – and still less for sustaining them – now that penalties for leaving have been radically reduced, something which makes the issue of relationships fraught in hitherto unprecedented ways.

Anthony Giddens goes so far as to suggest that we inhabit a 'post-traditional' world, one in which all received judgements and arrangements are subject to challenge. Whatever one decides to consider – moral codes, gender roles, how to treat animals, even the purposes of education – this seems to be borne out. All such cannot be justified in any foundational sense, there simply is no 'right way'. But though these things can be reconstructed in, potentially at least, any number of ways, there can be no firm basis for this reconstruction, and a consequence is, unavoidably, consternation.

It is important to note that this is no mere cerebral matter, a theoretical reconstruction without practical consequence. Quite the contrary, empirical developments continuously undermine one-time sureties, demonstrating time and again the ongoing reconstruction that marks human activity, and accordingly throw doubt in our way by compelling us to realize that there is no clear direction that we should be taking. For example, it has been demonstrated that 'nature' is 'invented', whether this be in the case of educating people to see the 'untamed' beauty and romance of the landscape around the Lake District (Urry 1995), or to designate and protect 'wilderness' areas and their 'managed scenery' such as in Dartmoor, or to radically recast, through irrigation and agribusiness especially, the terrain of Southern California to suit human needs (McNaghton and Urry 1996; Reisner 1987). Bill McKibben's (1990) depiction of the 'end of nature' underscores this point: human agency everywhere conceives, creates, and reconceives and recreates the 'natural' environment, even in circumstances where it may be experienced as 'sublime' (Nye 1994).

Again, previous types of work – ways that were tied closely to class, community, leisure pursuits and general outlooks on life – which provided what appeared to be fixed identities and perceptions, have been radically altered in recent decades. One thinks here, for instance, of the many studies of the English working class which describe the surety of life forms and social relationships, even in rumbustuous and trying circumstances such as the dangers of hard manual work and bitter political conflict, that were found in these domains (see for example, Dennis *et al.* 1956; Klein 1965; Hill 1976; Williamson 1982). The rapid destruction of many of these ways of life, and with this the reconstruction of the physical and social environment, is instructive. For example, in Durham County for most of the past two centuries at least 20 per cent of all households were occupied with a single occupation, coal mining, and in many villages the majority of the populace worked in the collieries (Bulmer 1978; Emery 1992). As recently as 1950 there were 127 collieries in this small county and well in excess of 100,000 colliers; today there are only one or two pits remaining and the occupation of miner rare (Durham County Local History Society 1992). And with the diminution have gone the long characteristic, indeed defining, spoil heaps, winding gear and coal stores associated with the pits, either levelled and planted with conifers or removed altogether. Even many of the colliery villages have been done away with, their houses demolished and people relocated to larger urban locations.

More acutely than before, it is possible to appreciate how these ways of life were socially created, how the physical as well as the social environment, and even the characters of the participants themselves, were products of a particular time and place, albeit that for many of the members these carried with them a surety, even an inevitabilism, as regards day to day conduct, attitudes and ethics.[4] The topography of the region has been transformed, and with it has come a redefinition of once assured notions of masculinity, community and even of human worth.[5] Today, with work patterns radically altered – part-time employment, feminization of the workforce, a sharp shift away from manual labour, high levels of uncertainty about the continuity of an occupational category let alone of a particular job (miners were well used to unemployment, though only recently have they come to realize that the occupation itself was vulnerable to extinction) – and with work itself being questioned, then it is hard to be unaware of the fluidity of our ways of life, and of work especially. And yet it is just this appreciation of the social character of all work which yet undermines confidence in forwarding suggestions for changes in the structure of work today. For if all is malleable, then why have work at all?

The city is a locus of much of this. Ever since, and still before, Simmel, the urban life has been regarded as the premier arena of possibilities, the place where most obviously new relationships could be forged, new institutions created, even new identities taken on (see Raban 1991). Balzac's Paris and Dickens' London pulsate with the energy of these themes. Not surprisingly, then, cities are the places where most commonly we encounter expressions of greatest confidence and panache – in their architecture, in music, in politics, in style . . . This exuberance has been manifest recently in an energetic urban entrepreneurialism which is dedicated to the thorough re-invention and re-imagination of cities, key elements of contemporary urban regeneration. In this way proposals are forwarded which suggest the 're-branding' of city imagery, to which end central areas are redesigned and rebuilt, and consultants employed to discover apposite slogans, logos, sponsorships and the like. There can be no denying that the visual appearance, the practical experiences, and the representation of the city is transformed in these ways. And yet this entrepreneurialism takes place amidst widespread scepticism and uncertainty about what might constitute the 'good city'. No one doubts that the city can be remade, but the assurance of Charles-Edouard Le Corbusier, Georges Haussmann, or even T. Dan Smith that it be made in this way has long gone. The striking confidence in the possibilities of the city co-exists with apprehension and anxiety (Robins 1991).

Knowledge

Knowledge (and information more generally) is crucial to these developments. It has been increased knowledge that has enabled the assault on all 'traditions', since analysis, examination and comparison quickly lead to calls for reform, whether it be in production techniques, political arrangements, or marital relationships. This is, of course, a familiar theme of social history, of which mounting secularization as well as the growth of science and technology are major expressions.

Integrally related to this role of knowledge has been the spread of the means of its storage. Libraries, universities, as well as other repositories remain important, but information and communication networks, spurred on over recent decades by technological innovations, have been especially significant in accelerating long-developing trends (and these have also enormously influenced institutional structures such as universities and libraries). While ICTs are very important to modern-day work organization, politics, retailing and much else, I want here to emphasize the expansion of media, taken to encompass the entire range from radio, television, books, news-papers, magazines, to the 'information superhighway'. The enormous expansion of media, and the resultant exposure to larger audiences and for greater periods of time, has been an inestimable factor in bringing home to a wide public that its own ways of life, however established they may be, are socially constructed.

To make this point is in no way to deny that the media are not skewed in favour of particular world views (most evidently Anglo- and US-centred perspectives) (Morley and Robins 1995). But it is to observe that media have played an important role in bringing into public consciousness an awareness of alternative vistas, scientific devel-opments, other religions, different ways of conduct within one's own society, varieties of musical traditions, even of once exclusive areas of life (the male locker room, the mortician's working life, the police station) (see Meyrowitz 1985; Scannell 1996). This too has been an important constituent of creating, in the present epoch, a ready awareness that the world does not have to be as it is. In addition, it is always necessary to remind ourselves that media audiences are not passive recipients of that which they receive – they cast a jaundiced eye on what they see and make sense of it in terms of their own reflections, experiences, circumstances and attitudes (Morley 1992). Part of this audience alertness is its recognition that the media themselves are social constructs: viewers do not regard television as a neutral 'window on the world' since they are well aware, not least from the media themselves, that television content – like the world which it inflects – is fabricated, composed by editors, journalists, writers, actors and the rest of the media retinue.

It has been this increased knowledge and the knowledgeability that accompanies it which allows us to appreciate that the world is changeable and to change it, but there has been a paradoxical corollary of this trend. One premise of the Enlightenment (the movement which underpins the spread of knowledge) was that knowledge would give control over the environment and over ourselves – and this it has done with a vengeance; but another, linked, premise was that Enlightenment would bring forth certainty about what to do – and this it assuredly has not delivered. Quite the contrary, the more knowledge there has been generated and disseminated, the more has there been induced a profound and often incapacitating scepticism as regards how we might best arrange our affairs. It is recognized that these can, and indeed that they will, be re-arranged, but confidence about how they ought to be constructed has evaporated.

For some commentators this is the stuff of the 'postmodern condition'. Zygmunt Bauman especially has written extensively on the ambivalences, ambiguities, even chaos of these postmodern times. To Bauman (1991), modernity's driving impulse was towards certainty, towards the 'one best way' of doing things, which manifested itself in an imperative to organize, to arrange, to repress dissent and unruliness, all to ensure maximum conformity to assured ends. Modernist politicians, intellectuals, and a host of corporate and especially state agencies strove to achieve stability and integration amongst their subjects, disciplining and schooling people in how to behave, what to think, even how to think of themselves (as citizens of a particular nation, with a particular history and culture and associated attitudes and ways of life). To be sure, modernist politicians differed as regards goals, most obviously in terms of the debate between collectivism and individualism, yet opponents shared a mentality of confidence in their alternative futures (and much else besides, such as faith in science and belief in progress).

Postmodernism represents the breakdown of this ethos and its practices. The postmodern era is characterized by doubt, scepticism, hesitancy, and even unruliness. The days of control by way of state machinery, socialization into conformity by complaisant media, or acquiescence to authoritative intellectuals, are gone. Nowadays people are left largely to their own devices, choosing their lifestyles by consumption in whatever they happen to prefer, only the 'seduction' of the marketplace sustaining any (shaky) order postmodernism maintains. Bauman identifies in postmodernism a radical removal of constraint. For instance, he reminds us of Freud's modernist insistence that civilization requires that people repress their deepest impulses and appetites, an orderly life demanding that we place inhibitions on personal freedom. Bauman argues that postmodernism supersedes Freud (Bauman 1997a). Today freedom is an overriding priority and there is reluctance to admit any attempt to limit individuals from fulfilling their self-directed desires. Indeed, nowadays people are expected to choose for themselves how they wish to behave. Collective moral constraint has been radically reduced, and freedom to choose has become a hallmark of the postmodern world. This extends from the decoration of one's home to preferences in music ('experts' who might dare to advise are routinely dismissed for their presumptuousness and partiality, both by alternative 'experts' and a sceptical public) (Bauman 1987).

The downside is that the removal of constraints has generated and exacerbated a nagging sense of insecurity. If one is not to be constrained by the national interest (and who believes all the trumpetry of nationalism today?), nor even by a need to conform to the norms of one's neighbours, then how is one to live? Worst of all, as I have observed earlier in this chapter, and as Bauman underlines, the question is no longer answerable by looking to an external reference point in hopes of identifying foundations, be it in religion, politics, aesthetics or even close neighbours.

This poses especial dilemmas for identity. If one might say that in the past identity was very much a received matter, something developed by a combination of family relationships, class, community, regional and national factors which were encultured in the individual, then for Zygmunt Bauman (1995) postmodern identity is crucially about the individual choosing his or her identity, choosing this freely – but choosing alone and without guidance. There is now a widespread perception that even the self is subject to invention and re-invention, that one can 'be anything you want to be'.

This does not mean that one simply becomes whatever might appeal to one's fancy, since there are clearly limitations, especially material, that intrude upon one being what one dreams,[6] but there is, nonetheless, validity in Anthony Giddens' (1991) argument that the key reference point for the self is what he terms self-reflexivity. As such identity is created and sustained by ongoing reflection on one's biography, by making continuous choices, large and small (this hair style rather than that one, this job rather than another, this body shape rather than that one, a child now or never . . .), in light of this reflexivity, and identity is cohered – if cohered at all – by an inchoate (and slippery) notion of being 'true to oneself'. The self, in short, is inescapably a matter of 'virtuality', with all the uncertainty, choice and lack of guidance this entails (Ignatieff 1984). This is of course an unsettling business, which is only to be expected in the 'risk society' (Beck 1992) that is the accompaniment of the 'virtual life'.

Fundamentalisms

It has become customary to argue that a wide variety of social phenomena, which have in common only the assertion of foundational tenets, reveal a search for certainty at a time – which will continue – when living with upheaval and turmoil, and scepticism and disenchantment, is inescapable. Encountering a world which presents no stability, and which acknowledges no boundaries, substantive or moral, then fundamentalisms are explained largely as reactions to the ensuing disconcertion. In these terms all movements which resort to foundational principles are regarded as fundamentalist. As such, the term can encompass a huge range, from calls for a return to 'family values', continued faith in the socialist ideals of equality and common ownership, claims that there are natural limits to growth, declarations that Jesus Saves, to exhortations that trust should be placed in scientific progress.

The 'virtual life' puts all such claims into question: each and all are challengeable and are so challenged as a matter of routine. Intellectual as well as practical acts of deconstruction – alternative family forms, the Soviet collapse, the confounding of neo-Malthusians, countering Christianity with Islam (and any number of other religions and religious sects), the distrust of science manifested in reactions to BSE, nuclear power accidents, global warming – remind us, time and again, that everything is contingent, is a matter of human construction and reconstruction, thereby an issue of choice rather than of foundational truth. And the doubt is never-ending, since the doubters are themselves doubted – thus families *per se* are attacked as oppressive, the failings of capitalism are set against collectivism's demise, increased longevity reintroduces the risks of surplus populations, global warming is questioned as a concept . . .

Such 'virtuality' sets the head swimming – and it is accordingly understandable if people respond by reaching out for certainties. Amongst the most discussed expressions of fundamentalism in recent years have been religious and quasi-religious revivalisms, especially in the Islamic regions, and ethnic nationalism which have affected just about every part of the world. However, even if they provide their devotees with some assurance, and even if they contribute to serious political unrest, such fundamentalisms are unlikely to prevail. They are too readily deconstructed, undermined by inside and outside critics, and challenged by alternative practices, while, in an era of global communications, much of this is too easily made available to all and sundry, to allow fundamentalisms to maintain their hold over the long term. Thus different groups within Islam itself, alternative religious movements, and accounts of the development of Islam with reference to events outside the Koran, combine to subvert its foundations. Similarly, ethnic nationalism is routinely debunked by journalists and historians who reveal miscegenation to be commonplace, who delineate movable frontiers, and document the creation of myths which support nationalist traditions (see for example, Ignatieff 1993, 1998). As has been observed, in searching for the sureties of 'roots', all people find is that they are capable of tracing extraordinarily variegated, complex and unclear 'routes' which they and their forebears have taken.

The imperative of sociability

We find ourselves in a world of infinite possibilities, but one which is unaccompanied by confidence about what we may want to create. Everything is possible, yet the value of anything we might create is doubted. Of course, there are some who have no particular difficulties with this. Jean Baudrillard, for example, famously concluded that the real had been abolished and that, in consequence, there was no meaning in anything and no point in trying to find any. This being so, one might just as well revel in the sensation (but not the meaning) of the 'hyper-real' (Baudrillard 1983).

However, most people find disconcerting, and even disturbing, a 'virtual life' which lacks criteria for being. It appears to me that 'communitarianism', in one or other of its guises, addresses this concern. The movement associated with Amitai Etzioni (1995) in the United States is but the most direct expression of a continuum of views which share a conviction that people must have a sense of belonging to a wider group in order to live satisfactorily. Emile Durkheim long ago insisted that, without absorption into the values of the group, individuals experience anomie with costs paid in terms of social dislocation and personal unhappiness (crime, suicide, depression and so forth) that were formidable. Etzioni, updating Durkheim, proclaims a need for society to take precedence over individual desires. This is offered in large part as a defence of social stability, but communitarianism is also motivated by the conviction that a fulfilled life can only be experienced in a context of meaningful social relationships, that it is only in our attachments to others that we may experience a genuine sense of self and self-fulfilment.

There is merit in this argument, but before I expand on it we had best highlight major shortcomings of communitarianism and its variants. First, its advocates have the hubris to presume to know what society, as well as the people in it, require for the best. Second, communitarianism proceeds to attempt to impose this vision on those people whom it considers legitimate members of the society while excluding those whom it considers to be outsiders. Third, communitarianism finds it hard to tolerate differences within society since it operates with a homogenizing concept of common beliefs and ways of life. Underpinning all these positions is the presupposition that society is clearly bounded. In other words, communitarianism works with an idea of community which it presents as embodying the appropriate way of life, it allows to be included only certain types of people, and it endeavours to discipline those who do not conform to its tenets. These seem to me the failings of almost all versions of community, from locally-based tribes to nationalisms. And for reasons to which I have already alluded, all such are challenged empirically and conceptually, for instance by the heterogeneity of multi-ethnicity and multi-culturalism within most nation states which subvert unifying definitions, by the astonishing geographical mobility of populations which encourage the fluidity of cultures, and by the ready demonstration that communal ideologies are mythic.

The root difficulty with communitarianism is that it seeks for order and stability, and it seeks this by conceiving a set of values and behaviours which must constrain and impose conformity on citizens. But there can be neither a foundational justification for these assumptions, nor can they be found in practice in anything beyond the most small-scale or authoritarian social orders. The allegedly uniform 'communal' values and habits of a society are neither to be found in practice and, were they to be found, would not be justifiable on any firm grounds. Furthermore, there is much to be said for the argument that it is precisely in the encounters between different ways of life, values and beliefs that the most vibrant, creative and rewarding cultures are found, since community, where strong, is constraining, inhibiting and unconducive to experiment (consider, for instance, the paucity of creative thought coming from fixed rural cultures).

But I am reluctant to dismiss communitarianism out of hand, because at the least it draws attention to a crucial issue. This is that life cannot be lived alone, that it is in sociability where we find our identities, and where we may forge our ways of life. No matter how individualistic and even fragmented may be our sense of identity, it remains the case that we know ourselves in relation to other people and organizations. Furthermore, sociability requires, at a minimum, some shared, even if inchoate, conception of the rights and obligations of fellow human beings. After all, the ideal of equality between citizens which undergirds all democratic politics and which acknowledges the worthiness of each individual, was nurtured in one undeveloped yet vital form of communitarianism – the nation state. From another direction, Norbert Elias (1994) spent much of his very long life delineating the expressions of this through time, charting a tendency towards greater acknowledgement of the recognition that others ought to be treated with courtesy and civility, restraint towards one's fellows being expressive of the 'civilizing process'.

It seems to me that the need for some communal bonds such as civility, neighbourliness and acknowledgement of fellow citizens is a requisite of all human life. This is so even in the most individualistic of societies, the United States of America. The insightful ethnography provided by Robert Bellah and his colleagues testifies to this need: although a 'first order language' in the United States is one of insular individualism, close behind is a complicated and inadequately articulated 'second order language' which draws upon biblical and republican traditions to stress friendship, voluntary commitments, and the worthiness of neighbours and colleagues. This is so because individualism, while it is 'theoretically imaginable (is) performatively impossible' (Bellah *et al.* 1985: 154).

I would go still further, to insist that it is a mistake to counterpose self-identity and communal ties, since an over-emphasis on self can actually subvert a sense of individuality. As citizens people may feel a strong sense of responsibility, involvement, and even empowerment, forces which promote self-awareness and identity. When the self is but a reflection of the things one owns, then this identity is 'narcissistic' in that it mirrors to oneself what one consumes. But it is thereby shallow, passive and resigned to wider forces. Christopher Lasch (1978, 1984, 1995) pointed out that it might seem highly autonomous in so far as it prioritizes 'me', but an identity which reflects only what the market expresses is weak and dependent, not to be compared to a self identity which holds to a sense of collective moral purpose – civic virtue – while having also a strong ethos of self-responsibility. It is something of this towards which Benjamin Barber (1995) draws attention in his study, Jihad versus McWorld. Barber observes that McWorld's market practices privilege collections of separate individuals who are similar only in terms of consumer preferences, and he suggests that this is antipathetic to 'strong democracy' since the latter requires, at the least, some form of communal bonds that reach beyond the shopping trolley. 'Consumers speak the elementary rhetoric of "me"', writes Barber, 'citizens invent the common language of "we"' (Barber 1995: 243).

Manuel Castells' (1997) focus on collective identities, forged in social movements such as feminism and environmental concerns, points also to what I would call the imperative of sociability. Other than for the most privileged elites – and perhaps even they are not exempt – the reflexive self, individually directed and conceived, is infeasible. While Giddens observes that reflexivity involves social relationships, there is in his concept a prioritization of an autonomous and self-engrossed individual which underestimates the contribution of wider groups to one's identity. This does not mean that one's self identity is reducible to that of the collectivity, but rather it reminds us that it is in engagement with others that one creates a meaningful life, and that the terms of this engagement require some commonality of bonds.

The constraints of the constructed

If we can establish that sociability, and with this some communal bonds and behaviour, are requisite of identity, then we are left with the question of the content of identity, and even what might be the arrangement of the wider society, at a time when anything is possible but nothing is foundationally justifiable. Anthony Giddens has been noticeably positive about this, stressing that the capability of overcoming all absolutes has presented the opportunity – and even necessity in a 'virtual society' – of making active choices about much that once had to be accepted because it presented itself as a 'fact of life' (or of nature, destiny, or fate). Everywhere now life has become 'disembedded' and as such it is increasingly recognized as something that is designed and chosen (Giddens 1990).

Giddens has emphasized the potential of 'dialogistic democracy', where actors may decide, through discussion and debate, appropriate balances of autonomy and social solidarity (or self and society, or localism and cosmopolitanism) in creating the lifestyles and identities they might wish to have. Such a position acknowledges that there can be no return, for example, to a 'my country right or wrong' ethic which required people to resign themselves to a particular national identity. We are too self-knowing neither to recognize that this is infeasible in a multi-cultural society, nor to be ignorant of the fact that the old nationalism was a flawed and fabricated phenomenon. But we can appreciate that the nation still has resonance, not least because it helps give shape to important elements of ourselves and it mitigates against excessive fragmentation. The task is, then, to construct a notion of the nation which may come to terms with a heightened cosmopolitanism both inside and outside the United Kingdom, yet which may also allow diverse members to feel that they belong, even inside a country which is highly disparate, and even frictional, in terms of cultural composition (Giddens 1994c; 1997).

I find a good deal of this argument persuasive. It is, of course, at the heart of today's 'third way' politics that prioritizes the ongoing negotiation of arrangements, and thereby the possibilities of political action being significant (Giddens 1998). But I do want to make an important qualification. My objection is to the ready consequences of anti-foundationalist thinking, which has it that, everything being constructed, then anything can be reconstructed (and, of course, that what is constructed is constantly being reconstructed). But what this ignores are the inhibitions to change, and the pressures to conform, that can accompany particular arrangements. This is, in my view, manifestly the situation as regards much of the 'virtual society'. For instance, the financial networks that have been established around the globe are undoubtedly socially designed and maintained day to day by human actions, but the prospects of radically changing them are severely restricted, not least because, were a threat made to commence precisely that in a given region, then currency outflows would be massive and immediate and, as such, enormously constraining on those with ambitions to remake relationships. Instances of the exercise of these constraints have been a regular occurrence throughout the 1990s. The triumph of capitalism and the increased imbrication of world markets, while it has induced enormous instability, has ushered change overwhelmingly in one direction – and that towards the intensification of capitalist practices. As such it has massively denuded the capacity of governments to shape their national economies other than in ways which appease business confidence (and often, let it be noted, this pressure is applied to already pro-market economies which are regarded as not quite in order by investors, dealers and advisers such as the World Bank and International Monetary Fund). Examples abound: from the speculative attacks on the Exchange Rate Mechanism in Europe between 1992 and 1993 which had Britain, Ireland, Italy and Spain reeling in turn, decisive pressure brought on Mexico in 1994–95, the run on East Asian economies during 1997, to huge disinvestment out of Russia in 1998. Such actions as these take place electronically, millions and even billions of units of currency moving as bits of information, but they leave behind closed factories, leaps in unemployment, diminished pensions, and spiralling inflation (Strange 1998).

Currency and financial flows are a familiar, if exemplary, example, but, when we reflect on the constraints emanating from human designs, we might consider too the price of participation in credit card services. On an individual level possession of a Visa or Mastercard can make life much easier. Certainly the banking networks mean travel and shopping are facilitated. But we also need to be aware that a corollary of participation in these networks is a willingness to be surveilled as a matter of routine and of being capable of maintaining a positive credit rating with the banks. This too is a disciplining factor. Moreover, it is the case that credit card facilities are becoming the norm in advanced capitalist societies. As they do so, not only are the excluded disadvantaged and stigmatized, but so too does it begin to become obligatory for all of us to be a part of the surveilled and credit-worthy population. In such ways, while the credit system is certainly humanly designed, it is extraordinarily difficult to imagine it being radically amended once it is widely adopted.

Let us now turn to the recent history of capitalism. To be sure, capitalism has many variants, and pure laissez-faire manifests itself but rarely in practice. And there are no natural foundations of capitalism: it requires, as it always did, constant human inventiveness and ingenuity. However, the fact is that, though there are certainly varieties of capitalism in the world today which reflect different cultures and historical circumstances, we cannot be blind either to the global extension of capitalism or to the constants that have accompanied its victory over recalcitrant or non-conformist practices. The extension of capitalism throughout the world places prodigious limits on the actions of even large states (Sklair 1991). This, of course, does not mean that all national endeavours are ineffective (Hirst and Thompson 1996), but it does acknowledge that, in the words of historian Eric Hobsbawm (1994: 572), by the end of the millennium '(t)he world economy was an increasingly powerful and uncontrolled engine' which demands conformity to its strictures. Tony Giddens (1998: 43–4) himself concedes that '(n)o-one any longer has any alternative to capitalism'. Amongst these constraints, whatever variations capitalism may be willing to tolerate, are a profound intolerance of proposals for radically different ways of life, to which the destruction of the peasantry – throughout recorded history, and well into the twentieth century, by far the majority of the human race – over the span of a few score years bears testimony (Worsley 1984).

In addition, capitalism insists that conditions are made propitious for its functioning, and these include opportunities to make profitable investment, the availability of wage labour, the unhindered movement of capital, encouragement of private ownership, the legitimacy of price signals from markets, legal support for contract and property relations, and the spread of ability to pay criteria as the basis for access to goods and services. Here and there one may come across situations that do not quite conform to these principles – tax rates might be marginally higher in one country compared to another, one state more interventionist than a neighbour – but, not least courtesy of the information networks which bind together the world nowadays, what variation is tolerable is subject to examination, definition and action by currency dealers, investment managers and finance brokers.

Moreover, while capitalism imposes its preconditions on those places it encounters, it is simultaneously a system which exacerbates and accelerates change and upheaval. In this way it serves to unsettle and disturb, to instil a sense of transience and even of 'creative chaos' (Castells 1996: 147) wherever it may operate (Harvey 1989). This observation, that capitalism, though undeniably a human construct is a dynamic, even frenetic system, leads me to my final point. As we have seen, much is now made of the choices that allegedly stem from recognition that everything is socially constructed. However, it is the case that globalized capitalism imposes inestimable constraints on human action all over the world. Moreover, integral to this constraint is an insistence that people change in accordance with capital's imperatives. Looked at in this light, the changeableness of life today looks much less a matter of human choices and much more the imposition of the imperatives of participation in capitalist relations.

In the advanced capitalist societies we are well aware of this, since it is something instanced in Britain during the early 1980s when recession rapidly cut swathes through the manufacturing sector (destroying one in three of all jobs in that realm), and ever since the 'collapse in tenure' has been a feature of life for a great many more (Sennett 1998). However, it is a very much more compelling experience of those – the vast majority of human beings – who inhabit the poorer regions of the planet. The 'virtual life' extends easily enough into areas such as India, China, Latin America and Africa. All are reached and touched in some way or another by the informational networks of global capitalism, by satellite television programming, business organization from afar, by economic analysts in New York or London. The key difference, however, is that, to adopt the terminology of Manuel Castells (1996: ch. 5), these people are interacted upon rather than interacting in the information age. They are, in short, recipients of decisions made by those with access to and privileges of capital, education, knowledge and, of course, technology on the global networks.

It is sobering to note that, at the end of the millennium, many nations may be regarded as irrelevant to global capitalism, perhaps only worth noticing when transnational media organizations arrive to report on the latest 'natural disaster', usually famine or internecine war, and to capture on video starving women and children or fleeing refugees. The peoples of Sub-Saharan Africa (about 500 million souls), for example, have neither the resources to appeal as markets nor other qualities that might appeal to investors of capital (Castells 1998: ch. 3). Other nations might have some attraction, perhaps to supply raw materials or to serve as cheap labour for corporations such as Nike or Adidas, but all are subject to decisions made for them by more powerful forces.

I would end with this observation: the World Health Organization calculated in 1994 that fully 20 per cent of the world's population lived in 'extreme poverty' (WHO 1994). That alone totals one billion people living at or below a 'dollar a day', but the WHO added that one-third of all the world's children are undernourished and that over 12 million under fives die annually, almost all from poverty-related illnesses. The United Nations has also calculated that by 1980 20 per cent of the world's people received over 80 per cent of total income, while the poorest fifth got 1.5 per cent (Athanasiou 1997: 53). Of the world's six billion people, four billion live in conditions that the vast majority of the remaining two would find abysmal. The divisions continue to worsen in favour of the already affluent.

As we have seen in this chapter, it is now conventional in social theory to argue that nothing happens for foundational reasons, that everything is an outcome of human construction and choice. With this, analysts may rush to record the manifold choices people make as regards their body shape, sexuality, intimate relationships, cultural pleasures, and even their lifestyle politics. Heterogeneity and difference are the liberatory keynotes of such refrains. Up to a point, one can only agree with such colleagues in this current orthodoxy. But they ought also be reminded that, as a result of complex but knowable processes, huge numbers of the world's population, our fellow human beings, have no choice but to endure conditions hard to imagine by those of us inhabiting the affluent nations. The 'virtual society' undoubtedly brings options to a good many people, and this must bring some satisfaction to those unhappy with their physique, occupation or partners. Because things don't have to be the way they are, many nowadays can change their circumstances, and accordingly they do so. Good luck to such people. But I would also insist that this world that has been created, for the bulk of humanity, means subordination to the strictures of capital, to constraining and systemic forces that are vastly more pervasive – intensively and extensively – than any previous empire. Talk of choice in a 'virtual society' to the world's majority is hollow indeed.

Notes

1 There is a wealth of such writing. A couple of recent examples of the genre are Michael L. Dertouzos, *What Will Be: How the New World of Information Will Change Our Lives*, London: Piatkus Books, 1997; Wilson Dizard, *Meganet: How the Global Communications Network Will Connect Everyone on Earth*, Boulder CO: Westview Press, 1997.

2 There is also some irony in the fact that Gibson's novel, while situated in an informationalized realm, has a plot, action and characterization very familiar to readers of the detective thriller genre, especially Raymond Chandler.

3 Actually, the malleability of sexuality is currently overstated. To concede that sexuality, and human nature more generally, is socially shaped is not to agree that it is entirely socially constructed. See Ted Benton, 'Biology and social theory in the environmental debate', in Michael Redclift and Ted Benton (eds), *Social Theory and the Global Environment*, London: Routledge, 1994, pp. 28–50; Peter Singer, 'Darwin for the Left', *Prospect* (31) June 1998: 26–30.

4 See, for instance, the biography of a local miner, Joe Robinson, *Tommy Turnbull: A Miner's Life*, Newcastle: TUPS Books, 1996; the writing of Sid Chaplin, *The Leaping Lad and Other Stories*, London: Longman, 1970; and the social history of Huw Beynon and Terry Austrin, *Masters and Servants: Class and Patronage in the Making of a Labour Organization*, London: Rivers Oram, 1994, especially chs 5–8.

5 On this see the recent study of the village of Horden on the East Durham coast by Mark Hudson, *Coming Back Brockens: A Year in a Mining Village*, London: Cape, 1994; cf. Raphael Samuel, 'North and South', *London Review of Books* 22 June 1995: 3–7.

6 See Bauman's acute observations on today's workless poor, who dream of consuming even while they lack resources to consume (and accordingly, in these aspirations, affront the rest of us who revel in the seductions of consumerism). Zygmunt Bauman, *Work, Consumption and the New Poor*, Buckingham: Open University Press, 1998.

15 'Have you got the customer's permission?'

Category management and circuits of knowledge in the UK food business

Ian Cook, Philip Crang and Mark Thorpe[1]

Introduction

This chapter emerges from our concern with exploring how commodity culture, and in particular that commodity culture associated with foods, is intertwined with the production, circulation and usage of geographical knowledge (Cook and Crang 1996). This has been the focus for a two-year, multi-locale, multi-method study of food retailing and consumption in North London, through which we have looked closely at the ways in which knowledges of foods and their geographies are drawn upon by, produced by, and contested between consumers, retailers and manufacturers. Practically, our own knowledges about these processes have been primarily constructed out of corporate interviews with retailing, manufacturing and associated food industry workers in four product sectors – spices, pasta, bread and chicken – and semi-structured interview series with twelve households of north London consumers – based around activities such as an accompanied shopping trip, household food inventories and food diaries. Elsewhere, we have begun to write through this material from the perspective of these consumers as they engage with increasingly globalised systems of provision, outlining both the ambivalent structuring of consumer knowledge of food origins around both a 'need to know and an impulse to forget' (Cook *et al.* 1998) and more particularly discussing the kinds of imaginative geographies being mobilised within food consumption (Cook *et al.* 1999a). Here, we want to complement this focus on consumer knowledges through an attention to commercial knowledges, or as Frank Mort terms them, 'commercial epistemologies' (Mort 1996: 89). More specifically, we will be outlining the knowledge economies of what in the food trade is termed 'Category Management' (hereafter referred to as 'CM').[2]

Within published trade accounts, CM is positioned as a key part of 'Efficient Consumer Response' (hereafter ECR), which is a wider but still emergent programme emphasising a stronger orientation to 'the consumer' within the food industry. ECR has been defined as having three main elements:

> *Category management*: managing product assortments, promotions and introductions to best meet consumer demands. *Product replenishment*: optimising the flow of products through the entire supply chain from ingredients into consumer shopping baskets. *Enabling technologies*: utilising key technologies to facilitate the implementation of ECR concepts.
>
> (Anon 1996: 1)

Therefore, under the rubric of CM and its stated concerns with understanding and then meeting consumer demands, both retailers and manufacturers have been constructing new combinations of quantitative and qualitative knowledges about the ways in which people shop for and think about food. At least according to the rather utopian version reported in the trade press, these knowledges should then be capitalised upon to: (a) re-structure store layouts to reflect the categorisations of products used by consumers; (b) re-organise management cultures/structures into multi-disciplinary category-centred teams; (c) re-structure relationships between retailers and manufacturers away from mutual suspicion and competition and towards trust and co-operation within categories; and (d) to thereby better satisfy consumer demands while, at the same time, developing more efficient and profitable relationships between retailers and manufacturers (Johnson and Pinnington 1997; McGrath 1997). In short, CM involves the restructuring of the social and economic geographies of food provision into a horizontal grid of vertically integrated 'categories' which themselves are taken from consumers' understandings of foods.

Building on Marianne Lien's assertion that 'the social production of foods for sale is closely linked to the social production of knowledge' (1997: 19) we want to consider critically CM's central concern with knowledge – with knowledge of consumers; with knowledge of product categories; with networks of knowledge exchange – as well as its status as knowledge itself (after all, like other products of management consultancy, CM is a piece of theoretical knowledge that itself is being bought and sold: see Thrift 1998a, 1998b). We also want to consider how this commercial emphasis on knowledge articulates with consumers' own knowledges as produced through culinary culture (although, as already mentioned, we will be leaving those consumers somewhat 'off stage' here). The chapter considers the commercial knowledges associated with Category Management in three main ways. First, we briefly highlight the political economies of knowledge production that are brought into play by CM, emphasising both the role that knowledges, or at least 'auras of knowledgability', play in the corporate doings of the food business, and the power relations that run through these capitalisations on knowledge (not least through the allocation by multiple retailers of so-called 'category captaincies' to particular manufacturers). Second, we examine the spatialities of provision–consumption relations assumed within and promoted by CM. Here, we make a distinction between the linear geographies of connection emphasised in the marketing discipline's promotional accounts of CM – in which knowledge flows back from consumers down systems of food provision – and the more circuitous geographies of knowledge apparent in CM practice.[3] Third, we turn to the epistemological differentiations – also called categorisations – involved in CM. Category Management is clearly more than a managerial theory of how to organise pre-existing product categories, but rather involves the active production of those categories themselves. Here we therefore critically consider the claims held out for CM that it can structure the food provision system according to the categorical understandings of consumers, and posit instead how categorisation is central to food providers' concerns with legibility, visibility and association.

Our approach is therefore to question an understanding of the economies of commercial knowledge framed around the logic of an unknown or only partly known world of consumption that it is commercially beneficial to gain knowledge about. In saying this we are not disputing that CM does indeed provoke ever more determined efforts to gain knowledge of consumer understandings and practices, efforts that intriguingly parallel those within academic consumption studies. What we do want to question, however, is the extent to which capitalisations upon this knowledge actually stem from their ability to accurately represent the consumer worlds. Instead, we point to a variety of other roles that knowledge can play: as tokens and auras that allow the claiming of commercial and professional authority; as a subject and product of interaction facilitating networks of trust and influence, of inclusion and exclusion; and as a structuring of a highly complex world of things, beliefs and actions that enables the production of some interpretations and associations rather than others. In summary, we argue that underlying the mimetic epistemological claims of CM – which frame it as a process of gaining knowledge about how 'the consumer' relates to foods and then acting on that knowledge – are a number of non-mimetic productivities for commercial knowledge.

Some political economies of categorising knowledge

We start with some observations on the political economy of knowledge promoted by CM, and its intertwining with the political economy of food provision and its multiple-retailer driven vertical integration. CM is a rubric of commercial organisation that places epistemological acts at its heart. According to the recurrent narratives of the grocery trade press, ECR – and with it CM – are said to have 'emerged as a global movement of the Grocery Industry in response to increasingly demanding consumers, shifting competitive environments and rapidly changing technologies' (Anon 1996: 1; Anon 1997b). To start with the last of these, what has been crucial is the barcode-scanning EPOS technology in supermarket checkouts and the subsequent intro-duction of customer loyalty cards. These mean that retailers now have the IT facilities to construct and (in some form at least) analyse huge amounts and new forms of data, which not only improve the efficiency of their 'product replenishment' activities but facilitate the surveillance of consumption. This data offers the potential to combine aggregate information on particular product lines with information on what goes with what in the trolleys and baskets of 'real' consumers, or at least their digital personae (which are also influenced by card sharing, receipt swapping etc.: McGrath 1997; Ody 1997). In so doing it acts as a coinage of epistemological authority for the large multiple retailers.

In this food literature, consumers themselves are portrayed as having become 'increasingly sophisticated and well-informed' (McGrath 1997: 7). As well as demand-ing good quality products and competitive prices, they are also believed to be looking for more interest and excitement in their shopping experiences, as well as an ever greater variety of products (the largest UK supermarket now has over 20,000 product lines). This poses problems for retailers in terms of rationalisation, and in terms of how to make sure that the New Product Development (NPD) needed to maintain or extend this variety is not too costly. According to one estimate, over 80 per cent of new product launches 'fail to sustain a market presence within two years' (McGrath 1997: 9). CM is therefore promoted as an approach which 'reduces inefficiencies and increases effectiveness by avoiding (the) redundant, improperly focused and often counter-productive efforts' regarding NPD (Anon 1996: 4). Of course, as the work of Alan Warde has emphasised (1997: esp. 168–73 and 191–6), we need to be careful in assuming in what way consumers actually engage with the variety offered by the food retail environment. His extensive data, chiming with our own intensive data, suggest that particular consumers do not necessarily have highly varied culinary regimes, even though an apparent variety of provision may be something they value in food retail outlets. What we can say, though, is that in emphasising 'variety' the trade press identifies an important element in the commercial auras which dominate contemporary UK food provision. And they signal how a prime impulse for CM is its positioning as a way of organising and producing this aura of variety; as, then, a way of knowing the world of goods that combines manageability with some sense of diversity and abundance.

To summarise, then, dominant trade accounts emphasise knowledge of food consumption as central to the contemporary food business. Food providers, it is said, need knowledge of consumers and the knowledges they have of foods. This knowledge is set up as a source of commercial efficiency, and hence power. More particularly, this knowledge is ideally to be produced within (and to be productive of) a collaborative political economy. Under CM the competitive buying and selling of branded products between retailers and manufacturers is supposedly downplayed, and the category itself is what has to do well. The category 'cannot be perceived as merely a collection of brands. It is an entity in itself', such that, in 'effect, the category becomes a business in itself' (Hutchins 1997: 179). Of course this category business embodies more complicated relations than those solely of trust and co-operation. One of the most controversial aspects of CM is that these new 'category businesses' are usually formed between retailers' category management teams and just one manufacturer within each category. These chosen manufacturers are referred to as 'category captains' and are responsible for 'taking charge of a particular category and providing in-depth analysis to the retailer' (Soars 1997: 26).

Being in charge of this analysis, however, by no means means that a category captain can simply pack the shelves with its own branded goods. Rather, in principle they have to 'guarantee impartiality' (Macdonald 1997: 28) as their charge 'is centred on maximising sales for the retailer. In so doing, . . . (they) expect to benefit too, although competitors may benefit equally' (McGrath 1997: 8). Thus, at least in the trade promotions of CM, category captains are charged with maximising the performance of the category rather than the individual brands which make it up. Thus, any 'biased approach would only result in driving away consumers and rejection of the category captain by the retailer' (Soars 1997: 26; Lewis 1997). Moreover, so the argument goes, because individual manufacturers are no longer being played off against one another by the retailers – having to constantly cut profit margins to remain or get on the supermarket shelf – CM 'can lead to reductions in price competition among brands within a category, hence benefiting all firms, whether they have adopted category management or not' (Hutchins 1997: 178). As well as the obvious financial implications of these changes, then, these are regularly talked about in the trade press as heralding a huge 'cultural' shift in the food business as personnel are entering into unfamiliar relationships with one another, largely related to the way that they share knowledge.

In practice, of course, it seems that CM's political economies of knowledge production are neither simply collaborative (a win–win scenario for all operating within a category) nor just a gloss for old forms of competition between manufacturers or between retailers and manufacturers. Rather, they signal new forms of competition based on inclusions and exclusions from 'the category', which in turn are legitimated by judgements of knowledgeability, and mediated by the thickness of relations established with retailers. An interview extract from our conversation with Fred,[4] the Managing Director of a small company manufacturing one range of 'regional' foods, illustrates this well:

> Category Management is potentially a really bad thing for people like us. If we were captained out of it, you know, we haven't got a thing to stand on. *Is that a verb now, 'to be captained?' [laughs].* Sorry, I made that up. . . . [But] that is happening . . . For us, we need to be able to react really quickly and to be so – it's a guerrilla tactic – we just need to be able to be moving the point of it all the time. That way we're quite hard to follow. . . . It won't be easy for someone to go after us and take [our company] out. We had a situation with one retailer recently where they said they wanted one supplier of (our regional) foods. They wanted to, they were into, supplier rationalisation and only wanted one supplier, kick everybody else out. Fortunately, we stayed in. We were told we were out in the first place and we managed to hang in there. And if we can be supplying another part of their store – the marinade buyer, and the relish buyer and the chutney buyer, and the frozen food buyer – they can't say, 'Our rationale is we want to rationalise [your company] out if it,' because we're in there too deeply. And that is one value of own label business in that, if you're in with them on that, you're an accredited supplier. You network up and down the account and you can try and fight for your brand. You know, you can be in there all the time and you're not going to be kicked out . . .

Rather than these thickening networks of collaboration simply replacing the kind of militaristic and territorial conceptions of the marketplace which Lien (1997) identified, these metaphors of commercial space seem to co-exist. The territorial metaphors are at work in that one can be 'in or out', one has to 'fight for shelf space' and retailer attention. But this is a battle in which the 'guerrilla tactics' of constant innovation and sophisticated networking can play an important role (for more of this, see Cook *et al.* 1999b).

Also at stake in these negotiations are the forms of knowledge that are given commercial legitimacy. As we have noted, a key premise of CM is that suppliers sell themselves to retailers not only in terms of their products and their qualities per se, but also in terms of the category knowledge they can provide (though obviously producers can influence retailers to concentrate on the former).[5] The latter has to be an 'objective' knowledge which covers entire categories. The kinds of knowledge constructed here are thus governed by an official economy which devalues both subjectivity and partiality. One effect of this is to devalue, at least in the context of category inclusions and exclusions, other less 'objective' knowledges. As Fred explained this:

> I was with a meeting of guys the other day, and they'd been de-listed on some lines because of these rationalisations. But it's very hard to argue because they've got all the information . . . The only thing is . . . the stores sell fantastically based

on the old premises they used to have ... and, umm, a gut feeling by the buyer, and there is a chance that all this is trying to be replaced with science. And I don't think you can. I mean, we as a company, umm, I'm very entrepreneurial and I've brought in a lot of people from big companies around me with big company skills and the science and stuff like that. And all that stuff's fine, and you need it, but when you stop flying by the seat of your pants, well, you, things aren't black and white. You know, a lot of it's got to be done on gut feeling. And if they lose that because of this [CM], you know they'll find it wasn't quite so wonderful.

Indeed one key result of the introduction of CM to the UK food business has been to promote the role of specialist data management companies as third parties within any CM set-up. These companies offer not only the specialist skills and equipment to 'deal with' complex masses of category data, but also the air of 'objectivity' in their analyses that can fend off criticisms and fears of 'bias' in favour of the branded products of category captains. Unsurprisingly, then, such companies are also the source of many of the articles promoting CM within the trade press. It is important to note, here, that trade press accounts of CM are not only about the food business, but texts in and of the food business. McGrath's (1997) report was published by the UK's *Institute for Grocery Distribution*, an independent research and training organisation which, among other things, runs courses on CM and produces expensive industry reports like hers. Many of the trade press articles have been written by people working for the market research and data management companies who could be that 'objective' third partner that readers are looking for: Dove and Hickman (1997) and Campbell (1997a, b, c) work for *IRI Infoscan* and, it is stated at the foot of Campbell's (1997a) article, this company's 'category management team offers consulting services, training courses, products and software tools' (p. 33); Dodgson (1997) is a 'distribution analyst at Datamonitor'; Anon (1996) was published by Coopers & Lybrand; Anon (1997a) seems to have been written by and for ACNielson; and Anon (1997b) was written by and for Ernst & Young.

The emphasis on knowledge of consumption and consumers in CM therefore does more than simply re-orientate the knowledge production activities of food providers towards consumption. It operates in ways that are not simply collaborative. Rather, it promotes a restructuring of commercial culture's epistemologies – in terms of both the sources and kinds of knowledges that are officially valued – and it tightens the logic of large retailer driven vertical integration.

From linear to circuitous geographies of knowledge

Linear geographies?

Claims that CM 'begins and ends with the consumer' (Noakes 1997: 6) run almost like a mantra through the trade press. It should by now be clear that we are sceptical of such claims, but they certainly have come to occupy a prime position in retailers' own narratives of their activities. Here, for example, is a detailed extract from our conversation with Jim, the buyer responsible for the cooking sauces category in one of the UK's 'big four' supermarkets:

It's dog eat dog out there at the moment. And, we're trying to grow our business. We're trying to get some customers back from [our main rivals] and other areas. And we've got to try any avenue to look at this. But [CM is] not an empty marketing thing. It's simply saying to the customer, 'Look. We're more likely to keep you, and keep you happy, and get other customers back in here if you feel that you're actually getting what you want.' . . . Everyone is the same in this. . . . There's that degree of arrogance there, that we actually know what's going on on our doorstep and all we need to know is the, sort of, macro level. Where, . . . at the same time we've realised that, as the years have gone by, that we don't actually know a great deal about the customers that we get through our door. *Was it almost, like, you trusted that you had a feeling for it beforehand, but now . . .?* Yeah. Well, the 'feeling for it' was that we were a bloody successful company! *Ha ha. I guess so. Yeah!* If you are successful, um, then you, you, you think, 'Why re-invent the wheel?' You know, why should we actually go out of the way to change what we're doing or, you know, look at that? In essence, we must be doing it right because we're successful. And I think, you know, that success is now, sort of, um, such that we're thinking, 'Well, actually', you know, 'come to think of it, we don't actually know a great deal about the specific categories and how they're shopped by the consumers and we need to know more.' And it's the retailer that understands his customer best which has the most durable long term future in the business. . . . We're being pulled in all sorts of directions: from the customer, from ourselves, from the suppliers, and the suppliers are obviously trying to pull us in a way which favours their own individual products rather than the overall category as such. But actually, more and more, I would say it is the customer who is winning . . . We've done a lot of research called 'usage and attitude tests' in terms of how they will expect to shop a fixture, what they would expect to see grouped together and that's even now being looked at in an operational way, to change our structure here, so that if it is perceived that pasta goes with pasta sauces, then pasta gets bought by the buyer that buys pasta sauces. [At the moment] whereas I'm the buyer of pasta sauces, I have a colleague in a different department who buys pasta. So, not only are we thinking of the way the fixture is displayed, but we're thinking of how the whole thing is bought and controlled. And that makes sense because, if the shopper is expecting that, by grouping products right the way down the line there will be efficiencies there.

The same kind of story was told to us by Louise, the 'bought-in' bread buyer working for the same company:

One of the catch-phrases of our directors is, 'Have you got the customer's permission?' It is true. The sentiments behind that are true – that you have to be thinking about your customers. You have to understand what they want and are you providing the products that they want? And I think the challenge in the bread industry is that the supply base is pretty unsophisticated and it's very production-led. Every baker that you will ever talk to is totally production led. And, from my position, I'm the person in between the customer and those producers. And it's trying to get the right answer as far as the customers are concerned.

The ideal version of CM promoted in the trade (press) is, therefore, characterised by a linear geography of knowledge. Consumers, it is said, understand foods in certain ways, and in their shopping, cooking and eating categorise them into knowable arrays of variety. Retailers and manufacturers research consumers, both quantitatively and qualitatively, to discover these categorisations and the understandings held about them. These categorisations then form the basis for the organisational geographies of food provision. As we have seen, 'vertical' category-based flows of knowledge and of goods are supposedly formed between retailers and their suppliers. Moreover, within these companies, these flows take place through a horizontal grid of categories in which occupational and professional situses are downplayed as organisational principles. Thus, as one trade press writer has explained how CM has affected the organisation of work in the retail sector, 'In the past, stores were aligned according to tasks: merchandising, buying, and so on. But now, often following management consultancy reviews, most have reorganised into category-based teams where each member is focused towards meeting the needs of the category's consumer' (Campbell 1997a: 32).

To be honest, this posited linear flow of knowledge from consumers to food providers ran counter to the expectation we had when we began our research. We had, in our proposal, put far more emphasis on the knowledge food providers were involved in producing for consumers to accompany and structure their food provision: forms of packaging, advice leaflets in store, magazine production and so on. In the spirit of Fine and Wright (1991), we were interested in the interactions of the food system and the information systems surrounding it, and in the ways in which food providers sought to shape the environments within which consumer choices were made. Take, for example, the standard histories told of the developing British taste for so-called 'foreign' foods. Here, a central role is usually accorded to 'ethnic' restaurants which have pre-disposed consumers to eat 'Italian', 'Indian' and 'Oriental' cuisines at home (see Cook *et al.* 2000). This is something which food providers for domestic consumption, having read those standard histories, are well aware of. Take what Owen, the marketing consultant working for a small 'ethnic' food importer/manufacturer, told us:

> You've probably read them, the ethnic foods reports from Mintel [e.g. 1996, 1997]. And they've said for a long time, the next eating experience is going to be Caribbean food. Now, which we all sort of go [with a deadpan voice] 'Whoopee, great'. But the problem is – and it's moving well, it's getting there, and we will be making conscious efforts to actually, sort of, raise our heads above the parapet a lot more than has happened in the past to say, 'Yeah we are the specialists', to the trade and to the consumer, 'These are the brands that will lead the way,' – but you know, the one thing that we do struggle with is, you know, whereas Indian food rode in on the back of thousands of Indian restaurants where people used to go out and experience and then go, 'Now I want to experience that at home,' that was the starting point. The same with Chinese food. But when you have a handful of Caribbean restaurants . . . that's really what's slowing the whole concept down a little bit. So, and we will be making bigger efforts to actually target the food service industry, not just Caribbean restaurants but the mainstream markets and restaurants and pub chains and whatever, to start educating them.

Our initial reaction to the co-existent emphasis on consumer-led CM was there-fore twofold. The first reaction was to see how many of the food provider activities we had thought of as knowledge provision also acted as forms of consumer research. For example, the recipe offers on products, or various clubs to which consumers can belong, can be seen to be as centrally concerned with establishing customer databases (both for actual information and as tokens of category knowledgeability), as to be about educating consumers about products or cuisines (see Cook *et al.* 1999b). Take, for instance, the following discussion with Fred, whose company manufactured high quality regional cooking sauces. We asked him about the strategies which his company used to encourage the demand for these products. His answer was long and complex:

> Education um that they're buying authentic products is, we haven't even scratched the surface because there's so much to do and you never stop doing it. I think the packaging is probably the starting point of it, because that's your 'first face' – if you like – and that's got to communicate that quality, authentic, with the right picture, the flavour's got to be good, the recipes, um, as well – although people don't want too complicated recipes generally – and then all the support type of thing, like a cookery club, a helpline, um, we've got a chuck wagon going around trying to sample people, now, you know, a 16 foot trailer that goes round with people standing around cooking, and things like that. *So, would that be outside supermarkets?* That's outside main supermarkets. Yeah. So, all that type of thing. We've just gone on the web, now, and we've got recipes on there and stuff like that. That's only just happened. So, it's just an ongoing thing. But mostly it's trying to get recipes in their hands and to taste it. And, when they taste it, they say, 'Ooh, that's special. It's different to what I normally eat.' Everything should be, everything we do should be aimed at getting people to put something in their mouth. And that's it. *So, how do these things, actually, I mean, where for instance would I come across one of these, one of these recipe cards? Would that be something in a supermarket, or . . .?* Well no, we've never managed to get those in the, those we used to give out at exhibitions, they were sent out to our cookery club members. We've had various ways of getting them out. We've had some neck-tag promotions. People could send off and get a discount on the sauce and got back recipe leaflets with it. A whole variety of mechanics um to try to get recipes to them. Best of all is point of sale in the supermarkets, you know, if you can put it on the shelf next to the products. Very few supermarkets want you to do that. They don't want the shelves littered with all sorts of brands trying to do their recipe ideas. So, very seldom does that work. *Right, so can you tell me a little bit about this cookery club?* Yeah. We started this cookery club really to try and um, the people who are already buying our products are obviously the start of loyal customers, we hope. And we want to try and encourage them to buy other products in the range and to use them more frequently. So this is our form of communicating directly with them through (the company's newsletter) as to 'here are recipe ideas', 'these are occasions when you might think you'll be eating our food', um, 'did you know?' all these kind of background information about (the source of our sauce). And it just gives us a bit of two-way between them and us, if you like. So um we've got about, I think it's about 15,000 names on that at the

moment and it's growing rapidly. In fact, most of our merchandising, most of our packaging now says, 'Join the cookery club', um [reading from a jar] 'For free membership of [the company's] cookery club, just send us your name and address'. And we get this coming in all the time. There must be 35 a day, 40, 50 a day. . . . These are just standard formats that everybody uses. *And what about these (other) kinds of booklets, as well? Who would, who would they be for?* They would go out to restaurants who were interested in using our products, and mostly from an enquiry about our products or giving them away at an exhibition.

What we see here, then, seems to be both a concern to 'educate' consumers into the tastes of these sauces, both by direct promotion and through shaping the wider provision environment, but also an effort to establish a verifiable consumer loyalty base that can be traded as an indication of market knowledge.

While our food trade interviewees often talked at length about their ability and efforts to shape consumers' tastes, they were also well aware that this was often undone in practices of consumption. They talked, in particular, about how they knew full well that consumers cooked across their carefully constructed categorisations. Jim, for one, was resigned to this fact:

> I think, in this business, you have to recognise that the usage of products is very much up to the customer. You can dictate so far. I know for a fact that my pasta sauces are being used as pizza toppings. I know they're being used as casseroles. . . . And it's why we need to get inside the heads of our customers. Sometimes we don't appreciate the way they are using products. . . . We shouldn't be dictating to customers, 'You use this product with this and this'. You can do that to help them, and some do pick up on that and buy into it. But a lot of customers say, 'Sod it, I want to use this product the way I want to'. And if we're not trying to show you can be flexible with products, then we're dictating to customers how to use it.

The vagueness of this account of consumers is itself intriguing. But the argument that consumers incorporate products into their own culinary repertoires is one we found to resonate strongly through our interview series work with households. In terms of both practice (what was done with foods) and meaning (the ways in which foods were understood and grouped) the same products were incorporated in differing ways by different consumers at different times. Having said this, though, these consumer understandings and practices were not totally unrelated to the categorisations displayed in store. This was pointed out by Fred as he talked about the difficulties in pitching new cooking sauce products:

> We launched with these jars, which is a concentrated recipe . . . And it's full of flavour in there, it really is. But with that, you can mix it with mayonnaise and it tastes really good, or with sour cream. Or you can eat it as it is, like a taco, like a hot ketchup, use it for salads. . . . You can make loads of lovely dishes out of it. Absolutely. Tons. But you can't tell everybody that on the label. And when they buy it, they think, 'Well, what equivalent Indian sauce is this?' You know, 'If this was [Indian] which of the Patak's range would it be? Is this a paste?' You know,

if it's versatile, 'Well, it's not really a sauce because I'm, of course, accustomed to sauces as cooking sauces.'

Food providers search, then, to match their own categories to those of consumers, and consumers operate in part through a conventionalised if changeable lexicon developed by food providers. So, how could we conceptualise this relationship between provider and consumer? As a one way flow of products and knowledges from the former to the latter? As a process in which knowledges about product categorisation and use are bounced back and forth between providers and consumers? Or as something else?

Circuitous geographies?

In a rather roundabout way, we have been trying to lay out a sense of the circuitous geographies of knowledge that characterise the relations between food providers and consumers. We have suggested that these seem to conform neither simply to the model of an information provision system shaping consumer knowledges and practices, nor to the 'CM ideal' of providers looking to replicate consumer knowledges. Here, we would like to reinforce this latter point through emphasising four elements of our discussion so far: first, that emphases on knowing and responding to 'the consumer' have no monopoly in commercial discourse; second, that this is brought to the fore when 'production-led' commercial practice (e.g. concerning the 'authenticity' of products) is itself seen to be valued by consumers; third, that knowledge about consumers is highly and multiply mediated, and actually directed as much if not more at the entity called 'the market' than at 'the consumer' *per se*; and fourth, that these commercial practices involve far more than just giving consumers what they want, in the form that they want it.

First, then, any emphasis derived from moves in CM towards providers moulding themselves to fit the categorisations unearthed by the research they undertake on consumers co-exists with a variety of other commercial discourses with very different spatialities of knowledge: on stimulating demand through promotional offers, on the education of consumers into new tastes, and on the structuring of consumer demand through supply. Here, for example, is an extract from our conversation with Louise the supermarket bread buyer:

> Our volumes don't tend to change enormously from week to week, so you can predict how much will sell in any given week. But, inevitably, things like speciality breads are a real impulse purchase. If they're there and in front of people, they'll probably buy them. But, if the store hasn't ordered enough, they won't sell them. . . . If it's a branded line . . . very often the supplier will give support for, there might be advertising on television or they might be doing some PR on the product. On many own label lines, we tend to do an intro-ductory offer, put it at a lower price for a limited period to get people in and trying it. We might want to do leaflets with it if it's something very different. But the introductory offer works very well because people will try something if they think they're getting a bargain. And hopefully they'll carry on buying it after then.

A second complication we need to offer to CM's linear account of commercial knowledge of consumption, is that a 'production-led' focus is not simply something that exists in opposition to a 'consumer-led' focus. Through key culinary discourses, such as those of 'authenticity' as well as those of novelty, an orientation to the concerns of production can itself be a quality valued by consumers. Certainly, for many small manufacturers unable to trade in large consumer databases, a looser appeal to consumers' desire for 'authenticity' was combined with an emphasis on a production-led focus (Cook *et al.* 2000), that is the spatiality of 'authenticity' in which producers reach consumers through 'diversion' rather than intention (Appadurai 1986; Crang 1996). To make that less opaque, the issue here is that in order to fulfil some consumers' desire for 'authenticity', food providers have to appear to pay little attention to consumers. Two illustrations may help here. The first concerns a small specialist 'Middle Eastern' bread company which was started in the early 1990s by two Lebanese brothers working as chefs and head waiters in a West London restaurant. Its General Manager told us:

> We believe what we are selling: we are selling an ethnic product to the main-stream market. We don't want to sell, we are not a company which has ambitions to produce ethnic products to [sell to] ethnic people. We are not in that game because we believe that the mainstream market is the main market and it's accepting. It's quite liberal. [Then he talks about a survey of 500 people where the results suggested that younger age groups were more open to these breads and, therefore, as this population grew so would his business]. . . . Because the market is so sophisticated, the only thing is to leave them behind. We can't keep on dropping the price. The only thing is to make difficult-to-do products, difficult-to-do breads. If you've got the basic bread, everybody is doing it. . . . We bring people for three months, six months and they will teach the boys how to do this, how to do that, based on the original recipe. So we have to because . . . otherwise we would be drifting away from the original recipe. . . . We bring people from Lebanon, from Syria, from Jordan. . . . But we also try to make them different: now we are working on pitta bread with a crust in order just to be different . . . pitta bread – it's not British – but, now, if we come up with a pitta bread with a crust, it might be that is what you call British in the long run.

The second is an extract from our interview with Fred where we asked about other foods that crossed such ethnic/regional categorisations:

> *One of the things that I've been doing as part of my bit of this project is looking at the histories of ingredients. And, because one of the things, this is the sort of thing that's important to me, anyway, is that the way in which, if you go into a supermarket, everything's, in the sauces section, it's divided up in different parts of the world. India there, China there, you know, the Americas or Mexican food there, you know. And it's like they seem to be treated completely separately, as if they're not connected to each other, these parts of the world. And I know that that's done because it makes it simpler for shoppers to shop the fixture. Because you can't start say, 'Actually, all of the cuisines completely overlap. They're hybrid.' . . . Yes. . . . If you follow the ingredients around, in fact, you can't make these distinctions at all. . . . No. That's right. . . . And you*

can read, you know, there are, at times, very interesting articles that come out. Like a flatbread, like a tortilla. You know, you've got the Chinese pancake, we eat Peking Duck in a Chinese restaurant. You've got chappatis from India. . . . You've got lots of flatbreads from North Africa. You've got the tortilla in Mexico. The pitta bread in Egypt. Right! You know, the whole thing of it. And they're all unleavened breads, you know, you use to roll around a filling to eat with. And even if you go to Africa, you find it. And um it suits us in many respects, as marketeers, to keep it niche. Yeah. To say, 'This is, you buy this when you buy Mexican,' and, 'You buy this when you buy Indian.' Equally, my tortillas work really well with Indian meals, like, to eat it instead of a chappati. We never have an Indian meal at home without having tortillas with it, which is great. So, I think it's nice for people to know about these things. . . . I think, foodies, yeah, it's fascinating for them. But, it would be, I don't think it's a, I don't think it would help the marketing of it . . . I think it would undermine our authenticity. If we wrote on it, 'Flour tortillas: wonderful chappati substitute,' or, 'Great with Indian meals,' I don't know if it would be a good thing. Part of me says we should try it and see what it does for the sales. And part of it makes me think, 'Would those people think, "Oh, well, I'd better buy [our major rival's] ones. That's the real Mexican"'.

So, at times, not demonstrating knowledge of what consumers do or might do with foods is itself an important element of commercial practice; because that practice is focused not only on the consumer but on positioning products and brands within the market.

Third, the image which providers develop of consumer 'needs' and 'understandings' is highly and multiply mediated. Despite the easy rhetorics of 'getting to know the customer', in practice the food trade personnel we talked to had as strong a sense as academic researchers of how hard it is to get to know what consumers think, feel and do. Masses of quantitative data, organised both by product lines and by customer types, were cited by the larger retailers and manufacturers, if not the smaller ones. But we found it far harder to get a sense of how this was actually used. Certainly, in an environment of CM development, it was a vital component of an 'aura of knowledgeability' that could be traded upon. But the alacrity with which interviewees seized upon our own work, and any mention of other qualitative or especially ethnographic work with consumers, suggested that they themselves were well aware of how this aura did not simply translate into an all-seeing God-like grasp of consumption. CM itself emphasises qualitative work and, in particular, focus groups, shuffle card analysis, accompanied shopping trips, individual interviews, shopper tracking, intercepted shopping, and kitchen cupboard inventories (Ody 1997; Soars 1997). One powerful reason for the introduction of these methods came from an early vision of CM where:

The new focus on data will be the 'why'. Why are consumers choosing a product or a category of products? Why are they shopping in this or that channel, or in one part or another of the store? Why do they hate or love the purchase experience itself?

(Molpus, in Owens 1997: 6, 70)

This new qualitative research is supposed to allow a bridging of the 'gap between the detailed transaction records showing combinations of products bought together, and the product oriented specialist sales knowledge that manufacturers add to the equation' (Ody 1997: 32). But in practice it too is supplemented by a variety of other 'research' practices undertaken by key personnel such as buyers, home economists and product development managers. There is much more going into understanding 'the consumer' than doing formalised market research. Asked how they kept in touch with what was going on in 'the market', the people we interviewed came up with a very wide range of sources. They talked about the more *specialist knowledges* they drew upon such as independent market reports produced by Mintel and Keynote; market reports produced by manufacturers such as *McCormick's Spice Report and Masterfoods' Cooking Sauce Report*; trade press articles such as those in *The Grocer*, or *SuperMarketing*; quantitative and qualitative market research which is 'bought in' – e.g. AGB data – commissioned, and/or done 'in-house'; sales figures from, for instance, EPOS returns; other feedback from stores through everyday contact with store personnel, letters from customers, and personal visits to stores; informal contacts between manufacturers and retailers through personal visits and telephone contacts; representations from pressure groups such as Christian Aid and the Farm Animal Welfare Network; and trial and error runs of NPDs on the shelf. It is also important to point out that buyers, home economists, marketing people, etc., were themselves also consumers and drew on more *popular knowledges* in their work. So, alongside the sources listed above to keep tabs on consumer trends, it was often also their job to pick up and respond to new ideas from, for example, TV chefs, cook books, travels to other places, restaurant visits, shopping trips to rival stores, and trips around other sections of their own stores.

To give a feel for how such a range of influences could come together to make sense of what 'the consumer' wants, take the following interview extracts from our conversation with Jim the cooking sauce buyer whose category was the pilot for his company's CM plans. Let us begin by reiterating his point about the pressures he felt from within the food business:

> We're being pulled in all sorts of directions, from the customer, from ourselves, from the suppliers, and the suppliers are obviously trying to pull us in a way which favours their own individual products rather than the overall category as such.

Although he emphasised here the interrelated importance of the three main players in CM discourse, his explanation of how he knew which NPDs to introduce to his shelf fixture and when to do so involved much more than the 'science' of CM. In practice, this seemed to give way to the 'savvy' of the experienced supermarket buyer:

> I think it's basically all sorts of things, not just one thing: the fact that you start to see recipes appearing from chefs; you start to see restaurants; you start to see market research which is based upon people travelling to those countries and bringing back those ideas; you start to see different areas of the business trying to latch on to that – produce that's a reflection of those countries – and therefore you can latch on to that. And you get to a point where it's on a roll. It's a train, very difficult to stop. At that point you latch on to it. It's very rare any retailer would jump in too early. You'd cut it dead. If you think about it, part of the reason

these areas grow is because of the mystique. If you jump in on Thai, stick it in the supermarket in a first week, you've killed off that mystique. Let it develop and nurture and grow, and then latch on to it in a way which isn't going to kill it dead. It's a slow build and that's what we try and do rather than steamroller in and knock all the inspiration out of it.

So, his aim was to get a grasp of a broader shift in commercialised culinary culture – or 'the market' – and then to time his intervention in it. It was not to just understand the consumer *per se*. Other aspects in the market were far more important; for example what his competitors were doing. Louise the bread buyer also emphasised the diversity of sources which she used for her new product ideas, as well as her inherently subjective and literally gut-based sense of what might sell:

> I've just been to New York on holiday. I was in all the delis looking at all the bread and getting quite excited about that. You are switched into having a look and seeing what's there. One of the buyers who works for me went to Italy for a week and has come back with some interesting ideas . . . [But] that's one of the most difficult things – getting hold of these people [i.e. representatives of small manufacturers]. And a lot of the smaller people think, 'I couldn't possibly speak to [that big supermarket chain]'. We actually go to regional centres and the local commerce board will set up for local people to come in and show us their products, which tends to be quite a good source of discovery. . . . You have to be able to try and recognise what will commercially sell and what won't. As a job as buyer, I used to buy cakes and I don't like cake at all. I never eat it and I haven't got a sweet tooth. But I had to do that job. You can find yourself in that position. I love bread. I've probably got more unusual tastes in bread than your average customer. So you have to be aware of what sells, and you know what sells.

Finally, not only is knowledge of the market a more multi-directional issue than simply knowing the consumer, but it also can have little to do with the consumer at all. Here, for example, is the wry response we got from Mike, the research and marketing co-ordinator for a national plant baker about bread marketing, and its dislocation from consumer understandings:

> There's a lot of confusion in the market from the consumer. There are a lot of different types of bread and I don't think the consumer really understands the difference between wholemeal and wheat germ, country grain and granary and soft grain. They're all just tinkering with a basic concept and changing it slightly. I think the consumer would only see white and brown, wholemeal and bread with bits in it, and the in-store bakery, and then rolls and baps and the cakey-type things like scones and tea cakes. They probably have a more simplistic view of the market. *Is that confusion in a way, I imagine certain people will benefit, presumably, for you, if there was more simplicity in the market . . .?* The more simplicity in the market, the less for our brand managers to do! A lot of it is marketing led, not consumer led. If you get a brand manager, this may be cynical, but brand managers will create niches or not exist . . . *despite market research?* . . . They work 52 weeks a year. They've got teams. They want bigger teams. They need to find a project

to do. Very often that means launching products. . . . Some work, some don't. *So, you're saying, basically, because of the way these things are organised here, the teams are set up, that you may get an over-production of . . .?* Basically, people will only eat so much bread, a level beyond which it's not really going to exceed. It's really just shifting them from one area to another . . . it's quite amusing, but that's what happens. You get all these different types of bread, and I don't think the consumer has the first clue about what half these things are about. They just know that they either like it or don't like it and continue to buy it . . . You can get locked in an ivory tower here. It's quite . . . a bit of a rude awakening sometimes when you do actually go to the consumer and start speaking to them. *What kinds of things do you find out?* I guess one of the things – it's prejudice and attitudes – you find out that the customer probably doesn't know a great deal about the market, doesn't know a great deal about bread. They're locked into prejudices and attitudes and that can affect the way that you market your product. . . . Maybe they've got strong views – 'I would never buy own label. I only ever buy brands,' or, 'I don't buy cheap bread, I only ever buy premium. But, equally, I want to get the best value for money so I buy the promotions pretty much'. 'I won't buy a particular brand because all the bread is the same anyway. So I'll buy Hovis one week and Allinsons the next week, Kingsmill one week, Hovis White the next.' You may think your brand is the most wonderful thing in the world but you find that customers don't. They might say, 'Bread is a commodity. It's all the same.'

Mike's job was to co-ordinate the findings of market research for his company's Category Management activities, and his company was also one of Louise's main suppliers. Earlier, we quoted her description of such plant bakers as 'unsophisticated and . . . production-led'. Echoing the rhetoric of CM, she argued that it was her job to understand what her customers wanted and to try to put pressure on manufacturers to get the 'right answer as far as the customers are concerned'. Yet, later in that interview, she echoed Mike's contradictory assertion that consumers didn't really know what they wanted and, by implication, neither could she:

> I'm quite interested in what you're doing. I did a Geography degree myself. It would be very useful if you had any feedback on why customers are buying certain products and what they think about them. Do they think there are products they would like that they can't buy? I think customers are led by what's available (rather) than saying, 'I want X.'

Our argument, in summary then, is that CM's claims to restructure commercial practice and knowledge simply in the image of consumer knowledge and practice oversimplifies both the commercial field and consumer culture. Their intersections are characterised far more by circuitous (and networked) relations of knowledge than by a simple flow through which providers come to understand 'the consumer'.

Categorisation, commercial culture and the consumer

In this short concluding section we want to hammer this point home by turning to consumers rather more directly. We have written up, and continue to write up, our household materials elsewhere, and space unfortunately precludes us from doing them justice here. However, we do want to emphasise that we are highly doubtful about the ability of commercial research to somehow grasp the categorisations used by consumers in such a way as to then establish organisational structures that can match these. We found consumers' categorisations of foods to be highly contextual and changeable and to be based on a number of different criteria (from state of preparation, to some sense of geographical origin, to intended use, to relationship to household financial and 'emotional' budgets, and so on). In a single household, for instance, pasta might be 'Italian' in one context – if a meal thought of as 'Italian' was being planned – and not at all 'Italian' in another. As an illustration we can take an extract from one of our interviews with Alexander and Laura, a married couple in their early forties. Here, they discussed their categorisation of such foods in a typically sparky manner:

Alexander: I wouldn't think, say, about cooking in an Italian style or cooking in a French style, though come to think of it, our favourite way of roasting chicken is a French style, but we think of it as roast chicken, not in terms of where it comes from [though, to complicate things, their food diary notes that they did buy a 'free range' chicken].

Laura: I disagree with you about Italian.

Alexander: Well, we don't think Italy when we're doing spaghetti.

Laura: Don't you?

Alexander: It's something favourite that we eat and you always put plenty of tomatoes in it, and what have you . . .

Laura: I guess you don't have such high standards if you are just cooking for us. But, if I had people around. I wouldn't put Italian food with a Greek salad. I'd do a proper salad.

Alexander: That's certainly true. But, in terms of standard cooking, I don't, umm, I don't think, 'I'm going to be in a Spanish mood tonight!'

Laura: But you said this morning, 'I feel like Chinese tonight.'

Alexander: Oh yeah [pause]. Urgh [pause]. But I didn't mean it like that. I'm not going to dress up or something.

As a result of these and other interviews, it is hard for us to see how the reality of consumers' categorisations can be legitimately used to resolve the hard commercial conflicts currently taking place over how to categorise and manage food provision. How, for instance, could a knowledge of Alexander and Laura's everyday (and often contradictory and inconsistent) categorisations of 'regional' or 'ethnic' foods inform trade debates about whether cooking sauces should be categorised by 'region of origin' or 'form of cooking'? On the shelves that he stocked, for instance, Jim told us:

We have a very clear objective which is to display goods by country of origin, to create an international foods area where you can have all Italian foods there, all Indian foods there. And the view from some manufacturers, predominantly because they're into some of the product areas outside of that, is that, 'No. The way people want to buy products is as a casserole, or a marinade.' So it's the preparation.

So what else does categorisation and Category Management do if it doesn't get providers into the heads of those consumers? In the end, what is at stake here is less the ability to replicate the categories of consumer culinary knowledge than: (a) the ability to claim to know, and therefore to act on behalf of, a 'virtual consumer' (see Marsden *et al.* 1998, 1999; Miller 1998); and (b) to thereby legitimate an 'organised array' of foods that suits particular interests within the provision system. We have tried to show how CM, like other managerialist discourses, is not only ideological but also plays a part in everyday work practices as rhetoric spoken by management professionals in ways which are 'not necessarily coterminous with organisational practice itself' (Thrift 1998a: 37). Thrift argues that this is because personnel faced with this 'new managerialism' are often ambivalent about it and that, while it has a place within business practice, this is often only in an incomplete, challenged, and leaky form. While the content, roots and routes of these discourses can be analysed through the trade press and the best-selling books of management gurus, however, he argues that very little is known about the ways in which such formalised business knowledges are consumed, including the possibilities that exist to resist and/or subvert them.

Despite this, Thrift suggests that such discourses have important roles to play because of the ways that managers (can) use them: to find some order to cling to in an ever changing business environment; to become empowered by as they provide new ways to frame the getting of results; and to gain esteem as managers may want to be seen as working in the most up to date ways and, perhaps, to have been involved in shaping those ways for their own firms. We believe that our analysis of Category Management in parts of the UK food business can shed some light on how managerialist discourses do (not) work in practice (see also Cook *et al.* 1999b). The main aims of CM seem to be to actively order foods so as to produce: *legibility* (ensuring the diversity is displayed in ways that facilitate easy purchase; and ensuring the divisions of responsibilities between category teams are operationally manageable); *associations* (within stores to promote cross-purchasing, so that one purchase can spark an impulse buy of another; and within systems of provision to promote 'economies of scope'); and *visibilities* (for example, of some product-lines and brands over others). Yet its practitioners seem to simultaneously work within and against the ordering logics of CM: through using the 'Science' of CM to feed into decisions still arrived at through the unscientific 'savvy' that can only come through experience and/or through engaging in 'guerrilla tactics' which draw on understanding of how CM works messily in practice, rather than neatly in theory. Our interviewees provided ample evidence that, while CM as neat theory was accepted and worked with, it was widely accepted as being wrong. Indeed, what was wrong with it seemed to have been accepted and worked with by all concerned.

Notes

1 This chapter has come from a larger project entitled 'Eating places: the provision and consumption of geographical food differentiations', which was funded by the ESRC (project number R000236404). We thank them for their support. It was presented in September 1998 at the ESRC 'Commercial Cultures' Seminar at the University of Sheffield. We thank the organisers and audience for their comments. Finally, we would like to express our appreciation to Peter Daniels and Nick Henry for their comments on our first draft and for their commendable feats of patience.

2 Here, we will be illustrating our findings through extracts from key interviews with people working in the cooking sauce and bread sectors. For a more wide ranging and detailed account of the project's take on CM, see Cook *et al.*, (1999b).

3 Here we are drawing on the 'circuits of culture' work produced by, for instance, Burgess (1990), du Gay *et al.* (1997) and Johnson (1986).

4 All real names, and some specific details, have been changed to preserve the anonymity of our interviewees and their companies. Quotations include the words of the interviewer in italics, and sequences of three full stops indicate inaudible words or minor editing.

5 One example is the butter producer Kerrygold for whom 'Promotions . . . play a key role in the absence of any extensive research it can peddle to the retailer . . . if the company offers "buy two get one free", the retailer will need to put more of the product on the shelves because demand will have risen' (Macdonald 1997: 29).

16 Being told and answering back

Knowledge, power and the new world of work

Jane Wills

Introduction

> For the last 100 years or so . . . we've assumed that there is one place where expertise should reside: with the 'expert' staffs at division, group, sector or corporate. And another, very different, place where 'the (mere) work gets done'. The new organizing regimen puts expertise back, close to the action – as it was in craft-oriented, pre-industrial revolution days . . . We are not, then, ignoring 'expertise' at all. We are simply shifting its locus, expanding its reach, giving it new respect – and acknowledging that everyone must be an expert in a fast-paced, fashionized world.
>
> (Peters 1992: 445)

In what Peters refers to as 'the new organizing regimen', companies are exhorted to recognise the knowledge workers have and to empower them to take decisions and play a full and productive role in 'their' firm. Over-turning the Taylorist divide between control and execution, the truly modern enterprise is argued to be one in which employees are knowledgeable partners, willing and able to play a full part in the life and success of the venture. In this context, a new set of managerial practices have come into being to tap and develop the knowledge of workers by fostering employee involvement and participation at work (see Blair 1995; Drucker 1993; Handy 1997; Kochan and Osterman 1994; Ramsay, 1990). Moreover, as the UK government's *Fairness at Work* white paper asserts, such measures are now firmly linked to economic performance:

The best modern companies, whether large or small, have some things in common:

- they seek to harness the talents of their employees in a relationship based on fairness and through a recognition that everybody involved in the business has an interest in its success;
- they ensure that everybody understands the business of that change is readily accepted and implemented, not feared;
- they set clear objectives for employees but also encourage them to exercise their initiative and to contribute their ideas to the development of the business; and
- they develop the workforce through training and work experience to respond to and lead change.

(Department of Trade and Industry 1998: 11)

Workers are now viewed as knowledgeable subjects and the declared aim of practices such as teamworking, quality circles and consultative forums is to tap employee knowledge in the service of greater productivity and improved working conditions. By drawing upon employee experiences of the processes of production and service, companies can make improvements in the way work takes place. Moreover, it is argued that workers will feel more involved and committed to their employment when they have a real influence on the ways things are done. At a time when competition is fuelling rapid workplace change, contented employees are more likely to adapt willingly to change and thus help the company meet new demands. Employee knowledge has become valorised as a vital economic resource, and tapping and developing this knowledge requires that new systems of corporate communication are put into place. Communication systems need to ensure that employees feel willing and have the space to share their knowledge and ideas about the labour process, technology and managerial practice. To be effective this communication needs to be two-way. Workers need business information to feel involved in the firm and so be more willing to provide information in the various forums provided. Managers need to talk to as much as listen to their staff if knowledge is to be effectively managed and developed at work.

In this chapter I scrutinise the ways in which information, communication, and knowledge are managed at work. Following a more detailed introduction to the wider issues involved, I draw upon original research into the development of European Works Councils to explore how information sharing takes place in practice. This research highlights some key questions about the rhetoric and reality of the management of knowledge at work.

Communicating at work

Contemporary businesses have to respond quickly to changes in the competitive environment, customer demands, technology and the geography of market opportunities. In order to survive, corporations are continually reconfiguring their products, production processes, organisational structures and the geography of production and service delivery, with major implications for those in employment (for an overview of these developments see Harvey 1989; Piore and Sabel 1984; Sayer and Walker 1992). Workers are being asked to make a greater contribution to the firm where they work at a time in which the majority feel threatened and insecure in the workplace (Allen and Henry 1997; Elliot and Atkinson 1998; Peck 1996; Reimer 1998).

In this context, corporations are faced by a paradox. In a climate of economic change, corporate success increasingly depends upon employee involvement, initiative and adaptability just at a time when change itself is likely to undermine employee commitment, loyalty and morale. Workers are less likely to feel motivated to work harder, innovate and adapt to the demands of the job when they feel threatened by reorganisation, redundancy or re-deployment. If companies are to get the most from their employees in this climate of change, they thus need to manage the way change happens and to pay careful attention to the ideas and emotions expressed by their staff. In short, a new business environment has given rise to the need for new systems of corporate governance and managerial practice, as Streeck explains:

With the new intensity of decision making characteristic of the new, flexible organizational structures, what firms needed was no longer just the experiential knowledge and the passive compliance of workforces, as potentially produced by traditional forms of – non-representative – consultation – but their *consensus*: their willingness to agree to continuous changes in rules and work procedures under high uncertainty, as well as their *involvement* and *commitment*; their willingness themselves to make decisions guided, not by bureaucratic rules or superiors, but by internalized organizational objectives. Consensus, involvement and commitment, however, require that workers develop an active *interest* in their work and 'their' firm – or rather that the interests they have in the firm be activated and redefined so as to make them contribute to the efficient organization of the productive process.

(Streek 1995: 331)

Building consensus, involvement and commitment at a time when workers can feel threatened by corporate change has involved a new approach to information exchange at work, and internal corporate communications have taken on much greater importance.

Many companies have come to recognise the need to manage information at work, ensuring that workers are armed with the means to understand change, to propose new ways of doing things and co-operate in new developments, and even to see the need to retrain for new jobs within and beyond their existing workplace. As Blakstad and Cooper explain:

It is no more than a decade or so since most companies came to terms with external communications, and the skills needed to handle the media or to promote their company and its products or services to outside audiences. In the 1990s, they are coming for the first time to appreciate that there is an equal and opposite discipline involved in internal communications.

(Blakstad and Cooper 1995: vii)

New systems of written, electronic and face-to-face communication have thus been devised to foster greater information exchange, and wider understanding, at work (*IRS Management Review* 1997). In most cases, such systems have been combined with new approaches to workplace organisation such as team working, multi-skilling and total quality systems, all of which are designed to foster employee involvement in the day-to-day management of their work (Ramsay 1990; Wills and Lincoln 1999; and for the latest Workplace Employee Relations Survey (WERS) data exploring such developments in the UK see Cully *et al.* 1999; Millward *et al.* 2000).

Increasingly then, many companies have tried to change their managerial style to foster participation, to devolve decision making and to share information. The 'new' workplace is argued to have a greater emphasis on those doing the work, and by re-uniting control with execution, work is to be more fulfilling for those on the front-line. In a situation of mutual gain, workers have more interesting, and better skilled, jobs, while managers are able to yield greater returns through employee involvement and adaptation to change on the back of improved workplace relations (Kochan and Osterman 1994). As Weiler suggests, governments, employers, unions and workers can all reap enormous gains from such an approach:

Our priority ought to be to devise arrangements which will enable the firm to tap the insights and ingenuity of the workforce in improving the efficiency of its operations and the quality of its product in a fast-changing and highly competitive marketplace; and which will at the same time allow employees to experience a sense of accomplishment and satisfaction from making an active and valuable contribution to the success of an enterprise in which they invest much of their working life.

(Weiler 1990: 192)

Yet in practice, concerns are being raised about the operation of such new managerial techniques. A growing body of empirical research points to a persistent gap between the rhetoric and reality of much employee involvement and workplace information exchange (Blyton and Turnbull 1992; Cunningham *et al.* 1996; Lewchuk and Robertson 1997; Milkman 1997; Parker and Slaughter 1988; Keenoy 1990; Purcell 1993). While companies document their support for employee involvement and improved information exchange, there is considerable evidence that workers actually have *less* control and discretion than they did in the past. Indeed, as far as information management is concerned, critics have suggested that:

- information flow has tended to be top-down rather than two-way, or bottom-up
- such information flow has been selective and senior managers have been reluctant to share sensitive and competitive information with their employees
- line managers have been particularly hostile to sharing information with, and listening to, their subordinates
- the views of employees have been collected but then largely ignored
- employees fear that the information they provide will be used to undermine their interests at work
- employees have not been given the training needed to understand financial information and business accounts which might allow them to play an equal part in debating corporate plans and decision making

In short, it would seem that in many cases, new practices of information exchange have simply been superimposed over long established relations of mistrust between managers and their employees. Indeed, in many workplaces, information has long been a source of political power at work, and those who have access to it have little real inducement to share it with those below them in the corporate hierarchy. While managers want their employees to understand the reasons for change, to co-operate in workplace reorganisation and to contribute to improving the products and processes of work, there has been little evidence that they are willing to listen to, or act upon employee views and opinions about their work. Without a genuine partnership, information flow will thus tend to reflect existing relations of power, and both managers and workers will use their knowledge as a source of influence over the other.

In her study of workplace change at the GM plant in Linden, New Jersey, for example, Milkman (1997) found that new corporate initiatives for employee involvement and greater information exchange were not realised in practice due to local managerial resistance:

What has stood in the way of change is not resistance from workers, who are ready to try anything in their desperation to escape the traditional system. Rather, management has proved itself unable (or perhaps unwilling) to implement even relatively modest changes like those announced with such fanfare at GM-Linden in the late 1980s.

(Milkman 1997: 17)

Despite a recognised need for change and employee willingness to play a new role, the shop-floor managers at the Linden plant were unable to deliver a new approach to work organisation. Moreover, when workers lose faith in new managerial systems, they are unlikely to volunteer their opinions and ideas for workplace improvements especially if they believe that such information is likely to increase the rate of work or threaten their jobs in the long term.

In the context of long established relations of mistrust, genuine partnerships and open information flow are very difficult to achieve. Moreover, it is significant that many pioneering agreements in workplace participation have tended to involve long and complex negotiations between managers and trade union representatives who are entrusted to ensure that change will not threaten workers' interests (see Scott 1988; and for the examples of Blue Circle Cement, Rover and some American employee owned firms, see TUC, undated and Wills and Lincoln 1999). The successful implementation of employee involvement initiatives and procedures for information exchange is thus largely dependent upon the nature of existing workplace relationships, union policy and managerial determination to change. In practice, genuine partnership and information exchange are dependent upon both 'sides' of the employment relation taking a new approach to the other – and without relations of trust any transformation will be superficial at best. Without trust, employee knowledge is not likely to be mobilised in the cause of the business, and managers will not trust employees with the information they need to feel involved in the first place. In the rest of this chapter such developments are further explored in the context of the communication that takes place at European Works Councils (EWCs) in UK-owned firms.

Councils of hope or despair?

Works councils are proving to be an important part of the armoury of new workplace developments, particularly as they focus on information exchange. As Rogers and Streeck suggest, by facilitating the flow of information and consultation between managers and workers, works councils can play a role in improving efficiency and productivity in the long term:

By reducing information asymmetries between managers and workers, consultation can lead to more efficient labor contracts. By lowering the costs of information to both parties, it can facilitate adjustment to changed circumstances. By increasing trust between managers and workers, it can increase their willingness to engage in cooperative ventures, and with it increase the rewards that accrue to cooperation.

(Rogers and Streeck 1995: 4)

A key feature of the post-World War II industrial relations landscape in Continental Europe, works councils (and particularly those involving co-determination) have been associated with managed workplace change, employee participation and long termism in business practice (Streeck 1995; Thelan 1993; Wever 1995). Partly as a result of the need for improved corporate information systems as outlined above, and in response to the European Works Councils Directive, these institutions have now begun to take root within Anglo-American capitalism.[1]

As an example of recent developments, the retailer Sainsbury's have decided to implement a local and national system of works councils to share information and consult with their 127,500 staff in the UK. An employee survey implemented during February 1996 indicated that less than 50 per cent felt respected and supported by management or that their ideas were listened to, and in response, a nation-wide system of works councils was put into place (*IRS Employment Trends* 1996; see also, *Managing Best Practice* 1998). Organised on a site-by-site basis, this involves elected representatives meeting with local shop stewards and managers. Over 400 of these forums meet at least once a quarter and they are designed to discuss business performance, working practices, equal opportunities, corporate plans and ideas for improvements. In addition, the group has a national works council which involves representatives from across the UK, from the two trade unions active in the company and from the company board. For the first time then, a large company like Sainsbury's has established a representative system for sharing corporate information with staff while listening to their views and opinions – and there is growing evidence that a considerable number of other UK-owned firms are following suite (*Managing Best Practice* 1998).[2]

The emerging lessons of European Works Councils

The European Works Council Directive, agreed in 1994, has proved the key impetus to such developments in the UK. This Directive requires that all companies which employ at least 1,000 workers in the European Economic Area (EEA), of whom at least 150 are in each of two separate member states, establish a mechanism for information sharing and consultation with their employees. It is estimated that this Directive covers at least 1,200 large and medium sized companies across Europe, employing about 15 million workers and eventually involving 30,000 EWC representatives (TUC 1997).[3] As indicated in Figure 16.1, the EWC Directive is a framework which sets the parameters for employee representatives and managers to negotiate a system for information sharing and consultation at a European scale (for more detail about the background to this development see Hall 1992, 1994; Hall *et al.* 1994; Gold and Hall 1994).

EWCs generally meet once a year for the purposes of information sharing and consultation, providing an opportunity for senior managers to outline corporate developments to employee representatives from across the European Union (EU). The brief of these meetings can be enormous, as implied by the specification given in the annex of the Directive:

> This meeting shall relate in particular to the structure, economic and financial situation, the probable development of the business and of production and

---

Article 1 – Objective

1. *The purpose of this Directive is to improve the right to information and to consultation of employees in Community-scale undertakings and Community-scale group of undertakings*
[in article 2(1) Community-scale undertaking is defined as having at least 1,000 employees within the member states and at least 150 employees in each of at least two member states and in article 2(2) these employment thresholds are to be based on the average number of employees, including part timers, employed during the previous two years]

2. *To that end, a European Works Council or a procedure for informing and consulting employees shall be established in every Community-scale undertaking and every Community-scale group of undertakings, where requested in the manner laid down in Article 5 (1) . . .*
[in article 2(f) consultation means the exchange of views and establishment of dialogue between employees' representatives and central management or any appropriate level of management]

The EWC Directive provides a **mechanism for management and employee representatives to negotiate the composition, powers and operation of the EWC in their particular case.**
In essence the Directive:

- defines which companies and other organisations are covered by the directive;
- outlines a procedure for negotiating an agreement to set up an EWC (involving a special negotiating body or SNB);
- lays down some of the issues which must be covered in an agreement; and
- provides a basic structure and brief for an EWC if negotiations cannot be successfully concluded within three years, or if managers do not respond to a request for negotiations within six months (the annex).

Other matters covered include confidentiality, the protection of employee representatives, compliance with the directive, links with other directives, and the circumstances when agreements were already in force before 22 September 1996 (article 13 agreements).

Figure 16.1 Key points in the European Works Council Directive
Source: TUC 1997: 6 and 61–73.

sales, the situation and probable trend of employment, investments, and substantial changes concerning organisation, introduction of new working methods or production processes, transfers of production, mergers, cutbacks or closures of undertakings, establishments or important parts thereof, and collective redundancies.

(TUC 1997: 72)

Moreover, the annex makes specific reference to exceptional situations when employee representatives may need to be informed and consulted outside the annual meeting:

Where there are exceptional circumstances affecting the employees' interests to a considerable extent, particularly in the event of relocations, the closure of establishments or undertakings or collective redundancies, the select committee, or where no such committee exists the European works council shall have the right to be informed. It shall have the right to meet, at its request, the central management, or any other more appropriate level of management within the Community-scale undertaking or group of undertakings having it own powers of decision, so as to be informed and consulted on measures significantly affecting employees' interests . . . This meeting shall not affect the prerogatives of the central management.

(TUC *op. cit.*: 72)

About 580 EWC agreements had been signed by May 2000 (European Works Councils Bulletin, May/June 2000: 2). There is considerable variation between the documents and the practices that have come into being and each EWC is shaped by the specificities of the company, the partners to the negotiations and the national legislation which applies to it (see Carley *et al.* 1996; Rivest 1996). From their exhaustive analysis of 386 voluntary agreements made before 22 September 1996, Marchington and his colleagues point to such diversity arguing that:

On the one hand are those EWCs whose potential appears to be confined to a largely formal or symbolic existence based on an annual meeting. In such EWCs there is little or no independent contact and no cooperation amongst employee members between meetings. On the other are those EWCs which exhibit the potential to develop an active role, in which there is continuing activity on the employee side between meetings and ongoing contact with management.

(Marginson *et al.* 1998: 81)

There is a broad continuum of agreements and activity that range from those that are formal annual events to those that generate ongoing contact between employee representatives and managers during the year.

Quantitative data suggest that managers in some of the 240 UK-owned firms that meet the terms of the directive are using the EWCs as part of their armoury of corporate communications.[4] A survey of all the UK-owned firms which are affected by the EWC Directive has indicated that as many as 88 per cent of the managers in firms with EWCs saw a key advantage being the opportunity to exchange views with representatives of their employees. In addition, 63 per cent valued the opportunity to get management views over to employees and 56 per cent emphasised the value of hearing the voice of their staff. When asked about the effect of such an information sharing and consultation body on employee relations, 71 per cent said it fostered identification with the corporate mission, 64 per cent that it widened understanding of the processes of management and 50 per cent that it improved communication between staff at different levels of the corporate hierarchy (for full details of this survey see Wills 1999).

Such data are further reinforced by qualitative research conducted to explore the development and implications of the EWCs at five UK-owned firms between 1996 and 1999.[5] In each case, senior managers used the EWC as a mechanism to impart their message to employee representatives from across Europe, asking them for questions at the end of each talk. Moreover, such information is generally welcomed by UK-based employee representatives who have traditionally had little opportunity to gain detailed business information about the company for which they work. EWCs have exposed employee representatives to business strategy, corporate philosophy, competitive pressures and the geographical reach and diversity of their employer. As two of the representatives in Company B explained:

> I think it gives us all a far more global picture because you can get very insular and very provincial, and I think it did enable me, certainly, and I hope others following on, to have more of an understanding of what the issues are and what the developments are within the company and its aspirations and so on, and the reasons for things . . . I thought it was very interesting.
>
> (Employee representative, Company B)

> I realised that there's a lot more to the company than our factory here. I realised the extent of employment [Company B] have in the . . . industry. You just dealt with your own area before. From talking to shop stewards from other countries you get the sense that you are not on your own. There are other people out there. It does make you realise that our Division is only part of it.
>
> (Employee representative, Company B)

Moreover, one representative from a service sector subsidiary in Company C argued that the EWC was a means to undermine the introspective nature of their corporate culture:

> It has always surprised me that [the subsidiary of Company C] is a very introspective organisation, so despite being members of this huge conglomerate, it has never felt as though we were part of anything other than the centre here. This [the EWC] was an opportunity to go out and look at it for myself, as well as meeting people from other parts of the organisation.
>
> (Employee representative, Company C)

The opportunity for un-mediated communication between workplace representatives and senior managers was also highly valued. EWCs have given shop floor employees an un-mediated channel to the people with power in their corporation. And as this representative in Company A explained, this un-mediated contact between the top and the bottom of the corporate hierarchy can be beneficial both to employees and to those who employ them:

> My view is what it can achieve will be as a result of the opportunity of a broad spectrum of people to question and interrogate the senior board members. They can come armed with whatever questions they have in their own minds or that other people give them and the senior officers of the company are obliged to give them serious answers and if not, they can be cross-questioned . . . The other benefit I think is actually letting the senior management hear what people think because at the moment they rely on third or fourth hand information. They are told by their subordinates who are told by their subordinates and the whole thing gets watered down so you're telling them what they want to hear and not the way it is. The EWC is a way of letting the senior management know what are the

worries, the views, what the people around the company are actually thinking and what concerns them through a direct route rather than through a convoluted and highly filtered route.

(Employee representative, Company A)

Whereas the view from the shop floor is usually mediated by union officials or middle management, EWCs allow workers to speak directly to those at the top and there are genuine advantages to both workers and managers from this type of exchange. Moreover, a number of the representatives argued that the main benefit of meeting senior staff came during the informal contact time after the EWC meeting, when they could ask about their particular office or plant. As this representative explained:

If I want to know about my own plant I can talk to the managers. I knew there was new machinery coming in here and I can ask the managers informally about when it's coming. You learn more about your own workplace at the informal meetings than you do in the formal one.

(Employee representative, Company B)

Unlike union negotiating meetings when managers and union representatives feel they have to play a particular role, EWCs should also allow communication in a more relaxed manner – a fact that was recognised by one senior manager at Company B:

Hopefully the way it will affect things for the good is that people will understand each a bit better and certainly, they will have produced a better set of personal relationships which will make conversations on the hard issues a little more civilised and a little more capable of reaching some useful conclusions.

(Human Resources Director, Company B)

Such new types of relationship will take time to develop and in this regard it is significant that the employees who were most positive about this new level of contact tended to be the newer representatives who arrived through non-union channels. Untainted by the traditions and experiences of the past, these EWC delegates approached their task with less scepticism than their union colleagues, optimistic that new relations were now taking shape.

The limits to communication at EWCs

With their emphasis on information sharing, communication and consultation, EWCs are clearly fostering new forms of industrial relations in the UK. These aspects of the new institutions have been broadly welcomed by managers, unions and employees alike. Yet despite these positive experiences, a considerable number of the employee representatives interviewed raised doubts about these new institutions. Most particularly, representatives questioned the quality of the information received, the wider benefits of their relationships with senior managers and the lack of opportunities for real consultation. As is illustrated by the following quotations from union representatives in two of the companies studied, many EWC delegates were convinced they would only hear things that it was in management's interest to tell them:

I suppose it depends on what you can get them to talk about, if you could get them to say more it would be a bit more useful. But it's like doctors and lawyers and that – they go by the book – they're only saying what they want you to know.

(Employee representative, Company B)

If they want to do something they just tell you it's being done. You can ask questions about it, where they are coming from, what their aims are, and each time they tell you they want to be the biggest seller. . . . The information management can give you is very limited. A lot of it is very sensitive. I get that impression anyway. They just tell you what they want to tell you . . . it's very hard to change that.

(Employee representative, Company C)

This cynicism was largely reinforced by experiences at the EWC meetings. A number of delegates reported their frustration at not being able to consult over decisions, arguing that managers provided corporate information simply to meet the terms of the Directive without any real interest in hearing the voice of their staff. As these representatives argued:

They just say, 'Well this is what we are going to do.' They give us the statistics or whatever, but they don't say, 'What is your opinion?' All they do say is, 'If you have any questions . . .'

(Employee representative, Company B)

I think the annoying part is that you have got the information but it is not up for discussion. We just sit there, make notes and that is it. You can't do anything with it, the minds have already been made up, there is no evidence that what we've actually said has made any difference to the outcome or anything.

(Employee representative, Company E)

In this respect, a considerable number of the representatives argued that they wanted to play a greater role in discussions with managers at the EWC. Given their experience of working conditions they were frustrated that they did not have any real opportunity to contribute to the discussion or to shape managerial practice. As this union official explained, the EWCs would be more beneficial if they provided employees with the opportunity to discuss the problems and challenges facing Company B *before* managers decided to make closures, cuts or to implement change:

The important thing is that we want to know when something is happening, before the statutory time limit. We want to know when the profitability of any plant is threatened so it's possible that we can do things to help it change. We want the opportunity to have that discussion as soon as possible.

(Union official, Company B)

Indeed, over time, employee experiences at the EWCs tended to reinforce frustration at the lack of genuine two-way exchange, as these comments from employee representatives, at their fourth Company D EWC meeting, indicate:

> There's no room for discussion. You just get a lot of financial information and
> some answers to your pre-disclosed questions.
>
> (Employee representative, Company D)

> They have put a lot into it, but I believe we have to get more involved in decision
> making with management. Before they finalise their future plans they should
> discuss with us first.
>
> (Employee representative, Company D)

> More debate is needed. [Company D] should answer the questions given in much
> more detail, and we should be trained to a higher level so we can respond.
>
> (Employee representative, Company D)

In two of the companies studied, employee expectations of the EWC had been
further dashed as management had failed to discuss particular closures, dis-investments
and relocations that had taken place during the year. In both cases, managers argued
that the matter was a national issue, whilst employee representatives expected a transnational dimension to the consultation procedure. As this non-union representative
in Company C put it:

> There have been things that I think have happened at a strategic level that
> we haven't been consulted on, like the acquisition of another company which
> happened over the last 12 months in Europe. It's relevant to us but we weren't
> involved in it. I don't suppose we would have been able to add anything but it
> would add credence to the process if we had been involved. I think they've got
> some work to do on that.
>
> (Employee representative, Company C)

The lack of European level communication and consultation during the processes of
closure, acquisition, sale or relocation tended to undermine the credibility of the EWC
and to reduce the trust employees felt towards their senior staff.

Finally, information exchange was also limited by the forum itself and a number of
the employee representatives were fearful of saying the wrong thing, of exposing any
weaknesses in their own workplace organisation or of 'giving away' information to
those at the top of the business. As one representative – who had attended two EWC
meetings at the time of the interview – explained:

> What I don't like and I've made a point of mentioning it to people is that we do
> talk at the dinner table, on the night before and you feel that what you're talking
> about is being taken up and they have a meeting after the informal dinner . . . You
> can talk to them [managers] but I think you have to be very careful what you say.
> It's just sometimes you feel a bit uncomfortable because you know that when
> they leave there they're going into another room and they're going to discuss
> what you talked about on the table.
>
> (Employee representative, Company B)

Such sentiment bears testament to the profound mistrust that still pervades many relationships between British managers and workers, and for this representative at least, the criteria of success were to say as little as possible. While this reaction was not universal, it is clear that many of the UK representatives were reluctant to contribute to EWC meetings and felt rather unprepared for this type of exchange (see also Wills 2000, 2001a, 2001b).[6] Without additional training, and evidence of managerial good faith at this level, it is unlikely that many employee representatives will feel any more comfortable about their contact with senior staff.

In short, the lessons of EWCs in UK-owned firms suggest that while information exchange is welcomed by managers and employees alike, existing relationships mediate the extent of this exchange. Many of the workers interviewed for this research suggested that managers would only tell them information which was of benefit to the company, and a number had evidence that the EWC had been denied the opportunity to discuss particular transnational developments which should have been brought to the forum. It is clear that information remains a political tool in the workplace. Although current business practice stresses the value of sharing ideas and opinions with the workforce, the lessons of EWCs suggests that such developments will remain one-sided and partial in practice. Managers remain better at imparting information (however limited it is) than they are at receiving it. This is not to suggest that every company will be the same, and each of the companies studied had a different tradition of industrial relations, and a different EWC as a result. But on balance, information exchange reflected the existing balance of power between managers and those doing the work: senior staff gave the message and employees were simply expected to listen and then make the odd remark in response.

Concluding remarks

The contemporary business environment has demanded a new approach to corporate communications. In order to secure the commitment and the participation of their employees, managers have had to turn their attention to matters of communication, information exchange and knowledge management in the workplace. To ensure that employees make the greatest contribution to the firm while also understanding the reasons for change, managers have devised new systems for information exchange in the workplace. A new barrage of techniques have come into being to ensure that employees have scope to contribute their ideas to the way things are done while also understanding the business strategy behind what they do. Works councils have proved one way of filling this managerial requirement, allowing senior managers to impart their vision to employee representatives, while also hearing staff voices and opinions about corporate change. Stimulated by the European Works Councils Directive, increasing number of UK-owned firms are developing such institutions for the very first time.

Research into the operation of EWCs illustrates the ways in which corporate information is managed, and the degree to which employees feel involved in 'their' business through information exchange. It would appear that managers value the opportunity of EWCs to relay information about the corporate vision while employees benefit from having greater knowledge about their employer. Indeed, the EWCs are also valuable in providing managers and employees with a new un-mediated channel of dialogue, and they have allowed new relationships to develop across the traditional hierarchy of corporate life. Yet EWCs also bear testament to the political importance of information at work, and the flow of information remains very partial. Managers select the detail they want to present to employees and there is little opportunity for real dialogue at these events. Reflecting the imbalance of power between managers and those whom they employ, information tends to flow down rather better than it moves up the corporate chain and employees express their frustration at not being able to bring their knowledge to bear on the subjects discussed.

Despite talk of employee empowerment, participation and partnership, managers will only tell their employees what they need to know to understand business strategy and their role within it and there is little scope for employee intervention at work. Ironically, it would appear that the knowledge embodied in workers is not being utilised by businesses in the UK. Although the rhetoric of managerial textbooks implores managers to treat their employees as knowing subjects who can supply valuable ideas and information about the way in which the processes of work can be altered, employees have few opportunities to bring their knowledge to bear on the real decisions being made (an observation reinforced by the latest WERS data in the UK, see Cully *et al.* 1999; Millward *et al.* 2000). Employees value EWCs as a means of gathering information and deepening their knowledge about the firm for which they work, but these new institutions are not being used to exchange ideas in a way that might shape the decisions being made. This research suggests that traditional barriers and lines of authority are limiting the benefits of communication for business, and the separation of control and execution remains largely in place. It would appear that, as Hyman suggests, managers are continuing to battle with the tension between the need for employee discipline and any autonomy which might allow them to be creative at work:

> The function of labour control involves both the direction, surveillance and discipline of subordinates whose enthusiastic commitment to corporate objectives cannot be taken for granted; and the mobilisation of the discretion, initiative and diligence which coercive supervision, far from guaranteeing is likely to destroy . . . Shifting fashions in labour management stem from this inherent contradiction: solutions to the problem of discipline aggravate the problem of consent, and vice versa. Accordingly, pragmatism may well be the most rational management principle.
>
> (Hyman 1987: 42)

As the corporate landscape alters, so too, managers have to devise new ways to maximise the contribution each employee can make to the business. In the contemporary age, improved information exchange can play a role in that bargain, but it does not imply genuine empowerment and autonomy for those doing the work. In contrast to Peters' (1992) optimistic assertion with which this chapter began, the contemporary worker is unlikely to experience any unification of control and execution at work. Rather, they might be given a little more information about the reasons why their work matters, and the occasional opportunity to voice some concerns. Such developments are important advances, but they do not necessarily allow workers to contribute their knowledge to the processes of production and service. The rhetoric and reality of knowledge management are still a long way apart and employees still have few real opportunities to put their knowledge to work in the workplace.

Acknowledgements

This research would not have been possible without the generous assistance of senior managers at the five EWCs studied, and in particular, those at Companies A, B and C where longitudinal research has been possible. It has been a pleasure to meet the EWC representatives at the five firms involved in this research and I thank them for all their assistance and insights. Finally, I am extremely grateful to the Economic and Social Research Council (ESRC) for funding this research (Grant Number R000221873).

Notes

1   Until recently, such institutions have only been associated with small scale experiments in industrial democracy in Anglo-American capitalism. In the UK they were pioneered by philanthropic industrialists such as Cadbury and have been a feature of employee owned firms such as the John Lewis Partnership (see Williams 1931; Bradley and Gelb 1983).

2   In recent years one of the motivations for the establishment of works councils in the UK may also be the new Employment Relations Act, passed by the Labour Government in 1999. This law makes provision for union recognition if a majority of employees request it and works councils may be a way of trying to head off such demands.

3   The figure of 30,000 representatives is generated by assuming that each of the 1,200 EWCs will have 25 members. Although this is the average size of the EWCs that have been established, actual numbers vary enormously. Indeed, during interview, a representative from Directorate-General Five (DG V) argued that as many as 50,000 representatives may eventually be involved:

> It's a big operation: 2,000 companies, with an average of 35 people on each council. That means that 50,000 representatives will be confronted with Europe for the first time. Not as something on television or as tourists, but as a real process, by meeting their colleagues. We have never had anything like this at the European level, with the citizens being directly involved.

4   EWCs have become an addendum to other outlets for information management in the workplace, augmenting the role of in house noticeboards, newsletters and briefing meetings for staff.

5   Longitudinal research has been carried out to explore the evolution of EWCs at three UK-owned multinational firms (called Companies A, B and C). Commencing in 1996, this research has involved interviewing the managers, employee representatives and union officials on each EWC, and attending the annual meetings as a neutral observer (although confidentiality has precluded this in the case of Company C as it was involved in merger deliberations during this time). In addition, I have had the opportunity to conduct a training session with another EWC which generated some questionnaire data (Company D), and I have conducted a limited piece of research for the employee side of an EWC in a major service company in the UK (Company E). This qualitative data informs the argumentation presented in this chapter, and for reasons of confidentiality, the names of the companies and individuals involved have not been provided. As each company is, by definition, a major multinational, they tend to have operations in various parts of the UK and they include heavy industry, manufacturing, financial services and sales operations (for more detailed discussions of the case studies of Companies A and B, see Wills 2000 and 2001a).

6 It is significant that the representatives who had least difficultly in this environment were the professional financial services delegates from Companies C and E. A number of these representatives were used to public speaking, chairing meetings and dissecting information at work and the EWC was less of a challenge to them.

17 Epilogue

John R. Bryson, Peter W. Daniels,
Nick Henry and Jane Pollard

Reading the completed manuscript of an edited book prior to writing the conclusion encourages reflection on the strengths and weaknesses of the collection as well as providing a final opportunity to draw out the key themes that have emerged. By adopting an interdisciplinary approach, it is hoped that this book will have advanced understanding of the complex, evolving relationship between knowledge, space and economy. When we embarked on this project in 1997 it was apparent that the social sciences were beginning to develop theoretical and, to a lesser extent, methodological approaches to understanding the growing importance of knowledge in economy and society. One of the reviewers of the original proposal for the book even considered that knowledge was too new a topic to warrant an edited collection. With the benefit of hindsight, some of the five aims specified in the original proposal were extremely ambitious:

- To consolidate and encourage an interdisciplinary research agenda around the notion of knowledge, space and economy concentrating on different forms of economic knowledges.
- To identify the diversity of theoretical perspectives for understanding the nexus of knowledge, space, economy.
- To highlight the contested nature of knowledge.
- To provide a series of theoretically informed case studies that explore the significance of knowledge for individuals, organisations and nation states.
- To highlight the spatiality of economic knowledges.

This raises the obvious question as to what extent these aims have been realised in the final product. In relation to the first aim, the contributions from geographers, sociologists, economists and management experts clearly demonstrate the increasingly central role given to understanding dimensions of economic knowledge in each of these disciplines. Knowledge really does matter and is currently at the centre of a research whirlwind. Realistically, however, the end result is a book that does more to encourage an emergent interdisciplinary research agenda, by bringing together a number of diverse authors under one dust jacket, than it does to consolidate contemporary thinking.

Our second aim was to identify the diversity of theoretical perspectives engaging with the knowledge, space, economy nexus. This has been one of the most interesting and challenging aspects of compiling this collection. What has emerged is a set of chapters that do indeed reflect theoretical diversity. The eclectic mix of flavours includes political economy (both Marxist and neo-Schumpeterian), variants of institutional and evolutionary economics, material cultures, state theory and post-structuralist thinking (à la Foucault and Baudrillard amongst others), and actor network theory as part of a revitalised contribution from literatures on the sociology of scientific knowledge.

Third, the contributions in this volume make clear that knowledge is a difficult, slippery concept to define. These definitional issues around knowledge, and its shifting topologies, illustrate very well the contested nature of knowledge. Which knowledges matter, and when and where they matter, is an ongoing social process of recognition and prioritisation. For example, cognitive knowledge, as Allen reminds us right at the beginning, has often been valued more highly than other forms of knowledge because of its association with a select set of economic labours and specific economic rationales. This prioritisation is, in turn, closely related to the ability to identify, measure, codify and ultimately make tangible different forms of knowledge. And as Bell highlights, economic knowledges, or more aptly, knowledges that can be made economic, often end up riding roughshod over rationales that produce other 'useful knowledges' (which often support alternative visions of society).

That this contestation over useful knowledge occurs in space and place has become very clear in recent debates concerning innovative clusters and learning regions. Tucked inside these visions of suitable economic futures is a clear privileging of particular forms of knowledge (tacit and untraded) combined with an increasingly problematic association of these knowledges with 'the local'. While many argue that all knowledges are local, how this is achieved requires theoretical and policy reflection. Whilst the precise spatial architectures of knowledge that combine to produce 'an innovative cluster' have received a great deal of attention, more recent research is problematising the all too neat alliance between tacit–local and global–codified and their stereotypical couplings in 'sticky' high technology clusters. A growing body of research on the spatialities of economic knowledges, as highlighted in this book, suggests a much greater diversity to the knowledge, space, economy nexus. Finally, the contributors have revealed some of the spatialities of economic knowledges through a range of case studies. Organised in two parts, through the foci on organisations and individuals, these case studies demonstrate the different topologies of knowledge that constitute the contemporary economy.

While a start may have been made, a number of issues and questions about the relationship between knowledge, space and economy remain largely unanswered. Here we suggest a select few issues that are worthy of further reflection and research.

First, very little detailed comparative research has been conducted into knowledge creation and utilisation in different cultural environments. It is clear that advanced capitalism is knowledge intensive, but not at all clear whether different cultures and countries use different forms of knowledge or even the same knowledge in different ways. It is well known that there are different forms of capitalism and that some of these forms are country specific (Castells and Aoyama 1994). Much more research needs to be conducted into the relationship between knowledge and different cultural formations.

Second, the definitional problem implies that knowledge can be everything and nothing. The shift in research towards knowledge implies that the methodologies of the social sciences need to be revisited. The key question is whether it is possible to identify and understand the important knowledges that lie behind the competitiveness of a region, company, industrial sector or national economy. The danger is that current methodologies will identify the most visible forms, but perhaps overlook the more important, but not necessarily tacit, forms of knowledge. The case studies in this collection display a diversity of methodologies, but it would still be possible to argue that most of the case studies explore the simpler and more visible forms of knowledge.

Third, the current emphasis on knowledge needs to be placed in an historical context. It may be the case that twenty-first century capitalism is more knowledge intensive than earlier periods of capital accumulation. The importance of knowledge may, however, have been understated in earlier periods and overstated in the operation of contemporary economies. Research efforts need to explore the past if only to provide the context for the growing importance of knowledge in the global economy.

In sum, this edited collection interrogates the knowledge, space, economy nexus through the examination of different forms of economic knowledges and their construction and mobilisation in, and through, space and place. In so doing, the book brings together an eclectic, yet rich, blend of interdisciplinary work that is grappling with the complexities of contemporary economic life. This interdisciplinary collection has been driven by the editors' belief that knowledge matters and, in its different guises, is transforming the contours of economic organisation and activity. Nevertheless, the key question remains, if knowledge matters, how much does it matter? We trust that this set of ambitious, provocative and diverse chapters illustrates something of the exciting (social science) research agenda now taking shape around knowledge, space, economy.

Bibliography

Abler, R. (1971) Distance, intercommunications, and geography, *Proceedings of the Association of American Geographers* 3: 1–4.

Abo, T. (ed.) (1994) *Hybrid Factory: The Japanese Production System in the United States*, New York: Oxford University Press.

Aburish, S. K. (1998) *A Brutal Friendship: The West and the Arab Elite*, London: Indigo.

Adam, B. (1990) *Time and Social Theory*, Cambridge: Polity Press.

Adam, B. (1992) Modern times: The technology connection and its implications for social theory, *Time and Society* 1(2): 175–91.

Adam, B. (1993) Within and beyond the time economy of employment relations: conceptual issues pertinent to research on time and work, *Social Science Information* 32: 163–84.

Adam, B. (1995) *Timewatch*, Cambridge: Polity Press.

Adam, B. (1996) Re-vision: the centrality of time for an ecological social science perspective, in Lash, S., Szerszynski, B. and Wynne, B. (eds) *Risk, Environment and Modernity: Towards a New Ecology*, London: Sage, 84–103.

Adam, B., Geibler, K., Held, M., Kummerer, K. and Schneider, M. (1997) Time for the environment: the Tutzing time ecology project, *Time and Society* 6: 73–84.

Aitken, H. G. J. (1960) *Taylorism at Watertown Arsenal: Scientific Management in Action: 1908–1915*, Cambridge, MA: Harvard University Press.

Alder, P. (1990) Shared learning, *Management Science* 36: 938–57.

Alexander, A. F. and Pollard, J. S. (2000) Banks, grocery retailers and the changing nature of financial services retailing in Britain, *Journal of Retailing and Consumer Services* 7(3): 137–47.

Allen, G. C. (1981) *The Japanese Economy*, New York: St. Martins Press.

Allen, J. and Henry, N. (1997) Ulrich Beck's risk society at work: labour and employment in the contract services industries, *Transactions of the Institute of British Geographers* 22: 180–97.

Allen, T. J. (1977) *Managing the Flow of Technology*, Cambridge, MA: MIT Press.

Allen, T. J. and Cohen, S. (1969) Information in R&D laboratories, *Administrative Science Quarterly* 14: 12–19.

Allen, T. J. and Hauptman, O. (1987) The influence of communication technologies on organizational structure: a conceptual model for future research, *Communication Research* 14: 575–87.

Alper, J. and Natowicz, M. (1992) The allure of genetic explanations, *British Medical Journal* 305, September: 666.

Alpers, S. (1989) *The Art of Describing: Dutch Art in the Seventeenth Century*, London: Penguin.

Althusser, L. (1972) *Politics and History: Montesquieu, Rousseau, Hegel and Marx*, London: New Left Books.

Altvater, E. (1993) *The Future of the Market*, London: Verso.

Ambrose S. E. (1985) *Rise to Globalism: American Foreign Policy Since 1938* (4th edn), Harmondsworth: Penguin.

Amin, A. (ed.) (1994) *Post-Fordism: A Reader*, Oxford: Blackwell.

Amin, A. and Thrift, N. J. (1992) Neo-marshallian nodes in global networks, *International Journal of Urban and Regional Research* 16(4): 571–87.

Amin, A. and Thrift, N. J. (1994) Living in the global, in Amin, A. and Thrift, N. J. (eds) *Globalization, Institutions and Regional Development in Europe*, Oxford: Oxford University Press, 1–22.

Amin, A. and Thrift, N. J. (2000) What kind of economic theory for what kind of economic geography?, *Antipode* 32(1): 4–9.

Amsden, A. (1979) Taiwan's economic history: a case of étatisme and a challenge to dependency theory, *Modern China* 5(3): 341–80.

Amsden, A. (1989) *Asia's Next Giant: South Korea and Late Industrialisation*, New York: Oxford University Press.

Anderson, C. and Aldous, P. (1992) Still room for HUGO?, *Nature* 355, January: 4–5.

Anderson, P. F. (1989) On relativism and interpretivism with a prolegomenon to the why question, in Hirschman, E. C. (ed.) *Interpretative Consumer Research*, Provo, UT: Association for Consumer Research.

Anderson, W. T. (1997) *Evolution Isn't What it Used to Be: The Augmented Animal and the Whole Wired World*, London: W.H. Freeman.

Andersson, A. E. (1985) Creativity and regional development, *Papers of the Regional Science Association* 56: 5–20.

Anon (1996) *Category Management*, Utrecht: Coopers & Lybrand.

Anon (1997a) Category management: managing retailer supplier collaboration, *Checkout*, October: 12.

Anon (1997b) Category management: survey and recommendations from Ernst & Young, *Progressive Grocer*, Special Supplement, May.

Antonelli, C. (1995) *The Economics of Localized Technological Change and Industrial Dynamics*, Norwell, MA: Kluwer.

Aoki, K. (1998) Considering multiple and overlapping sovereignties: liberalism, libertarianism, national sovereignty, global intellectual property, and the Internet, *Indiana Journal of Global Legal Studies* 5(2): 443–74.

Aoki, M. (1986) Horizontal versus vertical information structure of the firm, *American Economic Review* 76: 971–83.

Appadurai, A. (1986) Introduction: commodities and the politics of value, in Appadurai, A. (ed.) *The Social Life of Things: Commodities in Cultural Perspective*, Cambridge: Cambridge University Press, 3–63.

Appelbaum, D. (1995) *The Stop*, Albany: SUNY.

Arcangeli, F. (1993) Local and global features of the learning process, in Humbert, M. (ed.) *The Impact of Globalisation on Europe's Firms and Industries*, London: Pinter, 34–7.

Arcaya, A. (1992) Why is time not included in modern theories of memory?, *Time and Society* 1(2): 301–14.

Arndt, J. (1985) The tyranny of paradigms: the case for pragmatic pluralism in marketing, in Dholakia, N. and Arndt, J. (eds) *Changing the Course of Marketing: Alternative Paradigms for Widening Marketing Theory*, Greenwich, CT: JAI Press.

Arora, A. and Gambardella, A. (1994) The changing technology of technological change: general and abstract knowledge and the division of innovative labour, *Research Policy* 23: 523–32.

Arrow, K. J. (1962) The economic implications of learning by doing, *Review of Economic Studies* 29: 155–73.

Arrow, K. J. (1974) *The Limits of Organization*, New York: Norton.

Arthur, C. (1998) £1bn for science in public–private scheme, *The Independent* 15 July: 7.

Asheim, B. T. (1996) Industrial districts as learning regions: a condition for prosperity, *European Planning Studies* 4: 379–400.

Asheim, B. T. and Isaksen, A. (1997) Location, agglomeration and innovation: towards regional innovation systems in Norway?, *European Planning Studies* 5: 299–330.

Ashford, M. (1998) *Con Tricks: The Shadowy World of Management Consultancy and How to Make it Work for You*, London: Simon and Schuster.

Aspery, R. B. (1975) *War in the Shadows*, New York: Doubleday.

Aston, B. and Williams, M. (1996) *Playing to Win: The Success of UK Motorsport Engineering*, London: Institute of Public Policy Research.

Athanasiou, T. (1997) *Slow Reckoning: The Ecology of a Divided Planet*, London: Secker and Warburg, 53.

Badaracco, J. L. (1991) *The Knowledge Link: How Firms Compete through Strategic Alliances*, Boston: Harvard Business School Press.

Baechler, J., Hall, J. A. and Mann, M. (eds) (1988) *Europe and the Rise of Capitalism*, Oxford: Blackwell.

Balmer, B. (1996) Managing mapping in the Human Genome Project, *Social Studies of Science* 26: 531–73.

BankBoston (1997) *MIT: The Impact of Innovation*, Boston: BankBoston.

Barber, B. R. (1995) *Jihad vs McWorld*, New York: Ballentine Books.

Barley, S. R. (1996) Technicians in the workplace: ethnographic evidence for bringing work into organization studies, *Administrative Science Quarterly* 41(3): 404–41.

Barnes, T. (1996) *Logics of Dislocation: Models, Metaphors and Meanings of Economic Space*, New York: Guilford Press.

Barnett, L. (1957) *The Universe and Dr Einstein*, NewYork: William Sloane.

Barthes, R. (1973) [original 1957] *Mythologies*, St Albans: Paladin.

Bataille, G. (1985) *Visions of Excess: Selected Writings, 1927–1939* (ed. A. Stoekl) Minneapolis: University of Minnesota Press.

Bataille, G. (1988) *The Accursed Share: An Essay in General Economy. Volume I: Consumption*, New York: Zone Books.

Baudrillard, J. (1968/1996) *The System of Objects*, trans. J. Benedict, London: Verso.

Baudrillard, J. (1975) *The Mirror of Production*, St Louis: Telos Press.

Baudrillard, J. (1981) *For a Critique of the Political Economy of the Sign*, St. Louis: Telos Press.

Baudrillard, J. (1983) *Simulations*, New York: Semiotext(e).

Baudrillard, J. (1987) *Forget Foucault*, New York: Semiotext(e).

Baudrillard, J. (1990a) *Cool Memories*, London: Verso.

Baudrillard, J. (1990b) *Fatal Strategies*, New York: Semiotext(e).

Baudrillard, J. (1990c) *Seduction*, London: Macmillan.

Baudrillard, J. (1993) *Symbolic Exchange and Death*, trans. Hamilton Grant, London and New Delhi: Sage.

Baudrillard, J. (1994) *The Illusion of the End*, Cambridge: Polity Press.

Baudrillard, J. (1998) *The Consumer Society: Myths and Structures*, London: Sage.

Bauman, Z. (1983) Industrialism, consumerism and power, *Theory, Culture and Society* 1(3): 32–43.

Bauman, Z. (1987) *Legislators and Interpreters: On Modernity, Post-Modernity and Intellectuals*, Cambridge: Polity Press.

Bauman, Z. (1988) *Freedom*, Milton Keynes: Open University Press.

Bauman, Z. (1991) *Modernity and Ambivalence*, Cambridge: Polity Press.

Bauman, Z. (1992a) *Intimations of Postmodernity*, London: Routledge.

Bauman, Z. (1992b) *Mortality, Immortality and Other Life Strategies*, Cambridge: Polity Press.

Bauman, Z. (1993a) The sweet scent of decomposition, in Rojek, C. and Turner, B. S. (eds) *Forget Baudrillard?*, London: Routledge, 22–46.

Bauman, Z. (1993b) *Postmodern Ethics*, Oxford: Blackwell.

Bauman, Z. (1995) *Life in Fragments: Essays in Postmodern Morality*, Oxford: Blackwell.

Bauman, Z. (1996) From pilgrim to tourist – or a short history of identity, in Hall, S. and du Gay, P. (eds) *Questions of Cultural Identity*, London: Sage, 18–36.

Bauman, Z. (1997a) *Postmodernity and Its Discontents*, Cambridge: Polity Press.

Bauman, Z. (1997b) The haunted house, *New Internationalist* April: 24–5.

Bauman, Z. (1998) *Work, Consumerism and the New Poor*, Buckingham: Open University Press.

Baumard, P. (1995) Organizations in the fog: an investigation into the dynamics of knowledge, in Moingen, B. and Edmondson, A. (eds) *Organizational Learning and Competitive Advantage*, London: Sage Publications, 74–91.

Becattini, G. and Rullani, E. (1996) Local systems and global connections: the role of knowledge, in Cossentino, F., Pyke, F. and Sengenberger, W. (eds) *Local and Regional Response to Global Pressure: The Case of Italy and Its Industrial Districts*, Geneva: International Institute of Labour Studies, 159–74.

Beck, U. (1992) *Risk Society: Towards a New Modernity*, London: Sage.

Beck, U. (1994a) The reinvention of politics: towards a theory of reflexive modernisation, in Beck, U., Giddens, A. and Lash, S. (eds) *Reflexive Modernisation: Politics, Tradition and Aesthetics in the Modern Social Order*, Cambridge: Polity Press, 1–55.

Beck. U. (1994b) Self-dissolution and self endangerment of industrial society: what does this mean?, in Beck, U., Giddens, A. and Lash, S. (eds) *Reflexive Modernisation: Politics, Tradition and Aesthetics in the Modern Social Order*, Cambridge: Polity Press, 174–83.

Beck, U., Giddens, A. and Lash, S. (1994) *Reflexive Modernisation: Politics, Tradition and Aesthetics in the Modern Social Order*, Cambridge: Polity Press.

Becker, G. (1965) A theory of the allocation of time, *Economic Journal* 75, September: 493–517.

Becker, G. and Michael, R. (1973) On the new theory of consumer behaviour, *Swedish Journal of Economics* 75, September: 493–517.

Becker, H. (1963) *Outsiders*, New York: Free Press.

Beier, A. L. (1985) *Masterless Men: The Vagrancy Problem in England, 1500–1640*, London: Methuen.

Bell, D. (1973) *The Coming of Post-Industrial Society: A Venture in Social Forecasting*, New York: Basic Books.

Bell, D. (1976) *The Cultural Contradictions of Capitalism*, New York: Basic Books.

Bell, D. (1979) The social framework of the information society, in Dertouzos, M. L. and Moses, J. (eds) *The Computer Age: A Twenty-Year View*, Cambridge, MA: MIT Press, 163–211.

Bellah, R., Madsen, R., Sullivan, W., Swidler, A. and Tipton, S. (1985) *Habits of the Heart: Individualism and Commitment in American Life*, Berkeley, CA: University of California Press, 154.

Belussi, F. (1996) Local systems, industrial districts and institutional networks: towards a new evolutionary paradigm of industrial economics?, *European Planning Studies* 4: 5–26.

Benhabib, S. (1992) *Situating the Subject*, London: Sage.

Beniger, J. R. (1986) *The Control Revolution: Technological and Economic Origins of the Information Society*, Cambridge, MA: Harvard University Press.

Benjamin, W. (1968) Theses on the philosophy of history, in *Illuminations*, New York: Schocken, 253–64.

Bergadaa, M. (1992) The role of time in the action of the consumer, *Journal of Consumer Research* 17, December: 289–302.

Berger, M. T. and Beeson, M. (1998) Lineages of liberalism and miracles of modernisation: the World Bank, the East Asian trajectory and the international development debate, *Third World Quarterly* 19(3): 487–504.

Berger, P. and Luckman, T. (1985) *The Social Construction of Reality: A Treatise in the Sociology of Knowledge*, Harmondsworth: Penguin Books.

Bergson, H. (1950) *Time and Free Will*, trans. A. Mitchell, London: Macmillan.

Berle A. A. and Means G. C. (1932) *The Modern Corporation and Private Property*, New York: Macmillan.

Bernstein, R. (1980) Philosophy in the conversation of mankind, *Review of Metaphysics* 32(4): 762.

Berry, L. (1979) The time buying consumer, *Journal of Retailing* 55 Winter: 58–69.

Bessant, J. and Buckingham, J. (1993) Innovation and organizational learning: the case of computer-aided production management, *British Journal of Management* 4: 219–34.

Best, M. H. (1990) *The New Competition: Institutions of Industrial Restructuring*, Oxford: Polity Press.

Bettman, J. (1979) *An Information Processing Theory of Consumer Choice*, Reading, MA: Addison Wesley.

Bhabha, H. (1994) *The Location of Culture*, London: Routledge.

Bierly, P. and Chakrabarti, A. (1996) Generic knowledge strategies in the US pharmaceutical industry, *Strategic Management Journal* 17 Winter Special Issue: 123–35.

Bijker, W. E., Hughes, T. P. and Pinch, T. (eds) (1990) *The Social Construction of Technological Systems*, Cambridge, MA: MIT Press.

Blackler, F. (1995) Knowledge, knowledge work and organizations: an overview and interpretation, *Organization Studies* 16: 1021–46.

Blair, M. M. (1995) *Ownership and Control: Rethinking Corporate Governance for the Twenty-First Century*, Washington DC: The Brookings Institute.

Blair, T. (1998) Foreword, in *Building the Knowledge Driven Economy*, London: Department of Trade and Industry.

Blakstad, M. and Cooper, A. (1995) *The Communicating Organisation*, London: Institute of Personnel and Development.

Bloomfield, B. P. and Best, A. (1994) Management consultants: systems development, power and the translation of problems, *The Sociological Review* 40: 532–60, also in Bryson and Daniels (1998a).

Blyton, P. and Turnbull, P. (eds) (1992) *Reassessing Human Resource Management*, London: Sage.

Boisot, M. H. (1995a) *Information Space: A Framework for Learning in Organizations, Institutions and Culture*, London: Routledge.

Boisot, M. H. (1995b) Is your firm a creative destroyer? Competitive learning and knowledge flows in the technological strategies of firms, *Research Policy* 24: 489–506.

Boisot, M. H. (1998) *Knowledge Assets: Securing Competitive Advantage in the Information Economy*, Oxford: Oxford University Press.

Bolland, E. J. and Hofer, C. W. (1998) *Future Firms: How America's High Technology Companies Work*, New York: Oxford University Press.

Bonara, E. A. and Revang, O. (1993) A framework for analyzing the storage and protection of knowledge in organizations, in Lorange, P., Chakravarthy, B., Roos, J. and Van de Ven, A. (eds) *Implementing Strategic Processes: Change, Learning and Co-operation*, Oxford: Basil Blackwell, 190–213.

Bourdieu, P. (1977) *Outline of a Theory of Practice*, Cambridge: Cambridge University Press.

Boyer, R. and Saillard, Y. (eds) (1995) *Théorie de la Régulation. L'État des Savoirs*, Paris: La Découverte.

Braczyk, H. J., Cooke, P. and Heidenreich, M. (eds) (1998) *Regional Innovation Systems*, London: UCL Press.

Bradley, K. and Gelb, A. (1983) *Worker Capitalism: The New Industrial Relations*, London: Heinemann Educational Books.

Braudel, F. (1967) *Civilisation, Matérielle et Capitalisme. XV–XVII siècle*, Paris: Armand Colin.

Braverman, H. (1974) *Labour and Monopoly Capitalism: The Degradation of Work in the Twentieth Century*, New York: Monthly Review Press.

Breheny, M. J. and McQuaid, R. W. (eds) (1987) *The Development of High Technology Industries: An International Survey*, London: Croom Helm.

Brennan, T. (1995) Why the time is out of joint. Marx's political economy without the subject, Part I, Strategies, *Journal of Theory, Culture and Politics* 9–10: 18–37.

Brownlie, D. *et al.* (eds) (1994) *Rethinking Marketing, European Journal of Marketing*, Special edition.

Bryson, J. R. (1997) Business service firms, service space and the management of change, *Entrepreneurship and Regional Development* 9: 93–111.

Bryson, J. R. and Daniels, P. W. (1998a) *Service Industries in the Global Economy: Service Theories and Service Employment*, Cheltenham: Edward Elgar.

Bryson, J. R. and Daniels, P. W. (1998b) Business Link, strong ties and the walls of silence: small and medium-sized enterprises and external business service expertise, *Environment and Planning C: Government and Policy* 16: 265–80.

Bryson, J. R. and Daniels, P. W. (1998c) Recipe knowledge and the four myths of knowledge-intensive producer service research: the knowledge which producer service professionals bring to their clients, *Services, Space, Society*, Working Paper 3, Service Sector Research Unit, Birmingham: University of Birmingham.

Bryson, J. R., Daniels, P. W. and Ingram, D. R. (1999) Methodological problems and Economic Geography: the case of business services, *Service Industries Journal* 19(4) 1–16.

Bryson, J. R., Henry, N., Keeble, D. and Martin, R. (1999) *The Economic Geography Reader: Producing and Consuming Global Capitalism*, Chichester: Wiley.

Bryson, J. R., McGuinness, M. and Ford, R. G. (1999) A Third Way or an old way? Almshouse charities and the rise of the post Welfare State, *Services, Space, Society*, Working Paper 7, Service Sector Research Unit, Birmingham: University of Birmingham.

Bryson, J. R., Wood, P. and Keeble, D. (1993) Business networks, small firm flexibility and regional development in UK business services, *Entrepreneurship and Regional Development* 5, 265–77.

Bulmer, M. (ed.) (1978) *Mining and Social Change: Durham County in the Twentieth Century*, London: Croom Helm.

Burgess, J. (1990) The production and consumption of environmental meanings in the mass media: a research agenda for the 1990s, *Transactions of the Institute of British Geographers* 15: 139–61.

Burns, T. and Stalker, G. M. (1994) *The Management of Innovation*, Oxford: Oxford University Press (1st edn 1961).

Business Week (1997) Silicon Valley, 25 August: 64–147.

Bygrave, M. (1998) From Wall Street to high street, *Guardian Weekend* 25 July: 24–9.

Bynum, W. and Nutton, V. (eds) (1991) Essays in the history of therapeutics, *Clio Medica*, Amsterdam: Rodopi, 22.

Callon, M. (1986) Some elements of a sociology of translation: domestication of the scallops and the fisherman of St Brieuc Bay, in Law, J. (ed.) *Power, Action and Belief*, London: Routledge and Kegan Paul, 196–233.

Camagni, R. (1991) Local milieu, uncertainty and innovation networks: towards a new dynamic theory of economic space, in Camagni, R. (ed.) *Innovation Networks: Spatial Perspectives*, London: Belhaven Press, 121–44.

Camagni, R. (1995) Global network and local milieu: towards a theory of economic space, in Conti, S., Malecki, E. J. and Oinas, P. (eds) *The Industrial Enterprise and Its Environment: Spatial Perspectives*, Aldershot: Avebury, 195–214.

Campbell, J. (1995) Mid-life crisis on Route 128, *Regional Review*, Federal Reserve Bank of Boston 5(3): 6–11.

Campbell, M. (1997a) Team work, *The Grocer* 12 July: 32–3.

Campbell, M. (1997b) In the know, *The Grocer* 6 September: 41–2.

Campbell, M. (1997c) Shifting sands, *The Grocer* 22 November: 36, 39.

Cantor, C. (1990) Orchestrating the Human Genome Project, *Science* 248, April: 49–51.

Cantor, C. (1998) How will the Human Genome Project improve our quality of life?, *Nature Biotechnology* 16: 212–13.

Carley, M., Geissler, S. and Krieger, H. (1996) *The Contents of Voluntary Agreements on European-level Information and Consultation: Preliminary Findings of an Analysis of 111 Agreements*, paper produced by the European Foundation, Dublin. For additional analysis of this data see the *European Works Council Bulletin* (1997) Numbers 6 and 7.

Carlsson, B. and Eliasson, G. (1994) The nature and importance of economic competence, *Industrial and Corporate Change* 3: 687–711.

Carrier, J. G. (1997) Introduction, in Carrier, J. G. (ed.) *Meanings of the Market: The Free Market in Western Culture*, Oxford: Berg, 1–67.

Carrier, J. G. and Miller, D. (1998) *Virtualism: A New Political Economy*, Oxford and New York: Berg.

Carroll, P. N. and Noble, D. W. (1977) *The Free and the Unfree: A New History of the United States*, Harmondsworth: Penguin Books.

Cassirer, E. (1957) *The Philosophy of Symbolic Forms*, (vols I–III), New Haven, CT and London: Yale University Press.

Cassirer, E. (1979) Language and art, in Verene, D. P. (ed.) *Symbol, Myth and Culture: Essays and Lectures of Ernst Cassirer, 1935–45*, Newhaven and London: Yale University Press.

Castells, M. (1989) *The Informational City*, Oxford: Blackwell.

Castells, M. (1992) Four Asian tigers with a dragon head: a comparative analysis of the state, economy and society in the Asian Pacific rim, in Appelbaum, R. and Henderson, J. (eds) *States and Development in the Asian Pacific Rim*, Newbury Park, CA: Sage, 33–70.

Castells, M. (1996) *The Information Age: Economy, Society and Culture. Volume I: The Rise of the Network Society*, Oxford: Blackwell.

Castells, M. (1997) *The Power of Identity*, Oxford: Blackwell.

Castells, M. (1998) *End of Millennium*, Oxford: Blackwell.

Castells, M. and Aoyama, Y. (1994) Paths toward the informational society: employment structure in G7 countries, 1920–1990, *International Labour Review* 133(1): 5–33.

Castells, M. and Hall, P. (1994) *Technopoles of the World*, London: Routledge.

Castells, M. and Tyson, L. (1988) High technology choices ahead: restructuring interdependence, in Sewell, J. W. and Tucker, S. (eds) *Growth, Exports, and Jobs in a Changing World Economy*, New Brunswick, NJ: Transaction Books.

Castells, M. et al. (1986) *Nuevas tecnologías, economía y sociedad en España*, 2 vols, Madrid: Alianza Editorial.

Chandler, A. D. (1977) *The Visible Hand: The Managerial Revolution in American Business*, Boston: Harvard University Press.

Chandra, B. and MacPherson, A. D. (1994) The characteristics of high-technology manufacturing firms in a declining industrial region: an empirical analysis from western New York, *Entrepreneurship and Regional Development* 6: 145–160.

Chen, K.-H. (1996) Cultural studies and the politics of internationalisation: an interview with Stuart Hall, in Morley, D. and Chen, K.-H. (eds) *Stuart Hall: Critical Dialogues in Cultural Studies*, London: Routledge, 392–408.

Chesbrough, H. W. and Teece, D. J. (1996) When is virtual virtuous? Organizing for innovation, *Harvard Business Review* 74(1): 65–73.

Chesnais, F. (1994) *La Mondialisation du Capital*, Paris: Syros.

Clark, G. L. and Wrigley, N. (1997) Exit, the firm and sunk costs: reconceptualizing the corporate geography of disinvestment and plant closure, *Progress in Human Geography* 21: 338–58.

Clarke, A. (1995) Population screening for genetic susceptibility to disease, *British Medical Journal* 311, July: 35–8.

Clarke, A., Harper, P. and Unsworth, P. (1995) Eugenics in China, *The Lancet* 346: 508.

Clarke, D. B. (1997) Consumption and the city, modern and postmodern, *International Journal of Urban and Regional Research* 21(2): 218–37.

Clarke, D. B. and Bradford, M. G. (1998) Public and private consumption and the city, *Urban Studies* 35(5–6): 865–88.

Clarke, D. B. and Doel, M. A. (1995) Transpolitical geography, *Geoforum* 25(4): 505–24.

Clarke, P. D., Gardener, E. P. M., Feeney, P. and Molyneux, P. (1988) The genesis of strategic marketing control in British retail banking, *International Journal of Bank Marketing* 6(2): 5–19.

Clegg, S., Boreham, P. and Dow, G. (1986) *Class, Politics and the Economy*, London: Routledge.

Cleeremans, A. and McClelland, J. L. (1991) Learning the structure of event sequences, *Journal of Experimental Psychology: General* 120(3): 235–53.

Clemens, E. S., Powell, W. W., McLlwaine, K. and Okamoto, D. (1995) Careers in print: books, journals, and scholarly reputations, *American Journal of Sociology* 101(2): 433–94.

Cohen, S. (1990) Corporate nationality can matter a lot, testimony before the US Congress Joint Economic Committee, September.

Cohen, S. (1993) *Academia and the Luster of Capital*, Minneapolis: University of Minnesota Press.

Cohen, S. *et al.* (1985) *Global Competition: The New Reality*, vol. III of John Young (chair), *Competitiveness, The Report of the President's Commission on Industrial Competitiveness*, Washington, DC: Government Printing Office, 1.

Cohen, W. M. and Levinthal, D. A. (1989) Innovation and learning: the two faces of R&D, *Economic Journal* 99: 569–96.

Cohen, W. M. and Levinthal, D. A. (1990) Absorptive capacity: a new perspective on learning and innovation, *Administrative Science Quarterly* 35: 128–52.

Cohen, W. M. and Levinthal, D. A. (1994) Fortune favors the prepared firm, *Management Science* 40: 227–51.

Coleman, B. (1995) The Indy 500 is a very British affair, *Wall Street Journal* 26 May: 138.

Collinge, C. J. (1996) *Spatial Articulation of the State: Reworking Social Relations and Social Regulation Theory*, Birmingham: Centre for Urban and Regional Studies.

Connor, K. R. and Prahalad, C. K. (1996) A resource-based theory of the firm: knowledge versus opportunism, *Organization Science* 7: 477–501.

Cook, I. and Crang, P. (1996) The world on a plate: culinary culture, displacement and geographical knowledges, *Journal of Material Culture* 1(2): 131–53.

Cook, I., Crang, P. and Thorpe, M. (1998) Biographies and geographies: consumer understandings of the origins of foods, *British Food Journal* 100(3): 162–7.

Cook, I., Crang, P. and Thorpe, M. (1999a) Eating into Britishness: multicultural imaginaries and the identity politics of food, in Roseneil, S. and Seymour, J. (eds) *Practising Identities: Power and Resistance*, Basingstoke: Macmillan, 223–48.

Cook, I., Crang, P. and Thorpe, M. (1999b) Constructing the consumer: category management and circuits of knowledge in the UK food industry, unpublished manuscript available at http://www.bham.ac.uk/geography/research/human/staff/iancook-consumer.htm

Cook, I., Crang, P. and Thorpe, M. (2000) Regions to be cheerful: culinary authenticity and its geographies, in Cook, I., Crouch, D., Naylor, S. and Ryan, J. (eds) *Cultural Turns/ Geographical Turns: Perspectives on Cultural Geography*, Harlow: Longman, 109–39.

Cooke, P. (1995) Keeping to the high road: learning, reflexivity and associative governance in regional economic development, in Cooke, P. (ed.) *The Rise of the Rustbelt*, London: UCL Press, 231–45.

Cooke, P. (1996a) Building a twenty-first century regional economy in Emilia-Romagna, *European Planning Studies* 4: 53–62.

Cooke, P. (1996b) The new wave of regional innovation networks: analysis, characteristics and strategy, *Small Business Economics* 8: 159–71.

Cooke, P. and Morgan, K. (1993) The network paradigm: new departures in corporate and regional development, *Environment and Planning D: Society and Space* 11: 543–64.

Cooke, P. and Morgan, K. (1998) *The Associational Economy: Firms, Regions, and Innovation*, Oxford: Blackwell.

Cooke, P., Gomez Uranga, M. and Extebarria, G. (1997) Regional innovation systems: institutional and organisational dimensions, *Research Policy* 26: 475–91.

Coombe, R. J. (1998) *The Cultural Life of Intellectual Properties*, Durham, NC: Duke University Press.

Coombs, R. and Metcalfe, S. (1998) Distributed capabilities and the governance of the firm, paper presented at the DRUID Conference, Copenhagen.

Cooper, R. and Law, J. (1995) Organisation distal and proximal views, in Bocharach, S., Gaghardi, P. and Mundell, B. (eds) *Research in the Sociology of Organisations*, Greenwich, CT: JAI Press.

Copley, F. B. (1923) *Fredrick W. Taylor, Father of Scientific Management*, 2 vols, New York: Harper and Brothers.

Corbridge, S. (1992) Discipline and punish: the New Right and the policing of the international debt crisis, *Geoforum* 23: 285–301.

Corrigan, P. and Sayer, D. (1985) *The Great Arch: English State Formation as Cultural Revolution*, Oxford: Blackwell.

Cosgrove, D. (1994) Contested global visions: one-world, whole-earth, and the Apollo space photographs, *Annals of the Association of American Geographers* 84: 270–94.

Courlet, C. and Pecqueur, B. (1991) Local industrial systems and externalities: an essay in typology, *Entrepreneurship and Regional Development* 3: 305–15.

Courlet, C. and Soulage, B. (1995) Industrial dynamics and territorial space, *Entrepreneurship and Regional Development* 7: 287–307.

Cover, L. B. (1971) *Anthropology for Our Times*, New York: Oxford Book Company.

Cowan, R. and Foray, D. (1997) The economics of codification and the diffusion of knowledge, *Industrial and Corporate Change* 6: 595–622.

Coyle, D. (1997) *The Weightless World: Strategies for Managing the Digital Economy*, Oxford: Capstone.

Crang, P. (1996) Displacement, consumption and identity, *Environment & Planning A* 28: 47–67.

Crang, P. (1997) Cultural turns and the (re)constitution of economic geography in Lee, R. and Wills, J. (eds) *Geographies of Economies*, London: Arnold, 3–15.

Cressey, P. and Scott, P. (1992) Employment, technology and industrial relations in the UK clearing banks: is the honeymoon over?, *New Technology, Work and Employment* 3: 83–96.

Cromie, S., Birley, S. and Callaghan, I. (1993) Community brokers: their role in the formation and development of new ventures, *Entrepreneurship and Regional Development* 5: 247–64.

Cully, M., Woodland, S., O'Reilly, A. and Dix, G. (1999) *Britain at Work: As Depicted by the 1998 Workplace Employee Relations Survey*, London: Routledge.

Cunningham, I., Hyman, J. and Baldry, C. (1996) Empowerment: the power to do what?, *Industrial Relations Journal* 27(2):143–54.

Dalby, S. (1990) *Creating the Second Cold War: The Discourse of Politics*, London: Pinter.

Daniels, P. W. (1982) *Service Industries: Growth and Location*, Cambridge: Cambridge University Press.

Daniels, P. W. (1993) *Service Industries in the World Economy*, Oxford: Blackwell.

Darnton, R. (1984) *The Great Cat Massacre and Other Episodes in French Cultural History*, London: Allen Lane.

David, P. (1989) *Computer and Dynamo: The Modern Productivity Paradox in Historical Perspective*, Stanford, CA: Stanford University Center For Economic Policy Research, Working Paper No.172.

David, P. A. and Foray, D. (1995) Accessing and expanding the science and technology knowledge base, *STI Review* 16: 13–68.

Davidsson, P. (1995) Culture, structure and regional levels of entrepreneurship, *Entrepreneurship and Regional Development* 7: 41–62.

Davies, G. (1994) What should time be, *European Journal of Marketing* 28(8/9): 100–13.

Dawkins, R. (1989) *The Selfish Gene*, Oxford: Oxford University Press.

Dawson, A. C. (1998) The intellectual commons: a rationale for regulation, *Prometheus* 16(3): 275–89.

Daz, T. K. (1991) Time: the hidden dimension in strategic planning, *Journal of Long Range Planning* 24(3): 49–57.

De Grazia, S. (1972): Time and work, in Yaker, H. *et al.* (eds) *The Future of Time*, New York: Archer Books.

Deitrich, S. and Beauregard, R. A. (1995) From front-runner to also-ran – the transformation of a once-dominant industrial region: Pennsylvania, USA, in Cooke, P. (ed.) *The Rise of the Rustbelt*, London: UCL Press, 52–71.

De Laet, M. and Mol, A. (2000) The Zimbabwe bush pump: mechanics of a fluid technology, *Social Studies of Science* 30: 225–63.

Deleuze, G. (1992) Postscript on the societies of control, *October* 59: 3–7.

Deleuze, G. (1993a) *The Fold: Leibniz and the Baroque*, London: The Athlone Press.

Deleuze, G. (1993b) *The Deleuze Reader* (ed. C. V. Boundas), New York: Columbia University Press.

Deleuze, G. and Guattari, F. (1988) *A Thousand Plateaus: Capitalism and Schizophrenia*, London: The Athlone Press.

Dennis, N., Henriques, F. and Slaughter, C. (1956) *Coal is Our Life: An Analysis of a Yorkshire Mining Community*, London: Tavistock Publications.

Department of Media, Culture and Sport (1998) *Creative Industries Mapping Document*, London: DMCS.

Department of Trade and Industry (1998) *Fairness at Work*, London: The Stationery Office.

Derrida, J. (1994) *Specters of Marx: The State of the Debt, the Work of Mourning, and the New International*, London: Verso.

Derrida, J. (1978) *Writing and Difference*, trans. A. Bass, Chicago: University of Chicago Press.

Derrida, J. (1986) Declarations of independence, *New Political Science* 15: 7–15.

De Solla Price, D. J. (1963) *Little Science, Big Science*, London: Macmillan.

Dicken, P. (1992) *Global Shift: The Internalisation of Economic Activity*, London: Paul Chapman.

Diffie, B. W. and Winius, G. D. (1977) *Foundations of the Portuguese Empire, 1415–1580*, Minneapolis: University of Minnesota Press.

DiMaggio, P. J. and Powell, W. W. (1983) The iron cage revisited: institutional isomorphism and collective rationality in organisational fields, *American Sociological Review* 48: 147–60.

Dodgshon, R. A. (1987) *The European Past: Social Evolution and Spatial Order*, London: Macmillan.

Dodgson, N. (1997) Managing the shape of change, *Value Retailing* 4 September.

Dosi, G. (1988) The nature of the innovative process, in Dosi, G., Freeman, C., Nelson, R., Silverberg, G. and Soete, L. (eds) *Technical Change and Economic Theory*, London: Pinter Publishers, 221–38.

Dosi, G. (1995) The contribution of economic theory to the understanding of a knowledge-based economy, *International Institute for Applied Systems Analysis* WP-95-56, Luxenburg A-2361.

Dosi, G. and Marengo, L. (1993) Some elements of an evolutionary theory of organizational competences, in England, R. W. (ed.) *Evolutionary Concepts in Contemporary Economics*, Ann Arbor, MI: University of Michigan Press, 157–78.

Dosi, G., Teece, D. J. and Winter, S. (1995) Toward a theory of corporate coherence: preliminary remarks, in Dosi, G., Gianetti, R. and Toninelli, P. A. (eds) *Technology and Enterprise in a Historical Perspective*, Oxford: Oxford University Press, 185–211.

Douglas, M. and Isherwood, B. (1978) *The World of Goods: Towards an Anthropology of Consumption*, London: Allen Lane.

Dove, B. and Hickman, C. (1997) More can mean less, *The Grocer* 11 October: 34–5.

Downs, L. (1990) Industrial decline, rationalization and equal pay: the Bedaux strike at Rover automobile company, *Social History* 15(1): 45–74.

Drucker, P. (1969) *The Age of Discontinuity*, New York: Harper and Row.

Drucker, P. (1993) *Post-capitalist Society*, Oxford: Butterworth-Heinemann.

Drucker, P. (1994) *Management*, Oxford: Butterworth-Heinemann.

Drucker, P. (1999) *Management Challenges for the 21st Century*, London: Butterworth-Heinemann.

Dubini, P. (1989) The influence of motivations and environment on business start-ups: some hints for public policies, *Journal of Business Venturing* 4: 11–26.

du Gay, P., Hall, S., Janes, L., Mackay, H. and Negus, K. (1997) *Doing Cultural Studies: The Story of the Sony Walkman*, London: Sage.

du Gay, P. and Salaman, G. (1992) The cult(ure) of the customer, *Journal of Management Studies* 29(5): 615–33.

Dumont, L. (1977) *From Mandeville to Marx: The Genesis and Triumph of Economic Ideology*, Chicago: University of Chicago Press.

Dumont, L. (1986) *Essays on Individualism: Modern Ideology in Anthropological Perspective*, Chicago: University of Chicago Press.

Dunbar, R. (1996) *Grooming, Gossip and the Evolution of Language*, London: Faber and Faber.

Dunsire, A. (1996) Tipping the balance: autopoiesis and governance, *Administration and Society* 28(3): 299–334.

Duplessis, R. S. (1997) *Transitions to Capitalism in Early Modern Europe*, Cambridge: Cambridge University Press.

Durham County Local History Society (1992) *An Historical Atlas of County Durham*, Hexham: Durham County Local History Society, 44.

Durkheim, E. (1912) *Les Formes Élémentaires de la Vie Religieuse*, Paris: F. Alcan.

Durkheim, E. (1964) *The Division of Labor in Society*, New York: The Free Press, first published in 1933.

Dymski, G. A. (1996) On Krugman's model of economic geography, *Geoforum* 20: 1–14.

Eagleton, T. (1990) *The Ideology of the Aesthetic*, Oxford: Basil Blackwell.

The Economist (1996) The property of the mind, 27 July: 57–9.

The Economist (1999) Knowing your place, 13 March: 130.

Edwards, B. (1998) Capitals of capital: a survey of financial centres, *The Economist*, 9 May.

Ehrenreich, B. and Ehrenreich, J. (1979) The professional–managerial class, in Walker, P. (ed.) *Between Labour and Capital*, Brighton: Harvester.

Elias, N. (1982) *The Civilizing Process. Volume I: State Formation and Civilization*, Oxford: Blackwell.

Elias, N. (1992) *Time: An Essay*, London: Basil Blackwell.

Elias, N. (1944 [1939]) *The Civilizing Process: The History of Manners and State Formation and Civilization*, trans. E. Jephcott, Oxford: Blackwell.

Eliasson, G. (1996) *Firm Objectives, Controls and Organization: The Use of Information and the Transfer of Knowledge within the Firm*, Dordrecht: Kluwer.

Elliot, L. and Atkinson, D. (1998) *The Age of Insecurity*, London: Verso.

Emery, N. (1992) *The Coalminers of Durham*, Stroud, Glos.: Alan Sutton Publishing.

Engel, J., Blackwell R. and Minard P. (1986) *Consumer Behaviour*, 5th edn, New York: Holt, Reinhart and Winston.

Engel, J., Blackwell, N. and Minard, P. (1990) *Consumer Behaviour*, 6th edn, Chicago: Dryden Press.

Engel, J., Blackwell, N. and Minard, P. (1995) *Consumer Behaviour*, 7th edn, Chicago: Dryden Press.

Engel, J., Kollat, D. and Blackwell, R. (1968) *Consumer Behaviour*, New York: Holt, Reinhart and Winston.

Erlich, D., Guttman, I., Schenback, P. and Mills, J. (1957) Postdecision exposure to relevant information, *Journal of Abnormal and Social Psychology*, 54: 98–102.

Ermarth, E. (1992) *Sequel to History: Postmodernism and the Crisis of Representational Time*, Princeton, NJ: Princeton University Press.

Estimé, M.-F., Drilhon, G. and Julien, P.-A. (1993) *Small and Medium-sized Enterprises: Technology and Competitiveness*, Paris: Organisation for Economic Co-operation and Development.

Ettlinger, N. and Patton, W. (1996) Shared performance: the proactive diffusion of competitiveness and industrial and local development, *Annals of the Association of American Geographers* 86: 286–305.

Etzioni, A. (1995) *The Spirit of Community: Rights, Responsibilities and the Communitarian Agenda*, London: Fontana.

Evans, P. (1995) *Embedded Autonomy: States and Industrial Transformation*, Princeton, NJ: Princeton University Press.

Faulkner, W. (1994) Conceptualizing knowledge used in innovation: a second look at the science–technology distinction and industrial innovation, *Science, Technology and Human Values* 19: 425–58.

Faulkner, W., Senker, J. and Velho, L. (1995) *Knowledge Frontiers: Public Sector Research and Industrial Innovation in Biotechnology, Engineering Ceramics, and Parallel Computing*, Oxford: Clarendon Press.

Featherstone, M. (1991) *Consumer Culture and Postmodernism*, London: Sage.

Fébvre, L. (1968) *Le Problème de l'Incroyance au XVIᵉ Siècle*, Paris: Gallimard.

Filipec, J. (1986) Society and concepts of social time, *International Social Science Journal* 107: 19–32.

Financial Times (1997) F1 probe says threat to relocate racing was groundless, 15 December: 1 and 18.

Fine, B. and Wright, J. (1991) *Digesting the Food and Information Systems*, Birkbeck Discussion Paper no. 7 (December).

Firat, A. (1994) Gender and consumption transcending the feminine?, in Costa, J. (ed.) *Gender Issues and Consumer Behaviour*, Thousand Oaks, CA and London: Sage.

Fleck, J. (1994) Learning by trying: the implementation of configurational technology, *Research Policy* 23: 637–52.

Fleck, J. and Tierney, M. (1991) The management of expertise: knowledge, power and the economics of expert labour, Edinburgh PICT Working Paper, No. 29, Edinburgh: Research Centre for Social Science, University of Edinburgh.

Fletcher, A. and Stevenson, J. (eds) (1985) *Order and Disorder in Early Modern England*, Cambridge: Cambridge University Press.

Flora, C. B. and Flora, J. L. (1993) Entrepreneurial social infrastructure: a necessary ingredient, *Annals of the American Academy of Political and Social Science* 529: 48–58.

Flora, J. L., Sharp, J. and Flora, C. B. (1997) Entrepreneurial social infrastructure and locally initiated economic development in the nonmetropolitan United States, *Sociological Quarterly* 38: 623–45.

Florida, R. (1995a) Toward the learning region, *Futures* 27(5): 527–36.

Florida, R. (1995b) The industrial transformation of the Great Lakes region, in P. Cooke (ed.) *The Rise of the Rustbelt*, London: UCL Press, 162–76.

Florida, R. (1996) Regional creative destruction: production organization, globalization, and the economic transformation of the Midwest, *Economic Geography* 72: 314–34.

Flynn, P. M. (1993) *Technology Life Cycles and Human Resources*, Lanham, MD: University Press of America.

Foray, D. and Lundvall, B.-Å. (1996) The knowledge-based economy: from economics of knowledge to the learning economy, *Employment and Growth in the Knowledge-Based Economy*, Paris: OECD, 11–32.

Foss, N. J. (1996a) Knowledge-based approaches to the theory of the firm: some critical comments, *Organization Science* 7: 470–6.

Foss, N. J. (1996b) More critical comments on knowledge-based theories of the firm, *Organization Science* 7: 519–23.

Foss, N. J. (1996c) Introduction: the emerging competence perspective, in Foss, N. J. and Knudsen, C. (eds) *Towards a Competence Theory of the Firm*, London: Routledge, 1–12.

Foucault, M. (1972) *The Archeology of Knowledge*, London: Tavistock Publications.

Foucault, M. (1979) *Discipline and Punish: The Birth of the Prison*, Harmondsworth: Penguin.

Foucault, M. (1980) *Power/Knowledge: Selected Interviews and Other Writings 1972–1977*, (ed.) C. Gordon, Hemel Hempstead: Harvester.

Foucault, M. (1981) The order of discourse, in Young, R. (ed.) *Untying the Text: A Post Structuralist Reader*, Boston and London: Routledge and Kegan Paul.

Foucault, M. (1982) The subject and power, in Dreyfus, H. L. and Rabinow, R. (eds) *Michel Foucault: Beyond Structuralism and Hermeneutics*, Chicago: University of Chicago Press.

Foucault, M. (1989) Space, knowledge and power, in Lotringer, S. (ed.) *Foucault Live: Collected Interviews, 1961–1984*, New York: Semiotext(e).

Foxall, G. (1986) The role of radical behaviourism in the explanation of consumer choice, *Advances in Consumer Choice* 13: 187–91.

Foxall, G. (1987) Consumer choice in behaviourism and consumer research: theoretical promise and empirical process, *International Journal of Research Marketing* 4: 111–29.

Foxall, G. (1990) *Consumer Psychology in Behavioural Perspective*, London: Routledge.

Fox Keller, E. (1996) The biological gaze, in Robertson, G., Mash, M., Tickner, L., Bird, J., Curtis, B. and Putman, T. (eds) *FutureNatural: Nature, Science and Culture*, London: Routledge, 107–21.

Fox Keller, E. and Longino, H. (eds) (1996) *Feminism and Science*, Oxford: Oxford University Press.

Freeman, C. (1982) *The Economics of Industrial Innovation*, London: Pinter, 2nd edn.

Frow, J. (1996) Information as gift and commodity, *New Left Review* 219: 89–108.

Fujimara, J. (1996) *Crafting Science: A Sociohistory of the Quest for the Genetics of Cancer*, Cambridge, MA: Harvard University Press.

Fukuyama, F. (1992) *The End of History and the Last Man*, London: Hamish Hamilton.

Gabriel, Y. and Lang, T. (1995) *The Unmanageable Consumer. Contemporary Consumption and its Fragmentations*, London: Sage.

Game, A. (1994) Time, space, memory, with reference to Bachelard, in Featherstone, M., Lash, S. and Robertson, R. (eds) *Global Modernities*, London: Sage.

Gereffi, G. (1993) *Global Production Systems and Third World Development*, Madison, WI: University of Wisconsin Global Studies Research Program, Working Paper Series, August.

Gertler, M. S. (1995) Being there: proximity, organization and culture in the development and adoption of advanced manufacturing technologies, *Economic Geography* 71: 1–26.

Gertler, M. S. (1997) Between the global and the local: the spatial limits to productive capital, in Cox, K. R. (ed.) *Spaces of Globalization*, New York: Guilford, 45–63.

Gibbons, M., Limoges, C., Nowotny, H., Schwartzman, S., Scott, P. and Trow, M. (1994) *The New Production of Knowledge*, London: Sage.

Gibbs, P. (1993) Time as a dimension of consumption in financial services, Bournemouth Working Paper Series, Bournemouth: Bournemouth University.

Gibson, W. (1984) *Necromancer*, London: Grafton Books.

Gibson-Graham, J. K. (1996) *The End of Capitalism (as We Knew it): A Feminist Critique of Political Economy*, Oxford: Blackwell.

Giddens, A. (1979) *The Central Problem in Social Theory*, London: Macmillan.

Giddens, A. (1981) *A Contemporary Critique of Historical Materialism, vol. 1, Power, Property and the State*, London: Macmillan.

Giddens, A. (1984) *The Constitution of Society*, Cambridge: Polity Press.

Giddens, A. (1990) *The Consequences of Modernity*, Cambridge: Polity Press.

Giddens, A. (1991) *Modernity and Self-Identity: Self and Society in the Late Modern Age*, Cambridge: Polity Press.

Giddens, A. (1992) *The Transformation of Intimacy: Sexuality, Love and Eroticism in Modern Societies*, Cambridge: Polity Press.

Giddens, A. (1994a) Risk, trust and reflexivity, in Beck, U., Giddens, A. and Lash, S. *Reflexive Modernisation: Politics, Tradition and Aesthetics in the Modern Social Order*, Cambridge: Polity Press, 184–97.

Giddens, A. (1994b) Living in a post-traditional society, in Beck, U., Giddens, A. and Lash, S. (eds) *Reflexive Modernisation: Politics, Tradition and Aesthetics in the Modern Social Order*, Cambridge: Polity Press, 56–109.

Giddens, A. (1994c) *Beyond Left and Right: The Future of Radical Politics*, Cambridge: Polity Press.

Giddens, A. (1997) Brave New World: the new context of politics, in *In Defence of Sociology: Essays, Interpretations and Rejoiners*, Cambridge: Polity Press, 224–39.

Giddens, A. (1998) *The Third Way: The Renewal of Social Democracy*, Cambridge: Polity Press.

Gill, J. and Whittle, S. (1992) Management by panacea: accounting for transience, *Journal of Management Studies* 30(2): 281–95.

Gillespie, R. (1993) *Manufacturing Knowledge: A History of the Hawthorne Experiments*, Cambridge: Cambridge University Press.

Gjesme, T. (1981) Some factors influencing perceived goal distance in time: a preliminary check, *Perceptual Motor Skills* 53 (Aug.): 175–82.

Glasmeier, A. K. (1988) Factors governing the development of high tech industry agglomerations: a tale of three cities, *Regional Studies* 22: 287–301.

Glasmeier, A. K. (1991) Technological discontinuities and flexible production networks: the case of Switzerland and the world watch industry, *Research Policy* 20: 469–85.

Glasmeier, A. K., Fuellhart, K., Feller, I. and Mark, M. M. (1998) The relevance of firm-learning theories to the design and evaluation of manufacturing modernization programs, *Economic Development Quarterly* 12: 107–24.

Gold, M. and Hall, M. (1994) Statutory Works Councils: the final countdown?, *Industrial Relations Journal* 25(3): 177–86.

Goodman, N. (1969) *Languages of Art*, Oxford: Oxford University Press.

Goodman, N. (1978) *Ways of Worldmaking*, Hassocks, Sussex: The Harvester Press.

Gore, C. (1995) Introduction: markets, citizenship and social exclusion, in Rodgers, D., Gore, C. and Figueiredo J. (eds) *Social Exclusion: Rhetoric, Reality and Responses*, Geneva: IILS/UNDP, 1–40.

Grabher, G. (1993a) *The Embedded Firm*, London: Routledge.

Grabher, G. (1993b) Rediscovering the social in the economics of interfirm relations, in Grabher, G. (ed.) *The Embedded Firm: On the Socio-economics of Industrial Networks*, London and New York: Routledge.

Grabher, G. (1995) The disembedded regional economy: the transformation of East German industrial complexes into Western enclaves, in Amin, A. and Thrift, N. (eds) *Globalization, Institutions and Regional Development in Europe*, Oxford: Oxford University Press.

Grabher, G. (1996) Adaption at the cost of adaptability? Restructuring the East German regional economy, in Grabher, G. and Stark, D. (eds) *Restructuring Networks in Post-Socialism: Legacies, Linkages and Localities*, Oxford: Oxford University Press.

Graham, R. (1981) The role of perception of time in consumer research, *Journal of Consumer Research*, 7: 335–42.

Granovetter, M. (1985) Economic action and social structure: the problem of embeddedness, *American Journal of Sociology*, 91(3): 481–510.

Granovetter, M. and Swedburg, R. (eds) (1992) *The Sociology of Economic Life*, Boulder, CO and Oxford: Westview Press.

Grant, R. M. (1991) The resource-based theory of competitive advantage: implications for strategy formulation *California Management Review* 34, Spring: 114–35.

Grant, R. M. (1996) Toward a knowledge-based theory of the firm, *Strategic Management Journal* 17, Winter Special Issue: 109–22.

Grant-Braham, B. (1997) Formula 1, sponsorship and television: an historical perspective, in Brown, B. J. H. (ed.) *Explorations in Motoring History*, Oxford: Oxbow Books.

Gray, J. (1995) *Enlightenment's Wake: Politics and Culture at the Close of the Modern Age*, London: Routledge.

Gray, J. (1998) Global utopias and clashing civilizations: misunderstanding the present, *International Affairs* 74(1): 149–64.

Gregory, D. (1994) *Geographical Imaginations*, Oxford: Blackwell.

Gregson, N. (1997) On duality and dualism: the case of structuration and time geography, in Bryant, C. and Jary, D. (eds) *Anthony Giddens: Critical Assessments*, London: Routledge.

Greiner, L. and Metzger, R. (1983) *Consulting to Management*, Englewood Cliffs, NJ: Prentice-Hall.

Gronomo, S. (1989) Concepts of time: some implications for consumer research, *Advances in Consumer Research* 16: 339–45.

Gross, B. (1987) Time scarcity: interdisciplinary perspectives and implications for consumer behaviour research, *Advances in Consumer Research* 14: 339–45.

Grosse, R. (1996) International technology transfer in services, *Journal of International Business Studies* 27: 781–800.

Grosvenor, J. (1989) *Management Consultants 1990*, Oxford: The Ivanhoe Press.

Grove-White, R. (1997) Environmental sustainability, time and uncertainty, *Time and Society* 6: 99–106.

Guardian, (1997) Motor racing in pole position for influence, 15 December: 19.

Guillen, M. F. (1994) *Models of Management: Work, Authority and Organisation in a Comparative Perspective*, Chicago: University of Chicago Press.

Gupta, A. K. and Govindarajan, V. (1995) Organizing knowledge flows within MNCs, *International Business Review* 3: 443–57.

Gupta, S. and Govindarajan, V. (1991) Knowledge flows and the structure of control in multinational corporations, *Academy of Management Review* 16: 768–92.

Gurvitch, G. (1964) *The Spectrum of Social Time*, Dordrecht: Reidel.

Gurvitch, G. (1990) Varieties of social time, in Hassard, J. (ed.) *The Sociology of Time*, Basingstoke: Macmillan.

Haber, S. (1964) *Efficiency and Uplift: Scientific Management in the Progressive Era 1890–1920*, Chicago: University of Chicago Press.

Habermas, J. (1976) *Legitimation Crisis*, London: Heinemann.

Habermas, J. (1978) *Knowledge and Human Interests*, London: Heinemann.

Habermas, J. (1983) Modernity: an incomplete project, in Foster, H. (ed.) *The Anti-Aesthetic: Essays on Postmodern Culture*, Seattle: Bay Press, 3–15.

Habermas, J. (1985) *The Philosophical Discourse of Modernity*, Cambridge: Polity Press.

Habermas, J. (1989) *The Structural Transformation of the Public Sphere: An Inquiry into a Category of Bourgeois Society*, Cambridge: Polity Press.

Hall, J. A. (1985) *Powers and Liberties: The Causes and Consequences of the Rise of the West*, Harmondsworth: Penguin.

Hall, M. (1992) Behind the European Works Councils Directive: the European Commission's legislative strategy, *British Journal of Industrial Relations* 30(4): 547–66.

Hall, M. (1994) Industrial relations and the social dimension of European integration: before and after Maastricht, in Hyman, R. and Ferner, A. (eds) *New Frontiers in European Industrial Relations*. Oxford: Blackwell, 281–311.

Hall, M., Carley, M., Gold, M., Marginson, K. and Sisson, K. (1994) *European Works Councils: Planning for the Directive*, London: Industrial Relations Services.

Hall, S. (1980) Encoding/decoding, in Hall, S. *et al.* (eds) *Culture, Media and Language*, London: Hutchinson.

Hall, S. (1996a) When was the post-colonial?, in Curti, L. and Chambers, I. (eds) *The Post-Colonial in Question*, London: Routledge.

Hall, S. (1996b) Introduction: who needs identity?, in Hall, S. and Du Gay, P. (eds) *Questions of Cultural Identity*, London: Sage, 1–17.

Hall, S. and Du Gay, P. (eds) (1996) *Questions of Cultural Identity*, London: Sage.

Hamel, G. and Prahalad, C. K. (1994) *Competing for the Future*, Boston: Harvard Business School Press.

Hamilton, M. (1994) *Race Without End: The Grind Behind the Glamour of the Sasol Jordan Grand Prix Team*, Patrick Stephens Ltd.

Hammer, M. (1990) Reengineering work: don't automate – obliterate, *Harvard Business Review* July/August: 104–12.

Hammer, M. and Champy, J. (1993) *Reengineering the Corporation: A Manifesto for Business Revolution*, London: Nicholas Brealey Publishing.

Hampden-Turner, C. (1994) *Corporate Culture: How to Generate Organisational Strength and Lasting Commercial Advantage*, London: Piatkus.

Handy, C. (1997) *The Hungry Spirit: Beyond Capitalism – a Quest for Purpose in the Modern World*, London: Hutchinson.

Hanlon, G. (1994) *The Commercialisation of Accountancy: Flexible Accumulation and the Transformation of the Service Class*, Aldershot: St Martins Press.

Hansen, N. (1992) Competition, trust and reciprocity in the development of innovative regional milieux, *Papers in Regional Science* 71: 95–105.

Hansen, N. (1993) Endogenous growth centers: lessons from rural Denmark, in Barkley, D. (ed.) *Economic Adaptation: Alternatives for Nonmetropolitan Areas*, Boulder, CO: Westview Press, 69–88.

Haraway, D. (1989) Situated knowledges: the science question in feminism and the privilege of partial perspective, *Feminist Studies* 14: 575–99.

Haraway, D. (1991a) Situated knowledges: the science question in feminism and the privilege of partial perspective, in Haraway, D. (ed.) *Simians, Cyborgs and Women: The Reinvention of Nature*, London: Free Association Books.

Haraway, D. (1991b) *Simians, Cyborgs and Women: The Reinvention of Nature*, London: Free Association Books.

Haraway, D. (1997) Modest_Witness@Second_Millennium.Female_Man©_Meets_Oncomouse™: Feminism and Technoscience, New York and London: Routledge.

Hardill, I., Fletcher, D. and Montagné-Villette, S. (1995) Small firms' distinctive capabilities and the socioeconomic milieu: findings from case studies in Le Choletais (France) and the East Midlands (UK), Entrepreneurship and Regional Development 7: 167–86.

Harper, P. (1995) Genetic testing, common diseases and health service provision, The Lancet 346: 1645–46.

Harris, N. (1987) The End of the Third World, Harmondsworth: Penguin.

Harrison, B. (1992) Industrial districts: old wine in new bottles?, Regional Studies 26: 469–83.

Hartmann, B. (1987) Reproductive Rights and Wrongs: The Global Politics of Population Control and Contraceptive Choice, Harper Row.

Hartshorne, R. (1927) Location as a factor in geography, Annals of the Association of American Geographers, 17(2): 92–9 reprinted in Smith, R. H. T., Taaffe, E. J. and King, L. J. (eds) (1968) Readings in Economic Geography, Chicago: Rand McNally, 23–7.

Harvard Business School (1994) Time Based Competition, Management Programs, Harvard: Harvard University Press.

Harvey, D. (1982) The Limits to Capital, Oxford: Blackwell.

Harvey, D. (1989, 1991) The Condition of Postmodernity: An Enquiry into the Origins of Cultural Change, Oxford: Blackwell.

Harvey, D. (1996) Justice, Nature and the Geography of Difference, Oxford: Blackwell.

Hassard, J. (1990) The Sociology of Time, Basingstoke: Macmillan.

Hazard, P. (1968a) The European Mind 1680–1715, Harmondsworth: Penguin.

Hazard, P. (1968b) European Thought in the Eighteenth Century, Harmondsworth: Penguin.

Hebb, D. (1993) Racing the reason why, British Cars April: 51–5.

Hedberg, R. (1981) How organisations learn and unlearn, in Nystrom, P. C. and Starbuck, W. H. (eds) Handbook of Organisation Design, Oxford: Oxford University Press, 3–27.

Heidegger, M. (1962) Being and Time (Sein und Zeit, 1926), trans J. Macquarre and E. Robinson, New York: Harper Row.

Heidegger, M. (1971) Poetry, Language, Thought, trans A. Hofstadter, New York: Harper Row.

Hekman, S. (1990) Gender and Knowledge: Elements of a Postmodern Feminism, Oxford: Polity Press.

Hekman, S. (1995) Moral Voices, Moral Selves: Carol Gilligan and Feminist Moral Theory, Oxford: Polity Press.

Hendrix, P. (1980) Subjective elements in the examination of time expenditure, in Wilkie, W. (ed.) Advances in Consumer Research 6: 38–44, Ann Arbor, MI: Association for Consumer Research.

Henry, N. (1999) In Pole Position: The British Motor Sport Cluster and Lessons for Europe's Motor Industry, London: Euromotor Reports Ltd.

Henry, N. and Pinch, S. (1997) A regional formula for success? The innovative region of Motor Sport Valley, Birmingham: School of Geography, University of Birmingham.

Henry, N. and Pinch, S. (2000) Spatialising knowledge: placing the knowledge community of Motor Sport Valley, Geoforum: 31(2): 191–209. Reprinted in Henry, N. and Pollard, J. (eds) Theme Issue Capitalising on Knowledge (forthcoming).

Henry, N., Pinch, S. and Russell, S. (1996) In pole position? Untraded interdependencies, new industrial spaces and the British motor sport industry, Area 28(1): 25–36.

Hepple, L. (1986) The revival of geopolitics, Political Geography Quarterly 5: 21–36.

Herbig, P. A. (1994) The Innovation Matrix: Culture and Structure Prerequisites to Innovation, New York: Quorum Books.

Herbig, P. and Golden, J. E. (1993) How to keep that innovative spirit alive: an examination of evolving technology hot spots, Technological Forecasting and Social Change 43: 75–90.

Hetherington, K. (1997a) *The Badlands of Modernity: Heterotopia and Social Ordering*, London: Routledge.

Hetherington, K. (1997b) Museum topology and the will to connect, *Journal of Material Culture* 2: 199–218.

Hetherington, K. (1999) Whither the world? The presence and absence of the globe, in Laretta, E. *et al.* (eds), *Time in the Making and Possible Futures*, Rio de Janeiro: UNESCO.

Hetherington, K. and Lee, N. (2000) Social order and the blank figure, *Environment and Planning D: Society and Space* 18: 169–84.

Hewitt, P. (1993) *About Time. The Revolution in Work and Family Life*, London: Rivers Oram Press.

Hill, C. (1940) *The English Revolution 1640*, London: Lawrence and Wishart.

Hill, S. (1976) *The Dockers: Class and Tradition in London*, London: Heinemann.

Hirsch, F. (1976) *The Social Limits to Growth*, Cambridge, MA: Harvard University Press.

Hirschman, E. (1987) Theoretical perspectives of time use implications for consumer behaviour research, *Research in Consumer Behaviour* 2: 55–81.

Hirschman, E. and Holbrook, M. (1992) *Postmodern Consumer Research*, London: Sage.

Hirst, P. and Thompson, G. (1996) *Globalization in Question: The International Economy and the Possibilities of Governance*, Cambridge: Polity Press.

Hirst, P. and Woolley, P. (1982) *Social Relations and Human Attributes*, London: Tavistock Publications.

Hirszowicz, M. (1980) *The Bureaucratic Leviathan: A Study in the Sociology of Communism*, Oxford: Martin Robertson.

Hobsbawm, E. (1994) *Age of Extremes: The Short Twentieth Century*, London: Michael Joseph, 572.

Hoch, S. and Loewenstein, G. (1991) Time-inconsistent preferences and consumer self-control, *Journal of Consumer Research* 17 (March): 492–507.

Hodgson, G. (1998a) The approach of institutional economics, *Journal of Economic Literature* 36: 166–92.

Hodgson, G. (1998b) Competence and contract in the theory of the firm, *Journal of Economic Behavior and Organization* 35: 179–201.

Hodgson, G. (1999), *Economics and Utopia: Why the Learning Economy is not the End of History*, London: Routledge.

Holland, J. H., Holyoak, K. J., Nisbet, R. E. and Thagard, P. R. (1987) *Induction: Process of Inference, Learning and Discovery*, Cambridge, MA: MIT Press.

Holman, R. (1981) The imagination of the future: a hidden concept in the study of time, in Monroe, K. B. (ed.) *Advances in Consumer Research* 8: 187–91, Ann Arbor MI: Association for Consumer Research.

Hooley, G. J. and Mann, S. J. (1988) The adoption of marketing by financial institutions in the UK, *Service Industries Journal* 8 (4): 488–500.

Hornik, J. (1984) Subjective vs objective time measures: a note on the perception of time in consumer behaviour, *Journal of Consumer Research* 11 (June): 615–18.

House of Commons Science and Technology Committee (1995) *Human Genetics: The Science and its Consequences*, House of Commons Science and Technology Committee.

Howard, J. and Sheth, J. (1969) *The Theory of Buyer Behaviour*, New York: John Wiley and Sons.

Howells, J. (1995) Going global: the use of ICT networks in research and development, *Research Policy* 24: 169–84.

Howells, J. (1996) Tacit knowledge, innovation and technology transfer, *Technology Analysis & Strategic Management* 8: 91–106.

Howells, J. (1998) International innovation and technology transfer within multinational firms, in Grieve Smith, J. and Michie, J. (eds) *Globalization, Growth and Governance*, Oxford: Oxford University Press, 50–70.

Howells, J. (1999) Regional systems of innovation?, in Archibugi, D., Howells, J. and Michie, J. (eds) *Innovation Systems in a Global Economy*, Cambridge: Cambridge University Press.

Howitt, P. (1997) On some problems in measuring knowledge-based growth, in Howitt, P. (ed.) *The Implications of Knowledge-Based Growth for Micro-Economic Policies*, Calgary: University of Calgary Press, 9–29. http://web.mit.edu/newsoffice/founders/

Hu, Y-S. (1995) The international transferability of the firm's advantages, *California Management Review* 37(4): 73–88.

Huczynski, A. A. (1996) *Management Gurus: What Makes Them and How to Become One*, London: Thompson Business Press.

Hudson, R. (2000) The learning economy, the learning firm and the learning region: a sympathetic critique of the limits to learning, in Hudson, R. *Production, Places and Environment*, Harlow: Prentice Hall.

Hughes, T. P. (1983) *Networks of Power: Electrification in Western Society, 1880–1930*, Baltimore: Johns Hopkins University Press.

Hume, P. and Jordanova, L. (1990) Introduction, in Hume, P. and Jordanova, L. (eds) *The Enlightenment and its Shadows*, London: Routledge, 1–15.

Hunt, S. (1983) General theories and the fundamental explanda of marketing, *Journal of Marketing* 47 (Fall): 9–17.

Huntington, S. P. (1996) *The Clash of Civilizations and the Remaking of World Order*, New York: Simon and Schuster.

Husserl, E. (1928) *The Phenomenology of Internal Time-Consciousness*, (ed.) M. Heidegger, trans J. Churchill, Bloomington, IN: Indiana University Press.

Hutchins, R. (1997) Category management in the food industry: a research agenda, *British Food Journal* 99(5): 177–80.

Hyman, R. (1987) Strategy or structure? Capital, labour and control, *Work, Employment, Society* 1: 25–55.

Hyman, S. (1961) *An Introduction to Management Consultancy*, London: Heinemann.

Ignatieff, M. (1984) *The Needs of Strangers*, London: Chatto and Windus, ch. 2.

Ignatieff, M. (1993) *Blood and Belonging: Journeys into the New Nationalism*, London: Vintage.

Ignatieff, M. (1998) *The Warrior's Honour: The Ethnic War and the Modern Conscience*, London: Chatto and Windus.

Illeris, S. (1994) Proximity between service producers and service users, *Tijdschrift voor Economische en Sociale Geografie* 85: 294–302.

Ingold, T. (1993) The temporality of landscape, *World Archaeology* 25: 152–74.

Innis, H. A. (1951) *The Bias of Communication*, Toronto: University of Toronto Press.

Iriye, A. (1997) *Cultural Internationalism and World Order*, Baltimore: Johns Hopkins University Press.

IRS Employment Trends (1996) Talking shop: J Sainsbury's new works councils, *IRS Employment Trends* 622, December: 13–16.

IRS Management Review (1997) Communication in the workplace, *IRS Management Review* 6 July.

Jackson P. and Thrift, N. (1995) Geographies of consumption, in Miller, D. (ed.) *Acknowledging Consumption: A Review of New Studies*, London: Routledge.

Jacoby, J., Szybillo, G. and Berning, C. (1976) Time and consumer behaviour: an interdisciplinary overview, *Journal of Consumer Research* 2: 320–39.

Jacques, R. (1996) *Manufacturing the Employee: Management Knowledge from the 19th to 21st Centuries*, London: Sage.

Jameson, F. (1984) *Postmodernism or the Cultural Logic of Late Capitalism*, London: Verso.

Jameson, F. (1991) *Postmodernism, or, the Cultural Logic of Late Capitalism*, London: Verso.

Jamison, A. (1996) The shaping of the global environmental agenda: the role of non-governmental organisations, in Lash, S., Szerszynski, B. and Wynne, B. (eds) *Risk, Environment and Modernity. Towards a New Ecology*, London: Sage, 224–45.

Jessop, B. (1983) Accumulation strategies, state forms, and hegemonic projects, *Kapitalistate* 10: 89–111.

Jessop, B. (1990) *State Theory: Putting the Capitalist State in its Place*, Cambridge: Polity Press.

Jessop, B. (1998) The rise of governance and the risks of failure: the case of economic development, *International Social Science Journal* 155: 29–45.

Jessop, B. (1999) Reflections on the (il)logics of globalisation, in K. Olds *et al.* (eds) *Globalisation and the Asia Pacific: Contested Territories*, London: Routledge, 19–38.

Jevnaker, B. H. (1997) Competing through mobile design expertise, in Eskelinen, H. (ed.) *Regional Specialisation and Local Environment – Learning and Competitiveness*, Copenhagen: NordREFO, 1997, 65–96.

Johannisson, B., Alexanderson, O., Nowicki, K. and Senneseth, K. (1994) Beyond anarchy and organization: entrepreneurs in contextual networks, *Entrepreneurship and Regional Development* 6: 329–56.

Johnson, B. T., Holmes, K. R and Kirkpatrick, M. (1998) Index of Economic Freedom, Executive Summary. http://www.heritage.org/heritage/index/execsummary

Johnson, C. (1982) *MITI and the Japanese Miracle*, Stanford, CA: Stanford University Press.

Johnson, C. (1985) The institutional foundations of Japanese industrial policy, *California Management Review* 27(4).

Johnson, M. and Pinnington, D. (1997) Turning swords into ploughshares: supporting a co-operative approach to category management between retailers and manufacturers, paper presented at the Marketing Week Conference, 'Using in-store marketing to gain a competitive advantage', 20 April.

Johnson, P. S. (1973) *Cooperative Research in Industry*, London: Martin Robertson.

Johnson, R. (1986) The story so far: and further transformations?, in Punter, D. (ed.) *Introduction to Contemporary Cultural Studies*, Harlow: Longman, 277–313.

Johnson, R. H. (1994) *Improbable Dangers: U.S. Conceptions of Threat in the Cold War and After*, Basingstoke and London: Macmillan.

Johnson-Laird, T. (1983) *Mental Models*, Cambridge: Cambridge University Press.

Kakar, S. (1970) *Frederick Taylor: A Study in Personality and Innovation*, Cambridge, MA: MIT Press

Kant, I. (1788/1956) *Critique of Pure Reason*, trans. N. Kemp Smith, London: Macmillan.

Kant, I. (1987) *Critique of Judgement*, trans. W. S. Pluhar, Indianapolis and Cambridge: Hackett Publicity Company.

Kanter, R. M. (1995) *World Class: Thriving Locally in the Global Economy*, New York: Simon and Schuster.

Karlsson, C. and Larsson, J. (1993) A macro-view of the Gnosjö entrepreneurial spirit, *Entrepreneurship and Regional Development* 5: 117–39.

Katsh, M. E. (1995) Rights, camera, action: cyberspatial settings and the First Amendment, *Yale Law Journal* 194: 1,717.

Kay, L. (1993) *The Molecular Vision of Life: Caltech, The Rockefeller Foundation and the Rise of the New Biology*, Oxford: Oxford University Press.

Keeble, D., Bryson, J.R. and Wood, P. (1992) Entrepreneurship and flexibility in business services: the rise of small management consultancy and market research firms in the UK, in Caley, K., Chell, F., Chittenden, C. and Mason, C. (eds) *Small Enterprise Development: Policy and Practice in Action*, London: Paul Chapman, 43–58.

Keeble, D., Bryson, J. R. and Wood, P. (1994) *Pathfinders of Enterprise: The Creation, Growth and Dynamics of Small Management Consultancies in Britain*, Business Services Research Monograph, Series 1, No. 2, Milton Keynes: Small Business Research Trust.

Keenoy, T. (1990) Human Resource Management: rhetoric, reality and contradiction, *International Journal of Human Resource Management* 1(3): 363–84.

Kelly, K. (1994) *Out of Control: The New Biology of Machines*, London: Fourth Estate, 2.

Kelly, K. (1998) *New Rules for the New Economy*, London: Fourth Estate.

Kennan, G. [Mr. X] (1947) The sources of Soviet conduct, *Foreign Affairs* 25: 566–82.

Kennedy, D. (1996) Imperial history and post-colonial theory, *Journal of Imperial and Commonwealth History* 24: 345–63.

Kerfoot, D. and Knights, D. (1993) Management, masculinity and manipulation: from paternalism to corporate strategy in financial services in Britain, *Journal of Management Studies* 30: 659–78.

Kerr, A., Cunningham-Burley, S. and Amos, A. (1998a) Eugenics and the new genetics in Britain: examining contemporary professional accounts, *Science, Technology and Human Values* 23: 175–98.

Kerr, A., Cunningham-Burley, S. and Amos, A. (1998b) Drawing the line: an analysis of lay people's discussions about the new genetics, *Public Understanding of Science* 7: 113–33.

Kevles, D. (1997) Big science and big politics in the United States: reflections on the death of the SSC and the life of the Human Genome Project, *Historical Studies in the Physical and Biological Sciences* 27: 269–97.

Kharbanda, O. P. and Stallworthy, E. A. (1987) *Company Rescue: How to Manage a Company Turnaround*, London: Heinemann.

Kim, D. H. (1993) The link between individual and organisational learning, *Sloan Management Review* (Fall): 37–50.

Klein, J. (1965) *Samples from English Cultures: Vol. I*, London: Routledge.

Kline, S. J. and Rosenberg, N. (1986) An overview of innovation, in Landau, R. and Rosenberg, N. (eds) *The Positive Sum Strategy*, Washington, DC: National Academy Press, 275–305.

Knight, R. V. (1982) City development in advanced industrial societies, in Gappert, G. and Knight, R. V. (eds) *Cities in the 21st Century*, Beverly Hills, CA: Sage, 47–68.

Knight, R. V. (1995) Knowledge-based development: policy and planning implications for cities, *Urban Studies* 32: 225–60.

Knights, D. (1997) Fear of the Empty Space: The Threat of Postmodern Feminism, paper delivered to the 15 SCOS International Conference, Warsaw, 9–12 July.

Knights, D. and Morgan, G. (1993) Organisation theory and consumption in a post-modern era, *Organisation Studies* 14(2): 211–34.

Knights, D. and Odih, P. (1995) It's about time: the significance of gendered time for financial services consumption, *Time and Society* 14(2), June: 205–31.

Knights, D. and Odih, P. (1997) From here to eternity – a masculine project theorising time, space and risk within the classical, modern and postmodern epistemological regimes, in Rasmussen, H. (ed.) *Proceedings of the AOS, Accounting Time and Space Conference*, Copenhagen, 4–6 September.

Knights, D. and Sturdy, A. (1993) Marketing the Soul: From the Ideology of Consumption to Consumer Subjectivity, paper delivered at the 11th EGOS Colloquium, Paris, 6–8 July.

Knights, D., Sturdy, A. and Morgan, G. (1994) The consumer rules? An examination of the rhetoric and 'reality' of marketing in financial services, *European Journal of Marketing* 28(3): 42–54.

Knights, D. and Willmott, H. (1983) Dualism and domination: an analysis of Marxian, Weberian and existentialist perspectives, *The Australian and New Zealand Journal of Sociology* 19(1), March: 33–49.

Knorr-Cetina, K. (1981) *The Manufacture of Knowledge: An Essay on the Constructivist and Contextual Nature of Science*, Oxford: Pergamon Press.

Knox, P. and Agnew, J. (1994) *The Geography of the World Economy*, London: Edward Arnold.

Knox, P. and Taylor, P. J. (eds) (1995) *World Cities in a World-System*, New York: Oxford University Press.

Knudsen, C. (1996) The competence perspective: a historical view, in Foss, N. J. and Knudsen, C. (eds) *Towards a Competence Theory of the Firm*, London: Routledge, 13–37.

Kochan, T. A. and Osterman, P. (1994) *The Mutual Gains Enterprise: Forging a Winning Partnership among Labor, Management and Government*, Boston: Harvard Business School Press.

Kogut, B. and Zander, U. (1996) What firms do? Coordination, identity and learning, *Organization Science* 7 (5): 502–18.

Kopytoff, I. (1986) The cultural biography of things: commoditization as process, in Appadurai, A. (ed.) *The Social Life of Things: Commodities in Cultural Perspective*, Cambridge: Cambridge University Press, 64–91.

Krois, J. M. (1987) *Cassirer: Symbolic Forms and History*, Newhaven and London: Yale University Press.

Krugman, P. (1986) *Strategic Trade Policy and the New International Economics*, Cambridge, MA: MIT Press.

Krugman, P. (1991) *Geography and Trade*, Cambridge, MA: MIT Press.

Kubr, M. (ed.) (1986) *Management Consulting: A Guide to the Profession*, Geneva: International Labour Office.

Kuhn, T. (1970) *The Structure of Scientific Revolutions*, Chicago: University of Chicago Press.

Kundnani, A. (1998–9) Where do you want to go today? The rise of information capital, *Race and Class* 40(2–3): 49–72.

Lagendijk, A. and Cornford, J. (2000) Regional institutions and knowledge: tracking new forms of regional development policy, *Geoforum*.

Lamberton, D. (1994) Innovation and intellectual property, in Dodgson, M. and Rothwell, R. (eds) *The Handbook of Industrial Innovation*, Aldershot: Edward Elgar, 301–9.

Lamberton, D. (1997) The knowledge-based economy: a Sisyphus model, *Prometheus* 15: 73–81.

The Lancet (1995) editorial, Western eyes on China's eugenics law, 346: 131.

Langlois, R. N. and Robertson, P. L. (1995) *Firms, Markets and Economic Change: A Dynamic Theory of Business Institutions*, London: Routledge.

Lasch, C. (1978) *The Culture of Narcissism: American Life in an Age of Diminishing Expectations*, New York: Norton.

Lasch, C. (1984) *The Minimal Self: Psychic Survival in Troubled Times*, New York: Norton.

Lasch, C. (1995) *The Revolt of the Elites and the Betrayal of Democracy*, New York: Norton.

Lash, S., Szerszynski, B. and Wynne, B. (1996) *Risk, Environment and Modernity: Towards a New Ecology*, London: Sage.

Lash, S. and Urry, J. (1993) *The End of Organised Capitalism*, Cambridge: Polity Press.

Lash, S. and Urry, J. (1994) *Economies of Signs and Space*, London and New Delhi: Sage.

Latour, B. (1987) *Science in Action: How to Follow Scientists and Engineers Through Society*, Cambridge, MA: Harvard University Press and Milton Keynes: Open University Press.

Latour, B. (1990) Drawing things together, in Lynch, M. and Woolgar, S. (eds) *Representation in Scientific Practice*, Cambridge, MA: MIT Press, 19–68.

Latour, B. (1993) *We Have Never Been Modern*, New York and London: Harvester Wheatsheaf.

Lave, J. and Wenger, E. (1991) *Situated Learning: Legitimate Peripheral Participation*, Cambridge: Cambridge University Press.

Law, J. (1986) On the methods of long distance control: vessels, navigation and the Portuguese route to India, in Law, J. (ed.), *Power, Action and Belief: A New Sociology of Knowledge?* Sociological Review Monograph, 32, London: Routledge and Kegan Paul, 234–63.

Law, J. (1987), Technology and heterogeneous engineering: the case of the Portuguese expansion, in Bijker, W. E., Hughes, T. P. and Pinch, T. (eds) *The Social Construction of Technical Systems: New Directions in the Sociology and History of Technology*, Cambridge, MA: MIT Press, 111–34.

Law, J. (1992) Notes on the theory of the actor-network: ordering, strategy and heterogeneity, *Systems Practice* 5: 379–93.

Law, J. (1994) *Organizing Modernity*, Oxford: Blackwell.

Law, J. (1999) After ANT: complexity, naming and topology, in Law, J. and Hassard, J. (eds) *Actor Network Theory and After*, Oxford: Blackwell, 1–15.

Law, J. and Hassard, J. (eds) (1999) *Actor Network Theory and After*, Oxford and Keele: Blackwell and the Sociological Review.

Law, J. and Mol, A. (1995), Notes on materiality and sociality, *The Sociological Review* 43: 274–94.

Layton, E. T. (1974) Technology as knowledge, *Technology and Culture* 15: 30–41.

Leadbeater, C. (1999) *Living on Thin Air: The New Economy*, Harmondsworth: Viking.

Lee, F. C. and Has, H. (1997) A quantitative assessment of high-knowledge industries versus low-knowledge industries, in Howitt, P. (ed.) *The Implications of Knowledge-Based Growth for Micro-Economic Policies*, Calgary: University of Calgary Press, 39–78.

Lee, R. and Wills, J. (eds) (1997) *Geographies of Economies*, London: Arnold.

Lefebvre, H. (1991) *The Production of Space*, trans. D. Nicholson-Smith, Oxford: Basil Blackwell.

Leigh Star, S. (1995) The politics of formal representations: wizards, gurus and organizational complexity, in Leigh Star, S. (ed.) *Ecologies of Knowledge: Work and Politics in Science and Technology*, Albany: State University of New York Press, 88–118.

Leland, J. (1998) Blessed by the bull, *Newsweek* 27 April: 36–9.

Leonard-Barton, D. (1995) *Wellsprings of Knowledge: Building and Sustaining the Sources of Innovation*, Boston: Harvard Business School Press.

Levinthal, D. (1994) Surviving Schumpeterian environments: an evolutionary perspective, in Baum, J. A. C. and Singh, J. V. (eds) *Evolutionary Dynamics of Organizations*, New York: Oxford University Press, 167–78.

Lewchuk, W. and Robertson, D. (1997) Production without empowerment: work re-organisation from the perspective of motor vehicle workers, *Capital and Class* 63: 37–64.

Lewis, D. and Weigart, A. (1990) The structures and meanings of social-time, in Hassard, J. (ed.) *The Sociology of Time*, Basingstoke: Macmillan.

Lewis, J. (1997) Bringing everyone together, *In-Store Marketing* November: 26–7.

Lewis, J. and Weigart, A. (1986) Structures and meaning of social time, *Social Forces* 60: 433–62.

Lewis, R. (1997) Sharing professional knowledge: organizational memory, *International Journal of Continuing Education and Life-long Learning*, 7(2): 95–107.

Leyshon, A. and Pollard, J. S. (2000) Geographies of industrial convergence: the case of retail banking, *Transactions of the Institute of British Geographers*, 25: 203–20.

Leyshon, A. and Thrift, N. J. (1997) *Financial Infrastructure Withdrawal and Access to Financial Services*, End of Award Report to Economic and Social Research Council, March.

Leyshon, A. and Thrift, N. J. (1999) Lists come alive: electronic systems of knowledge and the rise of credit-scoring in retail banking, *Economy and Society*, 28, 434–66.

Leyshon, A., Thrift, N. J. and Justice, M. (1993) *A Reversal of Fortune: Financial Services and the South East of England*, Harlow: South East Economic Development Report.

Leyshon, A., Thrift, N. J. and Pratt J. (1998) Reading financial services: texts, consumers and financial literacy, *Environment and Planning D: Society and Space* 16: 29–55.

Liebskind, J. P. (1997) Keeping organizational secrets: protective institutional mechanisms and their costs, *Industrial and Corporate Change* 6: 623–63.

Lien, M. (1997) *Marketing and Modernity*, Oxford: Berg.

Light, J. (1997) The changing nature of nature, *Ecumene* 4: 181–95.

Littler, C. R. (1982) *The Development of the Labour Process in Capitalist Societies*, London: Heinemann Educational Books.

Livingstone, D. N. (1992a) Never shall ye make the crab walk straight: an inquiry into the scientific sources of American racial geography, in Driver, F. and Rose, G. (eds) *Nature and Science: Essays in the History of Geographical Knowledge*, Historical Geography Research Series, No. 28, 37–48.

Livingstone, D. N. (1992b) *The Geographical Tradition: Episodes in the History of a Contested Enterprise*, Oxford: Blackwell.

Livingstone, D. N. (1998) Reproduction, representation and authenticity: a re-reading, *Transactions of the Institute of British Geographers* 23: 13–19.

Loasby, B. J. (1998) The organisation of capabilities, *Journal of Economic Behavior and Organization*, 35: 139–60.

Lorenz, E. H. (1992) Trust, community, and cooperation: toward a theory of industrial districts, in Storper, M. and Scott, A. J. (eds) *Pathways to Industrialization and Regional Development*, London: Routledge, 195–204.

Love Brown, S. (1997) The Free Market as salvation from government: the anarcho-capitalist view, in Carrier, J. G. (ed.) *Meanings of the Market: The Free Market in Western Culture*, Oxford: Berg.

Lovejoy, A. O. (1961) *The Great Chain of Being: A Study of the History of an Idea*, Cambridge, MA: Harvard University Press.

Luke, T. W. (1994) Placing power/siting space: the politics of the global and the local in the new world order, *Environment and Planning D: Society and Space* 12(4): 613–28.

Lundvall, B-Å (1988) Innovation as an interactive process – from user-producer interaction to national systems of innovation, in Dosi, G., Freeman, C., Nelson, R., Silverberg, G. and Soete, L. (eds) *Technological Change and Economic Theory*, London: Pinter, 349–69.

Lundvall, B-Å. (1999) Technology policy in the learning economy, in Archibugi, D., Howells, J. and Michie, J. (eds) *Innovation Systems in a Global Economy*, Cambridge: Cambridge University Press.

Lundvall, B.-Å. and Johnson, B. (1994) The learning economy, *Journal of Industry Studies* 1: 23–42.

Luttwak E. (1990) From geopolitics to geoeconomics, *The National Interest* 20: 17–24.

Lyotard, J.-F. (1984) *The Postmodern Condition: A Report on Knowledge*, Manchester: Manchester University Press.

Lyotard, J.-F. (1986) *The Differend: Phrases in Dispute*, Manchester: Manchester University Press.

Lyotard, J.-F. and Thébaud, J.-L. (1984) *Just Gaming*, Manchester: Manchester University Press.

Mabey, C., Skinner, D. and Clark, T. (eds) (1998) *Experiencing Human Resource Management*, London: Sage.

McCracken, G. (1990) Culture and consumer behaviour, *Journal of the Market Research Society* 32(1): 3–11.

Macdonald, A. (1997) Keeping partners happy, *In-Store Marketing* November: 28–9.

Macdonald, S. (1996) Informal information flow and strategy in the international firm, *International Journal of Technology Management* 11: 219–32.

McGrath, M. (1997) *A Guide to Category Management*, Watford: Institute for Grocery Distribution.

McKelvey, W. (1982) *Organizational Systematics*, Berkeley, CA: University of California Press.

McKibben, B. (1990) *The End of Nature*, New York: Viking.

McLean, A. (1984) Myths, magic, and gobbledegook, a-rational aspects of the consultant's role, in Kakabadse, A. and Parker, C. (eds) *Power, Politics and Organisations: A Behavioural Science View*, New York: John Wiley, 147–67.

MacLeod, G. (2000) The learning region in an age of austerity: capitalizing on knowledge, entrepreneurialism and reflexive capitalism, *Geoforum*.

MacLeod, R. and Lewis, M. (1988) *Disease, Medicine and Empire: Perspectives on Western Medicine and the Expansion of Europe*, London: Routledge.

McLuhan, M. with Fiore, Q. (1967) *The Medium is the Message: An Inventory of Effects*, Harmondsworth: Penguin.

Macnaghten, P. and Urry, J. (1998) *Contested Natures*, London: Sage.

MacPherson, A. (1991) Interfirm information linkages in an economically disadvantaged region: an empirical perspective from metropolitan Buffalo, *Environment and Planning A* 23: 591–606.

Maillat, D. (1995) Territorial dynamic, innovative milieus and regional policy, *Entrepreneurship and Regional Development* 7: 157–65.

Maillat, D. (1998) Interactions between urban systems and localized productive systems: an approach to endogenous regional development in terms of innovative milieu, *European Planning Studies* 6: 117–29.

Maillat, D., Lecoq, B., Nemeti, F. and Pfister, M. (1995) Technology district and innovation: the case of the Swiss Jura Arc, *Regional Studies* 29: 251–63.

Mair, A. (1997) Strategic localization: the myth of the postnational enterprise, in Cox, K. R. (ed.) *Spaces of Globalization*, New York: Guilford, 64–88.

Malecki, E. J. (1987) The R&D location decision of the firm and creative regions, *Technovation* 6: 205–22.

Malecki, E. J. (1997) *Technology and Economic Development*, 2nd edn, London: Addison Wesley Longman.

Malecki, E. J. and Bradbury, S. L. (1992) R&D facilities and professional labour: labour force dynamics in high technology, *Regional Studies* 26: 123–36.

Malecki, E. J. and Oinas, P. (eds) (1999) *Making Connections: Technological Learning and Regional Economic Change*, Aldershot: Ashgate.

Malmberg, A., Sölvell, O. and Zander, I. (1996) Spatial clustering, local accumulation of knowledge and firm competitiveness, *Geografiska Annaler* 78B: 85–97.

Managing Best Practice (1998) Works councils, *Managing Best Practice* 49, July.

Manenti, Y. (1995) Capitalising on knowledge: organizational issues and constraints, *Journal of Computer Assisted Learning* 11: 225–30.

Mann, M. (1986) *The Sources of Social Power. Volume I: A History of Power from the Beginning to 1760 A.D.*, Cambridge: Cambridge University Press.

Mansfield, E. (1961) Technical change and the rate of imitation, *Econometrica* 29: 741–66.

Marginson, P., Gilman, M., Jacobi, O. and Krieger, H. (1998) *Negotiating European Works Councils: An analysis of Agreements Under Article 13*. Report prepared for the European Foundation for the Improvement of Living and Working Conditions and the European Commission.

Marglin, S. A. (1990) Towards a decolonization of the mind, in Marglin, F. A. and Marglin, S. A. (eds) *Dominating Knowledge: Development, Culture and Resistance*, Oxford: Clarendon Press.

Markham, C. (1987) *Practical Consulting*, London: Institute of Management Consultants and the Institute of Chartered Accountants.

Markusen, A. R. (1985) *Profit Cycles, Oligopoly, and Regional Development*, Cambridge, MA: MIT Press.

Markusen, A. R. (1996) Sticky places in slippery slopes: a typology of industrial districts, *Economic Geography* 72(3): 293–313.

Marsden, T., Flynn, A. and Harrison, M. (1999) *Consuming Interests: Retailing, Regulation and the Provision of Food Quality*, London: UCL Press.

Marsden, T., Harrison, M. and Flynn, A. (1998) Creating competitive space: exploring the social and political maintenance of retail power, *Environment and Planning A* 30: 481–98.

Marshall, A. (1899) *Elements of Economics of Industry*, London: Macmillan, 3rd edn (1932 reprint).

Marshall, A. (1909) *Principles of Economics*, London: Macmillan.

Marshall, A. (1975) *The Early Economic Writings of Alfred Marshall 1867–1890*, Whitaker, J. (ed.), London: Macmillan.

Marshall, J. N. and Richardson, R. (1996) The impact of telemediated services on corporate structure: the example of branchless retail banking in Britain, *Environment and Planning A* 28: 1843–58.

Marshall, J. N., Wood, P., Daniels, P. W., McKinnon, A., Bachtler, J., Damesick, P., Thrift, N., Gillespie, A., Green, A. and Leyshon, A. (1988) *Services and Uneven Development*, Oxford: Oxford University Press.

Martin, R. L. (1994a) Stateless monies, global financial integration and national economic autonomy: the end of geography?, in Corbridge, S., Martin, R. L. and Thrift, N. J. (eds) *Money, Space and Power*, Oxford: Blackwell.

Martin, R. L. (1994b) Economic theory and human geography, in Gregory, D., Martin, R. L. and Smith, G. (eds) *Human Geography: Society, Space and Social Science*, Basingstoke: Macmillan.

Martin, R. L. (1999a) The new geographical turn in economics: some critical reflections, *Cambridge Journal of Economics* 23: 65–91.

Martin, R. L. (1999b) The new economic geography of money , in Martin, R. L. (ed.) *Money and the Space Economy*, Chichester: Wiley.

Martin, R. L. and Sunley, P. (1996) Paul Krugman's geographical economics and its implications for regional theory: a critical assessment, *Economic Geography* 77: 259–92.

Marx, K. (1859) Preface to *Contribution to the Critique of Political Economy*.

Marx, K. (1976) *Capital, vol. I*, Harmondsworth: Penguin.

Marx, K. and Engels, F. (1948) *The Communist Manifesto*, reprinted in D. McLellan (ed.), *Karl Marx: Selected Writings*, Oxford: Oxford University Press, 1977, p. 224.

Maskell, P. (1999) Globalisation and industrial competitiveness: the process and consequences of ubiquitification, in Malecki, E. J. and Oinas, P. (eds) *Making Connections: Technological Learning and Regional Economic Change*, Aldershot: Ashgate, 35–59.

Maskell, P., Eskelinen, H., Hannibalsson, I., Malmberg, A. and Vatne, E. (1998) *Competitiveness, Localised Learning and Regional Development: Specialisation and Prosperity in Small Open Economies*, London: Routledge.

Maskell, P. and Malmberg, A. (1999) The competitiveness of firms and regions: ubiquitification and the importance of localized learning, *European Urban and Regional Studies* 6(1): 9–25.

Massachusetts Technology Collaborative (1997) *Index of the Massachusetts Innovation Economy*, Westborough, MA: Massachusetts Technology Collaborative.

Massey, D. (1995) Places and their pasts, *History Workshop* 39: 182–93.

Massey, D. (1999) Space-time, science and the relationships between physical geography and human geography, *Transactions of the Institute of British Geographers* 24(3): 261–76.

Massey, D., Quintas, P. and Wield, D. (1992) *High Tech Fantasies: Science Parks in Society, Science and Space*, London: Routledge.

Mead, G. (1959) *The Philosophy of the Present* (first published 1932), (ed.) Murphy, A. E. Preface by J. Dewey, La Salle: Open Court.

Menzies, H. (1998) Challenging capitalism in cyberspace, in McChesney, R. W., Meiksins Wood, E. and Bellamy, J. F. (eds) *Capitalism and the Information Age*, New York: Monthly Review Press, 87–98.

Merleau-Ponty, M. (1945) *The Phenomenology of Perception*, New York: Humanities Press.

Merton, R. K (1973) The Matthew effect in science, in Merton, R. K. and Storer, N. W. (eds)

The Sociology of Science: Theoretical and Empirical Investigations, Chicago: University of Chicago Press.

Messner, D. (1997) *The Network Society: Economic Development and International Competitiveness as Problems of Social Governance*, London: Cass.

Metcalfe, J. S. and De Liso, N. (1998) Innovation, capabilities and knowledge: the epistemic connection, in Coombs, R., Green, K., Richards, A. and Walsh, V. (eds) *Technological Change and Organization*, Cheltenham: Edward Elgar, 8–27.

Meyer, D. R. (1998) Formation of advanced technology districts: New England textile machinery and firearms, *Economic Geography* extra issue: 31–45.

Meyrowitz, J. (1985) *No Hiding Place: The Impact of Electronic Media on Social Behavior*, New York: Oxford University Press.

Micklethwait, J. (1997) Future perfect? A survey of Silicon Valley, *The Economist* 29 March.

Milkman, R. (1997) *Farewell to the Factory: Autoworkers in the Late Twentieth Century*, Berkeley, CA: University of California Press.

Miller, D. (1997) *Capitalism: an Ethnographic Approach*, Oxford: Berg.

Miller, D. (1998) Conclusion: a theory of virtualism, in Carrier, J. and Miller, D. (eds) *Virtualism: A New Political Economy*, Oxford: Berg, 187–215.

Miller, R. and Coté, M. (1987) *Growing the Next Silicon Valley*, Lexington, MA: Lexington Books.

Millward, N., Bryson, A. and Forth, J. (2000) *All Change at Work?*, London: Routledge.

Mintel (1996) *Ethnic Food Report* (March) London: Mintel Market Intelligence.

Mintel (1997) *Ethnic Food Report* (May) London: Mintel Market Intelligence.

Mol, A. (1998) Missing links, making links: the performance of some artheroscleroses, in Mol, A. and Berg, M. (eds) *Differences in Medicine: Unravelling Practices, Techniques and Bodies*, Durham, NC and London: Duke University Press, 144–65.

Mol, A. (2000) *The Body Multiple: Ontology in Medical Practice*, Durham, NC and London: Duke University Press, forthcoming.

Mol, A. and Law, J. (1994) Regions, networks and fluids: anaemia and social topology, *Social Studies of Science* 24: 641–71.

More, R. A. (1985) Barriers to innovation: intraorganizational dislocations, *Journal of Product Innovation Management* 3: 205–8.

Morello, G. (1988) Business requirements and future expectations in comparative bank services – the issue of time perceptions, Research for Financial Services Seminar, Milan: ESOMAR, 235–45.

Morgan, B. (1990) Transferring soft technology, in Robinson, R. D. (ed.) *The International Communication Technology*, New York: Taylor and Francis, 149–66.

Morgan, G. (1992) Marketing discourse and practice: towards a critical analysis, in Alvesson, M. and Willmott, H. (eds) *Critical Management Studies*, London: Sage.

Morgan, K. (1995) The learning region, institutions, innovation, and regional renewal, *Papers in Planning Research 157*.

Morgan, K. (1997) The learning region: institutions, innovation and regional renewal, *Regional Studies* 31: 491–503.

Morley, D. (1992) *Television, Audiences and Cultural Studies*, London: Routledge.

Morley, D. and Robins, K. (1995a) *Spaces of Identity: Global Media, Electronic Landscapes and Cultural Boundaries*, London: Routledge.

Morley, D. and Robins, K. (1995b) Tradition and translation: national culture in its global context, in Morley, D. and Robins, K., *Spaces of Identity, Global Media, Electronic Landscapes and Cultural Boundaries*, London: Routledge, 105–24.

Mort, F. (1988) Boys Own? Masculinity, style and popular culture, in Chapman, R. and Rutherford, J. (eds) *Male Order: Unwrapping Masculinity*, London: Lawrence and Wishart.

Mort, F. (1996) *Cultures of Consumption: Masculinities and Social Space in Late Twentieth-Century Britain*, London: Routledge.

Moulaert, F. and Djellal, F. (1995) Information technology consulting firms: economies of agglomeration from a wide-area perspective, *Urban Studies* 32: 105–22.

Moulaert, F. and Gallouj, C. (1993) The locational geography of advanced producer service firms: the limits of economies of agglomeration, *Service Industries Journal* 13: 91–106.

Moulaert, F., Scott, A. J. and Farcy, H. (1997) Producer services and the formation of urban space, in Moulaert, F. and Scott, A. J. (eds) *Cities, Enterprises and Society on the Eve of the 21st Century*, London: Pinter, 97–112.

Moulaert, F. and Tödtling, F. (eds) (1995) The geography of advanced producer services in Europe, *Progress in Planning* 43 (2–3): 89–274.

Moussouris, L. (1998) The higher education–economic development connection in Massachusetts: forging a critical linkage?, *Higher Education* 35: 91–112.

Muchembled, R. (1978) *Culture Populaire et Culture des Elites dans la France Moderne (XVᵉ–XVIIIᵉ Siècles)*, Paris: Flammaron.

Mulgan, G. (1997) *Connexity. How to Live in a Connected World*, London: Chatto and Windus.

Murdoch, J. (1997) Inhuman/nonhuman/human: actor-network theory and the prospects for a nondualistic and symmetrical perspective on nature and society, *Environment and Planning D: Society and Space* 15: 731–56.

Nadworny, M. J. (1955) *Scientific Management and the Unions, 1900–32*, Boston: Harvard University Press.

Nahapiet, J. and Ghoshal, S. (1998) Social capital, intellectual capital, and the organizational advantage, *Academy of Management Review* 23: 242–66.

Neef, D. (ed.) (1998) *The Knowledge Economy*, Boston: Butterworth-Heinemann.

Negroponte, N. (1995) *Being Digital*, London: Hodder and Stoughton.

Nelkin, D. (1994) Promotional metaphors and their popular appeal, *Public Understanding of Science* 3: 25–31.

Nelson, D. (1974) Scientific management, systematic management, and labor, 1880–1915, *Business History Review* 28: 479–500.

Nelson, R. R. (1980) Production sets, technological knowledge, and R&D: fragile and overworked constructs for analysis of productivity growth?, *American Economic Review* 70(2): 62–7.

Nelson, R. R. (1981) Research on productivity growth and productivity differences: dead ends and new departures, *Journal of Economic Literature* 19(3): 1029–64.

Nelson, R. R. (1988) Institutions supporting technical change in the United States, in Dosi, G. *et al. Technical Change and Economic Theory*, London: Pinter, 312–29.

Nelson, R. R. (1994) An agenda for formal growth theory, New York: Columbia University Department of Economics, unpublished paper (communicated by the author).

Nelson, R. R. (1995) Recent evolutionary theorizing about economic change, *Journal of Economic Literature* 33: 48–90.

Nelson, R. R. and Winter, S. G. (1982) *An Evolutionary Theory of Economic Change*, Cambridge, MA: Harvard University Press.

Nevis, E. C., di Bella, A. J. and Gould, J. M. (1995) Organizations as learning systems, *Sloan Management Review* 36(2): 73–85.

Nicosia, F., and Myer, R. (1976) Towards a sociology of consumption, *Journal of Consumer Research* 3 (September), 65–75.

Noakes, S. (1997) Consumer spice, *Logistics Manager* November/December, 6–7.

Nonaka, I. (1991) The knowledge-creating company, *Harvard Business Review* 69: 96–104.

Nonaka, I. (1994) A dynamic theory of organizational knowledge creation, *Organization Science* 5, 14–37.

Nonaka, I., Byosiere, P., Borucki, C. C. and Konno, N. (1994) Organisational knowledge creation theory: a first comprehensive test, *International Business Review* 3: 337–51.

Nonaka, I. and Takenchi, H. (1995a) *Theory of Organizational Knowledge Creation*, Oxford: Oxford University Press.

Nonaka, I. and Takeuchi, H. (1995b) *The Knowledge Creating Company: How Japanese Companies Create the Dynamics of Innovation*, Oxford: Oxford University Press.

Norton, B. and Smith, C. (1999) *Understanding Management Gurus*, London: Hodder and Stoughton.

Nowotny, H. (1988) From the future to the extended present time in social systems, in Kirsch, G., Nijkamp, P. and Zimmermann, K. (eds) *The Formation of Time Preferences in Multidisciplinary Perspectives*, Aldershot: Gower.

Nowotny, H. (1992) Time and social theory: towards a social theory of time, *Time and Society* 1: 421–54.

Nye, D. E. (1994) *American Technological Sublime*, Cambridge, MA: MIT Press.

O'Brien, R. (1992) *Global Financial Integration: The End of Geography*, London: Pinter Publishers.

Oakey, R. P. (1981) *High Technology Industry and Industrial Location*, Aldershot: Gower.

Odih, P. (1998) *Gendered Time and Financial Services Consumption*, PhD thesis, FSRC Manchester: UMIST, School of Management.

Odih, P. (1999) Gendered time in the age of deconstruction, *Time and Society* 8(1): 9–39.

Ody, P. (1997) Log on to your customers, *In-store Marketing* November: 32–3.

OECD (1995) *Industry and Technology: Scoreboard of Indicators 1995*, Paris: Organisation for Economic Co-operation and Development.

Ohmae, K. (1985) *Triad Power: The Global Shape of Coming Competition*, New York: Free Press.

O'hUallachain, B. (1998) Innovation and invention in the American urban system of technical advance, paper presented at the North American Meetings of the Regional Science Association International, Santa Fe, NM, November.

Oinas, P. (1995) Organisations and environment: linking industrial geography and organisation theory, in Conti, S., Malecki, E. J. and Oinas, P. (eds) *The Industrial Enterprise and Its Environment: Spatial Perspectives*, Aldershot: Avebury, 143–67.

Oinas, P. (1998) Competitive advantage and the role of regional culture: towards reconstructing a narrative, mimeo, Rotterdam: Economic Geography/Applied Economics, Erasmus University.

Oinas, P. and Malecki, E.J. (1999) Technology in space: tracing spatial systems of innovation, in Malecki, E. J. and Oinas, P. (eds) *Making Connections: Technological Learning and Regional Economic Change*, Aldershot: Ashgate, 7–33.

Oinas, P. and Virkkala, S. (1997) Learning, competitiveness and development: reflections on the contemporary discourse on learning regions, in Eskelinen, H. (ed.) *Regional Specialisation and Local Environment – Learning and Competitiveness*, Copenhagen: NordREFO, 263–77.

Origo, I. (1988) *The Merchant of Prato: Francesco di Marco Datini*, London: Penguin.

Orwell, G. (1954) *Nineteen Eighty-Four*, Harmondsworth: Penguin.

Osborne, P. (1995) *The Politics of Time: Modernity and the Avant-Garde*, London: Verso.

O'Shea, J. and Madigan, C. (1997) *Dangerous Company: The Consulting Powerhouses and the Businesses They Save and Ruin*, New York: Times Business/Random House.

O'Sullivan, P. (1982) Antidomino, *Political Geography Quarterly* 1: 1.

O'Tuathail, G. (1996) *Critical Geopolitics*, London: Routledge.

O'Tuathail, G. (1998) Introduction: thinking critically about geopolitics, in O'Tuathail, G., Dalby, S. and Routledge, P. (eds) *The Geopolitics Reader*, London and New York: Routledge, 1–12.

O'Tuathail, G., Dalby, S. and Routledge, P. (eds) (1998) *The Geopolitics Reader*, London and New York: Routledge.

Oudshoorn, N. (1996) A natural order of things? Reproductive sciences and the politics of othering, in Robertson, G., Mash, M., Tickner, L., Bird, J., Curtis, B. and Putman, T. (eds) *FutureNatural: Nature, Science and Culture*. London: Routledge, 122–32.

Owens, J. (1997) Category management boom foreseen, *SN* 1 September: 6 and 70.

Pahl, R. (1995) *After Success: Fin-de-Siècle Anxiety and Identity*, Cambridge: Polity Press.

Parker, M. and Slaughter, J. (1988) *Choosing Sides: Unions and the Team Concept*, Boston: South End Press.

Pavitt, K. (1991) Key characteristics of the large innovating firm, *British Journal of Management* 2: 41–50.

Peck, J. (1996) *Work-Place: The Social Regulation of Local Labour Markets*, London: Guildford.

Penrose, E. (1959) *The Theory of the Growth of the Firm*, Oxford: Basil Blackwell.

Perkin, H. (1996) *The Third Revolution: Professional Elites in the Modern World*, London: Routledge.

Peters, T. (1992) *Liberation Management: Necessary Disorganization for the Nanosecond Nineties*, London: Macmillan.

Pettigrew, A. (1985) *The Awakening Giant: Continuity and Change in Imperial Chemicals Industries*, Oxford: Blackwell.

Pfeffer, J. (1998) Seven practices of successful organizations, *California Management Review* 40(2): 96–124.

Philo, C. (compiler) (1991) *New Words, New Worlds: Reconceptualising Social and Cultural Geography*, Aberystwyth: Cambrian Printers.

Pinch, S. and Henry, N. (1999a) Discursive aspects of technological innovation: the case of the British motor sport industry, *Environment and Planning A* 31: 665–82.

Pinch, S. and Henry, N. (1999b) Paul Krugman's geographical economics, industrial clustering and the British motor sport industry, *Regional Studies* 33(9): 815–27.

Pinch, S. Henry, N. and Turner, D. (1997) In pole position: explaining the development of Britains Motor Sport Valley, in Brown, B. J. H. (ed.) *Explorations in Motoring History: The Proceedings of the First United Kingdom History of Motoring Conference 12th October 1996*, Oxford: Oxbow Books.

Plant, S. (1997) *Zeros and Ones: Digital Women and the New Technostructure*, London: Fourth Estate, 46.

Plotkin, H. (1994) *The Nature of Knowledge: Concerning Adaptations, Instinct and the Evolution of Intelligence*, London: Allen Lane.

Podolny, J. M. and Stuart, T. E. (1995) A role-based ecology of technological change, *American Journal of Sociology* 100(5): 1224–60.

Polanyi, K. (1957 [1944]) *The Great Transformation: The Political and Economic Origins of Our Time*, Boston: Beacon Press.

Polanyi, M. (1958) *Personal Knowledge: Towards a Post-Critical Philosophy*. London: Routledge & Kegan Paul.

Polanyi, M. (1966) The logic of tacit inference, *Philosophy* 41: 1–18.

Polanyi, M. (1967) *The Tacit Dimension*, London: Routledge and Kegan Paul.

Pollard, S. (1965) *The Genesis of Modern Management: A Study of the Industrial Revolution in Great Britain*, Cambridge MA: Harvard University Press.

Poole, S. (1997) *Vision, Race and Modernity: A Visual Economy of the Andean Image World*, Princeton, NJ: Princeton University Press.

Popkin, R. H. (1979) *The History of Scepticism from Erasmus to Spinoza*, Berkeley, CA: University of California Press.

Porat, M. (1977) *The Information Economy: Definition and Measurement*, Washington, DC: US Department of Commerce, Office of Telecommunications Publications 77–12 (1).

Porritt, J. (1998) Putting her awe in, *Times Higher Educational Supplement* 1 May: 29.

Porter, M. E. (1985) *Competitive Advantage*, New York: Free Press.

Porter, M. E. (1990) *The Competitive Advantage of Nations*, New York: Free Press.

Porter, M. E. and Millar V. E. (1985) How information gives you competitive advantage, reprinted in Porter, M. E. (1998) *On Competition*, Boston: Harvard Business School Press.

Porter, R. (1997) *The Greatest Benefit to Mankind: A Medical History of Humanity from Antiquity to the Present Day*, London: HarperCollins.

Postrel, V. I. (1997) Resilience vs. anticipation, *Forbes ASAP* 25 August, 57–61, 94.

Poulantzas, N. (1978) *State, Power, Socialism*, London: New Left Books.

Prahalad, C. and Hamel, G. (1990) The core competence of the corporation, *Harvard Business Review* 68: 79–91.

Pratt, D. J. (1998) Re-placing money: the evolution of branch banking in Britain, *Environment and Planning A* 30(12): 2211–26.

Pred, A. R. (1966) *The Spatial Dynamics of US Industrial Growth, 1800–1914*, Cambridge, MA: MIT Press.

Prins, G. (ed.) (1996) *Threats Without Enemies: Facing Environmental Insecurity*, London: Earthscan.

Proctor, J. (1998) Ethics in geography: giving moral form to the geographical imagination, *Area* 30: 8–18.

Purcell, J. (1993) The challenge of Human Resource Management for industrial relations research and practice, *The International Journal of Human Resource Management* 4: 511–27.

Quah, D. T. (1996) *Growth and Dematerialisation: Why Non-Stick Frying Pans Have Lost their Edge*, London: CEP London School of Economics.

Raban, J. (1991) *Soft City*, London: Fontana.

Rabinow, P. (1989) *French Modern: Norms and Forms of the Social Environment*, Cambridge, MA: MIT Press.

Ramsay, H. (1990) Re-inventing the wheel? A review of the development and performance of employee involvement, *Human Resource Management Journal* 1: 1–22.

Rattansi, A. (1997) Postcolonialism and its discontents, *Economy and Society* 26: 480–500.

Ravetz, J. R. (1971) *Scientific Knowledge and its Social Problems*, Oxford: Clarendon Press.

Ray, C., (1991) *Time, Space and Philosophy*, London: Routledge.

Reich, R. B. (1991) *The Work of Nations*, New York: Random House.

Reimer, S. (1998) Working in a risk society, *Transactions of the Institute of British Geographers* 23: 116–27.

Reisner, M. (1987) *Cadillac Desert: The American West and Its Disappearing Water*, New York: Penguin.

Richardson, G. B. (1972) The organization of industry, *Economic Journal* 82: 372–83.

Richardson, H. W. (1973) *Regional Growth Theory*, London: Macmillan.

Ridgeway J. (1997) Heritage on the Hill: the rights pre-eminent P.R. machine, *The Nation*, 22 December: 11–18.

Ring, P. S. (1992) Cooperating on tacit know-how assets, chapter presented at the First Annual Meeting of the International Federation of Scholarly Association of Management, Tokyo, September 1992.

Rist, G. (1997) *The History of Development: From Western Origins to Global Faith*, London and New York: Zed Books.

Rivest, C. (1996) Voluntary European Works Councils, *European Journal of Industrial Relations* 2(2): 235–53.

Roberts, E. B. (1991) *Entrepreneurs in High Technology: Lessons from MIT and Beyond*, New York: Oxford University Press.

Robins, K. (1991) Prisoners of the city: whatever could a postmodern city be?, *New Formations* 15(1): 1–22.

Robins, K. and Gillespie, A. (1992) Communication, organisation and territory, in Robins, K. (ed.) *Understanding Information: Business, Technology, and Geography*, London: Belhaven, 148–62.

Robins, K. and Webster, F. (1987) Information as capital: a critique of Daniel Bell, in Slack, J. and Fejes, F. (eds) *The Ideology of the Information Age*, Norwood, NJ: Ablex Publishing, 95–117.

Rogers, A. (1992) The boundaries of reason: the world, the homeland, and Edward Said, *Environment and Planning D: Society and Space* 10: 511–26.

Rogers, J. and Streeck, W. (1995) *Works Councils: Consultation, Representation and Co-operation in Industrial Relations*, Chicago: University of Chicago Press.

Rosegrant, S. and Lampe, D. R. (1992) *Route 128: Lessons from Boston's High-Tech Community*, New York: Basic Books.

Rosenberg, N. (1963) Technical change in the machine tool industry, 1840–1910, *Journal of Economic History* 23: 414–43.

Rosenberg, N. (1982) *Inside the Black Box: Technology and Economics*, Cambridge: Cambridge University Press.

Ruggles, R. L. (1997) Tools for knowledge management: an introduction, in Ruggles, R. L. (ed.) *Knowledge Management Tools: Resources for the Knowledge-Based Economy*, Boston: Butterworth-Heinemann, 1–8.

Rushkoff, D. (1996) *Playing the Future: How Kid's Culture Can Teach Us to Thrive in an Age of Chaos*, London: HarperCollins.

Sabel, C. F. (1992) Studied trust: building new forms of co-operation in a volatile economy, in Pyke, F. and Sengenberger, W. (eds) *Industrial Districts and Local Economic Regeneration*, Geneva: International Institute for Labour Studies, 215–50.

Sampson, P. (1994) Postmodernity, in Sampson, P., Samuel, V. and Sugden, C. (eds) *Faith and Modernity*, Oxford: Regnum, 29–57.

Samuelson R. J. (1998) Why we are all married to the market, *Newsweek* CXXXI, 17, 27 April: 32–6.

Sanchez, R. and Mahoney, J. T. (1996) Modularity, flexibility and knowledge management in product and organisation design, *Strategic Management Journal* 17, Winter Special Issue: 63–76.

Sassen, S. (1991) *The Global City: New York, London, Tokyo*, Princeton, NJ: Princeton University Press.

Sassen, S. (1996) *Losing Control? Sovereignty in an Age of Globalisation*, New York: Columbia University Press.

Saurin, J. (1993) Global environmental degradation, modernity and environmental knowledge, *Environmental Politics* 2(4): 46–64.

Saussure, F. de (1974) *Course in General Linguistics*, London: Fontana.

Saviotti, P. P. (1998) On the dynamics of appropriability of tacit and of codified knowledge, *Research Policy* 26: 843–56.

Saxenian, A. (1994) *Regional Advantage: Culture and Competition in Silicon Valley and Route 128*, Cambridge, MA: Harvard University Press.

Sayer, A. and Storper, M. (1997) Ethics unbound: for a normative turn in social theory, *Environment and Planning D: Society and Space* 15: 1–17.

Scannell, P. (1996) *Radio, Television and Modern Life: A Phenomenological Acccount*, Oxford: Blackwell.

Schiffman, L. and Kanuk, L. (1991) *Consumer Behaviour*, 4th edn, Englewood Cliffs NJ: Prentice Hall International.

Schiller, D. (1988) How to think about information, in Mosco, V. and Wasko, J. (eds) *The Political Economy of Information*, Madison, WI: University of Wisconsin Press, 27–44.

Schlegel, J. (1998) The ABCs of ADRs, *Individual Investor* 17(4) April: 61.

Schoenberger, E. (1994) Corporate strategy and corporate strategists: power, identity, and knowledge within the firm, *Environment and Planning A* 26: 435–51.

Schoenberger, E. (1997) *The Cultural Crisis of the Firm*, Oxford: Blackwell.

Schon, D. A. (1983) *The Reflective Practitioner: How Professionals think in Action*, New York: Basic Books.

Schon, D. A. (1991) *The Reflective Practitioner: How Professionals Think in Action*, Aldershot: Avebury.

Schumpeter, J. A. (1939) *Business Cycles: A Theoretical, Historical and Statistical Analysis of the Capitalist Process*, New York: McGraw-Hill.

Schur, E (1971) *Labelling Deviant Behaviour*, New York: Harper and Row.

Scott, A. (1994) *Willing Slaves? British Workers under Human Resource Management*, Cambridge: Cambridge University Press.

Scott, A. J. (1988) *Metropolis: From the Division of Labour to Urban Form*, Berkeley, CA: University of California Press.

Scott, J. (1997) *Corporate Business and Capitalist Classes*, Oxford: Oxford University Press.

Segal, Quince and Wicksteed (SQW) (1985) *The Cambridge Phenomenon: The Growth of High Technology Industry in a University Town*, SQW, Cambridge.

Seidman, S. (ed.) (1994) *The Postmodern Turn: New Perspectives on Social Theory*, Cambridge: Cambridge University Press.

Seidman, S. (1998) *Contested Knowledge: Social Theory in the Postmodern Era*, Oxford: Blackwell.

Senge, P. (1990) The leader's new work: building learning organisations, *Sloan Management Review* 32: 7–23.

Senker, J. (1995) Tacit knowledge and models of innovation, *Industrial and Corporate Change* 4: 425–47.

Sennett, R. (1998) *The Corrosion of Character: The Personal Consequences of Work in the New Capitalism*, New York: Norton.

Serres, M. (1988) Ambrosia and gold, in Bryson, N. (ed.) *Calligram, Essays in New Art History from France*, Cambridge: Cambridge University Press, 116–30.

Serres, M. (1991) *Rome: The Book of Foundations*, Stanford, CA: Stanford University Press.

Shakespeare, T. (1995) Disabled people and the new genetics, *Genethics News* 5: 8–11.

Shield, R. (1992) *Lifestyle Shopping: The Subject of Consumption*, London: Routledge.

Shilling, C. (1993) *The Body and Social Theory*, London: Sage.

Shiva, V. (1993) *Monocultures of the Mind: Perspectives on Biodiversity and Biotechnology*, London: Zed Books.

Shusterman, R. (1997) *Practicing Philosophy: Pragmatism and the Philosophical Life*, New York and London: Routledge.

Sidaway J. D. (1998) What is in a Gulf? From the arc of crisis to the Gulf War, in O'Tuathail, G. and Dalby, S. (eds) *Rethinking Geopolitics*, London and New York: Routledge, 224–39.

Simon, D. (1998) Rethinking (post) modernism, postcolonialism and posttraditionalism: South–North perspectives, *Environment and Planning D: Society and Space* 16: 219–45.

Sklair, L. (1991) *Sociology of the Global System*, Hemel Hempstead: Harvester.

Slater, D. (1997) Spatialities of power and postmodern ethics – rethinking geopolitical encounters, *Environment and Planning D: Society and Space* 15: 55–72.

Slaughter, S. (1993) Innovation and learning during implementation: a comparison of user and manufacturer innovations, *Research Policy* 22: 81–95.

Slouka, M. (1995) *War of the Worlds: Cyberspace and the High-Tech Assault on Reality*, London: Abacus, 30.

Smart, B. (1992) *Modern Conditions, Postmodern Controversies*, London: Routledge.

Smith, A. (1976 [1776]) *An Inquiry into the Nature and Causes of the Wealth of Nations, Volumes I and II*, Oxford: Clarendon Press.

Smith, D. (1997a) Geography and ethics: a moral turn?, *Progress in Human Geography* 21: 583 90.

Smith, D. (1997b) Back to the good life: towards an enlarged conception of social justice, *Environment and Planning D: Society and Space* 15: 19–35.

Soars, B. (1997) Focusing on the consumer, *In-Store Marketing* November: 25–6.

Sohn-Rethel, A. (1978) *Intellectual and Manual Labour*, London: Macmillan.

Sorokin, P. and Merton, R. (1990) The structures and meanings of social time, in Hassard, J. (ed.) *The Sociology of Time*, London: Macmillan.

Sparrow, J. (1996) *Knowledge in Organizations: Access to Thinking at Work*, London: Sage Publications.

Speed, R. and Smith G. (1992) Retail financial services segmentation, *Service Industries Journal* 12(3): 368–83.

Spender, J.-C. (1996a) Making knowledge the basis of a dynamic theory of the firm, *Strategic Management Journal* 17 (Winter): 45–62.

Spender, J.-C. (1996b) Competitive advantage from tacit knowledge? Unpacking the concept and its strategy implications in Moingen, B. and Edmondson, A. (eds) *Organizational Learning and Competitive Advantage*, London: Sage Publications, 56–73.

Spender, J.-C. and Grant, R. M. (1996) Knowledge and the firm: overview, *Strategic Management Journal* 17: 5–9.

Stahel, A. W. (1999) Time contradictions of capitalism, *Culture, Nature, Society* 10(1): 101–32.

Stallings, B. (1993) *The New International Context for Development*, Madison, WI: University of Wisconsin, Working Paper Series on the International Context of Development, no. 1.

Steinberg, J. (1998) Following freedom, *Individual Investor* 17 April, 4: 61.

Stern, P. and Fineberg, H. (eds) (1996) *Understanding Risk: Informing Decisions in a Democratic Society*, Washington, DC: National Academy Press.

Stiglitz, J. E. (1987) Learning to learn, localized learning and technological progress, in Dasgupta, P. and Stoneman, P. (eds) *Economic Policy and Industrial Performance*, Cambridge: Cambridge University Press, 125–53.

Stiglitz, J. and Weiss, A. (1981) Credit rationing in markets with imperfect information, *American Economic Review* 71: 393–410.

Stoneman, P. (1995) Introduction, in Stoneman, P. (ed.) *Handbook of Economics of Innovation and Technological Change*, Oxford: Basil Blackwell, 1–13.

Stork, J. (1997) Bahrain's crisis worsens, *Middle East Report* No. 204, 27, 3, (July–September): 33–5.

Storper, M. (1995) The resurgence of regional economies ten years later: the region as a nexus of untraded interdependencies, *Journal of European Urban and Regional Studies* 2: 191–221.

Storper, M. (1997a) *The Regional World: Territorial Development in a Global Economy*, New York and London: The Guilford Press.

Storper, M. (1997b) The spatial and temporal constitution of social action, in Bryant, C. and Jary, D. (eds) *Anthony Giddens: Critical Assessments*, London: Routledge.

Storper, M. (1998) *The Regional World*, New York: The Guilford Press.

Storper, M. and Christopherson, S. (1987) Flexible specialisation and regional industrial agglomerations: the case of the U.S. motion picture industry, *Annals of the Association of American Geographers* 77(1): 104–17.

Storper, M. and Salais, R. (1997) *Worlds of Production: The Action Frameworks of the Economy*, Cambridge, MA: Harvard University Press.

Storper, M. and Scott, A. J. (1995) The wealth of regions: market forces and policy imperatives in local and global context, *Futures* 27: 505–26.

Strange, S. (1998) *Mad Money*, Manchester: Manchester University Press.

Strange, S. (1999) *Mad Money: When Markets Outgrow Governments*, Ann Arbor, MI: University of Michigan Press.

Strathern, M. (1996) Enabling identity? Choice and the new reproductive technologies, in Hall, S. and Du Gay, P. (eds) *Questions of Cultural Identity*, London: Sage, 37–52.

Stratton, J. and Ang, I. (1996) On the impossibility of a global cultural studies. British cultural studies in an international frame, in Morley, D. and Chen, K.-H. (eds) *Stuart Hall: Critical Dialogues in Cultural Studies*, London: Routledge, 361–91.

Strauss, E. (1966) Human traces, in Strauss, E. (ed.) *Phenomenological Psychology*, New York: Basic Books.

Streeck, W. (1995) Works councils in Western Europe: from consultation to participation, in Rogers, J. and Streeck, W. (eds) *Works Councils: Consultation, Representation and Co-operation in Industrial Relations*, Chicago: University of Chicago Press, 313–48.

Sum, N.-L. (1998) Theorizing the development of East Asian Newly Industrializing Countries: a regulationist perspective, in Cook, I. *et al.* (eds) *Dynamic Asia*, Aldershot: Ashgate, 44–78.

Suzuki, D. (1989) *Genethics: The Ethics of Engineering Life*, London: Unwin Hyman.

Sweeney, G. P. (1987) *Innovation, Entrepreneurs and Regional Development*, New York: St Martins Press.

Sweeney, G. P. (1996) Learning efficiency, technological change and economic progress, *International Journal of Technology Management* 11: 5–27.

Szulanski, G. (1996) Exploring internal stickiness: impediments to the transfer of best practice within the firm, *Strategic Management Journal* 17 (Winter): 27–43.

Talbott S. (1992) Beware of the three-way split, *Time* 15 June: 39.

Taylor, F. W. (1934) *The Principles of Scientific Management*, New York: Harper and Brothers.

Taylor, M. C. (1993) *Nots*, Chicago: University of Chicago Press.

Taylor, M. J. (1975) Organisational growth, spatial interaction and location decision-making, *Regional Studies* 9: 213–23.

Taylor, P. (1996) Embedded statism and the social sciences: opening up to new spaces. *Environment and Planning A* 28: 1917–28.

Teece, D. J. (1982) Towards an economic theory of the multiproduct firm, *Journal of Economic Behaviour and Organization* 3: 39–63.

Teece, D. J. (1986) Profiting from technological innovation: implications for integration, collaboration, licensing and public policy, *Research Policy* 15: 285–305.

Teece, D. J. (1987) Capturing value from technological innovation: integration, strategic partnering and licensing decisions, in Brooks, H. and Guide, B. R. (eds) *Technology and Global Industry*, Washington, DC: National Academy Press.

Teece, D. J. and Pisano, G. (1994) The dynamic capabilities of firms: an introduction, *Industrial and Corporate Change* 3: 537–56.

Thelan, K. (1993) *Union of Parts: Labor Politics in Post-war Germany*, Ithaca and London: Cornell University Press.

Thomas, K. (1978) *Religion and the Decline of Magic*, Harmondsworth: Peregrine.

Thomas, R., Saren, M. and Ford, D. (1994) Technology assimilation in the firm: managerial perceptions and behaviour, *International Journal of Technology Management* 9 (2): 227–40.

Thompson, E. P. (1967) Time, work, discipline and industrial capitalism, *Past and Present* 38: 56–97.

Thompson, J. K. and Rehder, R. R. (1995) Nissan U.K.: a workers' paradox?, *Business Horizons* 38(1): 48–58.

Thrift, N. J. (1985) Flies and germs: a geography of knowledge, in Gregory, D. and Urry, J. (eds) *Social Relations and Spatial Structures*, London: Macmillan, 330–73.

Thrift, N. J. (1989) The perils of transition models, *Environment and Planning D: Society and Space* 7: 127–9.

Thrift, N. J. (1994) On the social and cultural determinants of international financial centres: the case of the City of London, in Corbridge, S., Martin, R. and Thrift, N. (eds) *Money, Space and Power*, Oxford: Basil Blackwell, 327–55.

Thrift, N. J. (1996a) *Spatial Formations*, London, Thousand Oaks and New Delhi: Sage.

Thrift, N. J. (1996b) The rise of soft capitalism, *Cultural Values* 1: 27–57.

Thrift, N. J. (1998a) The rise of soft capitalism, in Herod, A., O'Tuathail, G. and Roberts, S. (eds) *An Unruly World? Globalisation, Governance and Geography*, London: Routledge, 25–71.

Thrift, N. J. (1998b) Virtual capitalism: the globalisation of reflexive business knowledge, in Carrier, J. and Miller, D. (eds) *Virtualism: A New Political Economy*, Oxford: Berg, 161–86.

Thrift, N. J. (1999) Steps to an ecology of place, in Massey, D., Allen, J. and de Sarre, P., *Human Geography Today*, Cambridge: Polity Press.

Thrift, N. J. and Olds, K. (1996) Refiguring the 'economic' in economic geography, *Progress in Human Geography* 20(3): 311–37.

Thurley, K. and Wirdenuis, H. (1973) *Supervision: A Re-Appraisal*, London: Heinemann.

Tiles, M. (1996) A science of Mars or Venus?, in Fox Keller, E. and Longino, H. (eds) *Feminism and Science*, Oxford: Oxford University Press, 220–34.

Tisdall, P. (1982) *Agents of Change: The Development and Practice of Management Consultancy*, London: Heinemann.

Tödtling, F. (1994) The uneven landscape of innovation poles: local embeddedness and global networks, in Amin, A. and Thrift, N. J. (eds) *Globalization, Institutions and Regional Development in Europe*, Oxford: Oxford University Press, 68–90.

Toffler, A. (1980) *The Third Wave*, London: Collins.

Toffler, A. (1990) *Powershift: Knowledge, Wealth and Violence at the Edge of 21st Century*, New York: Bantam Books.

Tolliday, S. (1983) Militancy and organisation: women workers and trade unions in the motor trades in the 1930s, *Oral History* 48.

Törnqvist, G. (1983) Creativity and the renewal of everyday life, in Buttimer, A. (ed.) *Creativity and Context* (Lund Studies in Geography, Series B, no. 50), Lund: Gleerup, 91–112.

Tronto, J. (1993) *Moral Boundaries: A Political Argument for an Ethic of Care*. London: Routledge.

Truett Anderson, W. T. (1997) *Evolution Isn't What it Used To Be: The Augmented Animal and the Whole Wired World*, London: W. H. Freeman.

Tsering, Y. (1995) China aims to improve health of newborn by law, *British Medical Journal* 310: 1538.

Tsoukas, H. (1996) The firm as a distributed knowledge system: a constructionist approach, *Strategic Management Journal* 17 Winter Special Issue: 11–25.

TUC (1997) *European Works Councils: A TUC Guide for Trade Unionists*, London: Labour Research Department.

Turnbull, D. (1999), *On With the Motley: The Contingent Assemblage of Knowledge Spaces*, London: Harwood.

Turner, B. (1993) *Citizenship and Social Theory*, London: Sage.

Turpin, T., Garrett-Jones, S. and Rankin, N. (1996) Bricoleurs and boundary riders: managing basic research and innovation knowledge networks, *R&D Management* 26: 267–80.

Twitchen, A. (2000) The body's second skin: forming the protective community of Grand Prix motor racing, in *Organising Bodies*, Basingstoke: Macmillan.

Ullman, E. L. (1958) Regional development and the geography of concentration, *Papers and Proceedings of the Regional Science Association* 4: 179–198.

United Nations (1993) *Management Consulting: A Survey of the Industry and its Largest Firms*, United Nations Conference on Trade and Development.

Urlich, D., Jick, T. and von Glinow, M. A. (1993) High-impact learning: building and diffusing learning capability, *Organizational Dynamics* 22(2): 52–66.

Urry, J. (1995) *Consuming Places*, London: Routledge.

Urry, J. (2000) Mobile Sociology, *British Journal of Sociology*.

Urwick, L. and Brech, E. F. L. (1945) *The Making of Scientific Management, Vol 1: Thirteen Pioneers*, London: Management Publications Trust.

Urwick, L. and Brech, E. F. L. (1953) *The Making of Scientific Management, Vol 2: Management in British Industry*, London: Pitman and Sons.

Vaessen, P. and Keeble, D. (1995) Growth-oriented SMEs in unfavourable regional environments, *Regional Studies* 29: 489–505.

van der Pijl, K. (1998) *Transnational Classes and International Relations*, London: Routledge.

Vaughan, M. (1991) *Curing their Ills: Colonial Power and African Illness*, Cambridge: Polity Press.

Veltz, P. (1996) *Mondialisation de Villes et Territoires: l'Économie Archipel*, Paris: Presses Universitaires de France.

Venkatesan, J. and Anderson, B. (1985) Time budgets and consumer services, in Bloch, T. M., Block, G. D., Chan, V. A. and Zeitmal, A. M. (eds) *Services Marketing in a Changing Environment*.

Verene, D. P. (1979) *Symbol, Myth and Culture: Essays and Lectures of Ernst Cassirer, 1935–45*, Newhaven and London: Yale University Press.

Vestner C. (1998) Freedom markets, *Individual Investor* 17(4), April: 46–61.

Vincenti, W. G. (1984) Technological knowledge without science: the innovation of flush riveting in American airplanes, ca. 1930–ca. 1950, *Technology and Culture* 25: 540–76.

Vincenti, W. G. (1990) *What Engineers Know and How They Know It: Analytical Studies from Aeronautical History*, Baltimore, MD: Johns Hopkins University Press.

Von Glinow, M. A. (1988) *The New Professionals: Managing Todays High-Tech Employees*, Cambridge, MA: Ballinger.

Von Hippel, E. (1994) Sticky information and the locus of problem solving: implications for innovation, *Management Science* 40: 429–39.

Wade, R. and Veneroso, F. (1998) The gathering world slump and the battle over capital controls, *New Left Review* 231: 13–42.

Walker, R.A. (1995) Regulation and flexible specialisation as theories of capitalist development: challengers to Marx and Schumpeter?, in Liggett, D. C. and Perry, C. (eds) (1995) *Spatial Practices*, London: Sage.

Wallerstein, I, (1974a) *The Modern World System*, New York: Academic Press.

Wallerstein, I. (1974b), *The Modern World System 1: Capitalist Agriculture and the Origins of the European World Economy in the Sixteenth Century*, New York, San Francisco and London: Academic Press.

Warde, A. (1994) Consumption, identity-formation and uncertainty, *Sociology* 28(4): 877–98.

Warde, A. (1997) *Consumption, Food and Taste: Culinary Antinomies and Commodity Culture*, London: Sage.

Watson, J. (1968) *The Double Helix*, London: Atheneum Press.

Watson, J. (1990) The Human Genome Project: past, present and future, *Science* 248, April: 44–9.

Weiler, P. (1990) *Governing the Workplace: The Future of Labor and Employment Law*, Boston: Harvard Business School Press.

Wever, K. (1995) *Negotiating Competitiveness: Employment Relations and Organizational Innovation in Germany and the US*, Boston: Harvard Business School Press.

Whitworth, S. (1994) The International Planned Parenthood Federation, in *Feminism and International Relations. Towards a Political Economy of Gender in Interstate and Non-Governmental Organisations*, New York: St. Martins Press, 80–118.

Wilkes, T. (1993) *Perilous Knowledge: The Human Genome Project and its Implications*, London: Faber and Faber.

Williams, A. (1997) The postcolonial flaneur and other fellow-travellers: conceits for a narrative of redemption, *Third World Quarterly* 18: 821–41.

Williams, I. (1931) *The Firm of Cadbury*, London: Constable.

Williams, S. J. (1996) The limits of medicalisation? Modern medicine and the lay populace in late modernity, *Social Science and Medicine* 42: 1609–20.

Williamson, W. (1982) *Class, Culture and Community: A Biographical Study of Social Change in Mining*, London: Routledge.

Willke, H. (1997) *Supervision des Staates*, Frankfurt: Suhrkamp.

Wills, J. (1998) *Making the Best Of It? Managerial Attitudes Towards, and Experiences of European Works Councils in UK-owned Multi-national Firms*. Re-scaling Workplace Solidarity? The implications of European Works Councils for British Industrial Relations, Working Paper Three, Department of Geography, Southampton: University of Southampton.

Wills, J. (1999) Managers making the best of it: top down convergence through European Works Councils, *Human Resource Management Journal*, 9, 4: 19–38.

Wills, J. (2000) Great expectations: three years in the life of a European Works Council, *European Journal of Industrial Relations* 6: 83–105.

Wills, J. (2001a) Re-scaling trade union organisation: lessons from the European front-line, in Munck, R. (ed.) *Labour and Globalisation: Results and Prospects*. London: Macmillan.

Wills, J. (2001b) Uneven geographies of capital and labour: the lessons of European Works Councils, *Antipode* 33:3 and in Waterman, P. and Wills, J. (eds) *Space, Place and the New Labour Internationalisms*, Oxford: Blackwell.

Wills, J. and Lincoln, A. (1999) Filling the vacuum in new management practice? Lessons from American employee-owned firms, *Environment and Planning A* 30: 1497–512.

Wilson, D. and Holman, R. (1980) Economic theories of time in consumer behaviour, in Lamb, C. W. and Dunne, P. M. (eds) *Theoretical Developments in Marketing: Advances in Consumer Research*, Chicago: Chicago American Marketing Association, 265–8.

Wilson, J. (1999) *Successful Consultancy in a Week*, London: Hodder & Stoughton.

Wilson, M. I. (1995) The office farther back: business services, productivity, and the offshore back office, in Harker, P. T. (ed.) *The Service Productivity and Quality Challenge*, Dordrecht: Kluwer, 203–24.

Winter, S. (1987) Knowledge and competence as strategic assets, in Teece, D. (ed.) *The Competitive Challenge: Strategies for Industrial Innovation and Renewal*, Cambridge, MA: Ballinger, 159–84.

Wood, P., Bryson, J. and Rotherham, D. (1994) *Consultant Use and the Management of Change: Insights into Expertise*, Working Paper No. 3, Business Services Research Project, London: University College London.

Woolley, B. (1992) *Virtual Worlds: A Journey in Hype and Hyperreality*, Oxford: Blackwell.

World Bank (1997) *Private Capital Flows to Developing Countries*, Oxford: OUP/World Bank.

World Health Organization (1994) *Bridging the Gaps*, Geneva: WHO.

Worsley, P. (1984) *The Three Worlds: Culture and World Development*, London: Weidenfeld and Nicolson, ch. 2.

Wright, P. and Weitz, B. (1977) Time horizon effects on product evaluation strategies, *Journal of Marketing Research* 14 (November): 429–43.

Wynne, B. (1991) Knowledges in context, *Science, Technology and Human Values* 19: 1–17.

Wynne, B. (1995) The public understanding of science, in Jasanoff, S., Markel, G., Petersen, J. and Pinch, T. (eds) *Handbook of Science and Technology Studies*, London: Sage, 361–88.

Zelizer, V. A. (1989) The social meanings of money: special monies, *American Journal of Sociology* 95: 342–77.

Zelizer, V. A. (1994) *The Social Meaning of Money*, New York: Basic Books.

Zimbardo, P., Marshall, G. and Moslach, C. (1971) Liberating behaviour from time bound control, *Journal of Applied Social Psychology* 1: 305–23.

Zukin, S. (1995) *The Culture of Cities*, Oxford: Basil Blackwell.

Zukin, S. (1998) Urban lifestyles: diversity and standardization in spaces of consumption, *Urban Studies* 35(5–6): 825–39.

Index

Knowledge, Space, Economy

We are now living through a period of knowledge capitalism in which, as Manuel Castells puts it, 'the action of knowledge upon knowledge is the main source of productivity'. In the face of such transformation, the economic, social and institutional contours of contemporary capitalism are being reshaped. At the heart of this process is an emergent set of economies, regions, institutions and people central to the flows and translations of knowledge.

This book provides the first interdisciplinary review of the triad of knowledge, space, economy on entering the twenty-first century. Drawing on a variety of disciplinary backgrounds, the first part of the book comprises a set of statements by leading academics on the role of knowledge in capitalism. Thereafter, the remaining two parts of the book explore the landscape of knowledge capitalism through a series of analyses of knowledge in action within a range of economic, political and cutural contexts.

Bringing together contributions from across the social sciences, this book provides both a major theoretical statement on understanding the economic world and empirical exemplification of the power of knowledge in shaping the spaces and places of today's economy and society.

John R. Bryson is Senior Lecturer in Economic Geography, **Peter W. Daniels** is Professor of Geography and **Jane Pollard** is Lecturer in Economic Geography, all at the University of Birmingham. **Nick Henry** is Reader in Urban and Regional Studies at the University of Newcastle upon Tyne.